Nanotechnology in the Life Sciences

Series Editor

Ram Prasad
Department of Botany
Mahatma Gandhi Central University
Motihari, Bihar, India

Nano and biotechnology are two of the 21st century's most promising technologies. Nanotechnology is demarcated as the design, development, and application of materials and devices whose least functional make up is on a nanometer scale (1 to 100 nm). Meanwhile, biotechnology deals with metabolic and other physiological developments of biological subjects including microorganisms. These microbial processes have opened up new opportunities to explore novel applications, for example, the biosynthesis of metal nanomaterials, with the implication that these two technologies (i.e., thus nanobiotechnology) can play a vital role in developing and executing many valuable tools in the study of life. Nanotechnology is very diverse, ranging from extensions of conventional device physics to completely new approaches based upon molecular self-assembly, from developing new materials with dimensions on the nanoscale, to investigating whether we can directly control matters on/in the atomic scale level. This idea entails its application to diverse fields of science such as plant biology, organic chemistry, agriculture, the food industry, and more.

Nanobiotechnology offers a wide range of uses in medicine, agriculture, and the environment. Many diseases that do not have cures today may be cured by nanotechnology in the future. Use of nanotechnology in medical therapeutics needs adequate evaluation of its risk and safety factors. Scientists who are against the use of nanotechnology also agree that advancement in nanotechnology should continue because this field promises great benefits, but testing should be carried out to ensure its safety in people. It is possible that nanomedicine in the future will play a crucial role in the treatment of human and plant diseases, and also in the enhancement of normal human physiology and plant systems, respectively. If everything proceeds as expected, nanobiotechnology will, one day, become an inevitable part of our everyday life and will help save many lives.

More information about this series at http://www.springer.com/series/15921

Inamuddin • Abdullah M. Asiri
Editors

Advanced Nanotechnology and Application of Supercritical Fluids

Springer

Editors
Inamuddin
Chemistry Department
King Abdulaziz University
Jeddah, Saudi Arabia

Department of Applied Chemistry
Aligarh Muslim University
Aligarh, India

Abdullah M. Asiri
Chemistry Department
King Abdulaziz University
Jeddah, Saudi Arabia

ISSN 2523-8027 ISSN 2523-8035 (electronic)
Nanotechnology in the Life Sciences
ISBN 978-3-030-44983-4 ISBN 978-3-030-44984-1 (eBook)
https://doi.org/10.1007/978-3-030-44984-1

This Springer imprint is published by the registered company Springer Nature Switzerland AG
The registered company address is: Gewerbestrasse 11, 6330 Cham, Switzerland

Preface

At present, science is looking towards renewability and sustainability, which encompass the reduction in the use of toxic chemical substances and a sharp turn towards the applications of eco-friendly, chemically efficient products with an increasing trend in grasping the philosophy of green chemistry. Supercritical fluids are defined as fluids with properties (temperature and pressure) higher than their critical point. Supercritical fluids exhibit a wide range of chemical and physical properties, varying from gas-like to liquid-like behavior due to the variations in the ionic product and dielectric constant. The supercritical fluid marks one of the most important step-ups towards green chemistry to develop sustainability in chemical industries. Being a green medium, this proves as a potential substitute for the traditional toxic organic solvents and volatile organic compounds that are extremely hazardous to the environment. Owing to its inflammability, non-toxicity, minimum expenses, and easy availability, supercritical carbon dioxide makes itself highly valuable to modern industrial protocols. Thus, the utilization of eco-friendly supercritical fluid technologies to obtain a wide range of products from renewable resources is a topic at the forefront of innovation for a novel circular economy.

This book *Advanced Nanotechnology and Application of Supercritical Fluids* explores supercritical fluid technology applications for modern society. To keep up with developments in society, researchers have been studying modern technologies that use green supercritical fluid for processing a variety of raw materials. Therefore, this book presents a brief motivation to study fundamentals, properties, and nanotechnology applications of supercritical fluids. Topics discussed include Amazonian natural resources, extraction strategies, non-catalytic and catalytic remediation processes, biomass conversion, production of platform chemicals, biological, pharmaceutical, and industrial applications, and few more. This book targets undergraduates, postgraduates, researchers, engineers, R&D professionals, lecturers, and faculty who are working in academics and industrial technologies. Based on thematic topics, the book edition contains the following ten chapters.

Chapter 1 discusses the applications of supercritical fluid (SCF) technology, as a green solvent, in the various avenues of pharmaceutical product development. Recent exploitation of the technology in drug delivery, sterilization of pharmaceutical products, chiral separation, and bioactive extraction is discussed in detail. The occupational safety aspects of SCF are also discussed.

Chapter 2 presents the use of green solvents such as carbon dioxide, water, and ethanol in supercritical extraction technology, as an alternative to conventional methods that apply toxic solvents, aiming to obtain purer bioactive compounds from raw vegetable materials from the Amazon region, as well as the application of their extracts or isolated compounds in biological studies.

Chapter 3 presents the historical background, fundamentals, classification, reactor designs, industrial applications, and commercialization of supercritical water oxidation (SCWO) technology. Also, reaction pathways and kinetics for phenol oxidation in supercritical water and non-catalytic and catalytic SCWO of phenol in the petroleum industry and other wastewaters are discussed.

Chapter 4 provides an overview of recent advances in the production of platform chemicals (furfural, levulinic acid, 5-hydroxymethylfurfural (5-HMF), L-lysine, amino acids, and L-glutamic acid) using supercritical fluid technology. The mechanism, method, raw material, and process description are reviewed. Finally, the future outlook and perspectives of these platform chemicals are discussed.

Chapter 5 illustrates the chemistry of supercritical carbon dioxide and its chemical intricacies and plausible solutions for making it widely applicable in different chemical industries. The wide range of applications of this green medium in chemical reactions, reactive extraction procedures, nanoparticle synthesis, foam production, and hydrocarbon exploration is discussed.

Chapter 6 generalizes the types of phenolic compounds present in different plant parts and discusses various processing parameters needed for their extraction by supercritical fluid extraction.

Chapter 7 discusses the application of supercritical carbon dioxide in the extraction of a variety of biomolecules. The optimization of extraction parameters for obtaining high extraction yields of target molecules, including the development of mathematical models to describe the extraction process for improvement and development of scale-ups, is discussed.

Chapter 8 discusses the advances of green solvents like ionic liquids, switchable solvents, supercritical carbon dioxide, and subcritical and supercritical water in chemical-related areas. The main properties and advantages of green solvents over traditional solvents are highlighted to favor specific applications in a wide range of processes.

Chapter 9 discusses supercritical fluid extraction (SFE) with carbon dioxide (CO_2) as a valuable alternative technique in the rubber industry. A combination of these two offered many advantages such as sample recovery, maintenance of purity factor, high selectivity in products, and very short processing time. In this chapter, the role of SFE in rubber industries and the importance of the rubber industry in Malaysia are summarized.

Chapter 10 describes the potential of biorefinery approaches based on compressed fluid technologies and gives an overview of their applications to obtain high added-value products from food residues. Combinations of compressed fluid technologies are reviewed, always taking into account a green chemistry point of view.

Aligarh, India Inamuddin
Jeddah, Saudi Arabia Abdullah M. Asiri

Contents

Contributors

Ederson R. Abaide Laboratory of Agroindustrial Processes Engineering (LAPE), Federal University of Santa Maria (UFSM), Cachoeira do Sul, RS, Brazil

Norfahana Abd-Talib Universiti Kuala Lumpur, Branch Campus Malaysian Institute of Chemical and Engineering Technology (UniKL MICET), Alor Gajah, Melaka, Malaysia

Poonam Aggarwal Department of Food Science and Technology, Punjab Agricultural University, Ludhiana, Punjab, India

Akil Ahmad School of Industrial Technology, Universiti Sains Malaysia, Penang, Malaysia

G. Arun Drug Delivery Systems, Talla Padmavathi College of Pharmacy, Warangal, India

Jhonatas Rodrigues Barbosa Laboratório de Extração (LABEX), Programa de Pós-Graduação em Ciência e Tecnologia de Alimentos, Universidade Federal do Pará, Belém, Pará, Brazil

Fernanda Wariss Figueiredo Bezerra Laboratório de Extração (LABEX), Programa de Pós-Graduação em Ciência e Tecnologia de Alimentos, Universidade Federal do Pará, Belém, Pará, Brazil

Mónica Bueno Laboratory of Foodomics, Institute of Food Science Research (CIAL-CSIC), Madrid, Spain

Natascha Cheikhyoussef Research and Innovation, Ministry of Higher Education, Technology and Innovation, Windhoek, Namibia

Vânia Maria Borges Cunha Laboratório de Extração (LABEX), Programa de Pós-Graduação em Ciência e Tecnologia de Alimentos, Universidade Federal do Pará, Belém, Pará, Brazil

Wanessa Almeida da Costa Programa de Pós-Graduação em Engenharia de Recursos Naturais da Amazônia, Universidade Federal do Pará, Belém, Pará, Brazil

Jorddy Neves da Cruz Laboratório Adolpho Ducke, Museu Paraense Emílio Goeldi, Belém, Pará, Brazil

Marcilene Paiva da Silva Laboratório de Extração (LABEX), Programa de Pós-Graduação em Ciência e Tecnologia de Alimentos, Universidade Federal do Pará, Belém, Pará, Brazil

Raul Nunes de Carvalho Junior Laboratório de Extração (LABEX), Programa de Pós-Graduação em Ciência e Tecnologia de Alimentos, Universidade Federal do Pará, Belém, Pará, Brazil

Programa de Pós-Graduação em Engenharia de Recursos Naturais da Amazônia, Universidade Federal do Pará, Belém, Pará, Brazil

Mozaniel Santana de Oliveira Laboratório de Extração (LABEX), Programa de Pós-Graduação em Ciência e Tecnologia de Alimentos, Universidade Federal do Pará, Belém, Pará, Brazil

Priscila do Nascimento Bezerra Laboratório de Extração (LABEX), Programa de Pós-Graduação em Ciência e Tecnologia de Alimentos, Universidade Federal do Pará, Belém, Pará, Brazil

Crisleine P. Draszewski Department of Chemical Engineering, Federal University of Santa Maria (UFSM), Santa Maria, RS, Brazil

Santanu Ghosh Coal Geology and Organic Petrology Laboratory, Department of Applied Geology, Indian Institute of Technology (Indian School of Mines), Dhanbad, Jharkhand, India

Mohamad Nasir Mohamad Ibrahim School of Chemical Sciences, Universiti Sains Malaysia, Penang, Malaysia

Shahryar Jafarinejad Department of Chemical Engineering, College of Engineering, Tuskegee University, Tuskegee, AL, USA

Marcio A. Mazutti Department of Chemical Engineering, Federal University of Santa Maria (UFSM), Santa Maria, RS, Brazil

Nurabiyiah Mohamad Universiti Kuala Lumpur, Branch Campus Malaysian Institute of Chemical and Engineering Technology (UniKL MICET), Alor Gajah, Melaka, Malaysia

Siti Hamidah Mohd. Setapar Department of Chemical Processes, Malaysia-Japan International Institute of Technology, University Technology Malaysia, Kuala Lumpur, Selangor, Malaysia

Cristiano C. Muller Prati Donaduzzi & Cia. Ltda., Pharmaceutical Industry, Toledo, PR, Brazil

Khairul Faizal Pa'ee Universiti Kuala Lumpur, Branch Campus Malaysian Institute of Chemical and Engineering Technology (UniKL MICET), Alor Gajah, Melaka, Malaysia

Satyanarayan Pattnaik Drug Delivery Systems, Talla Padmavathi College of Pharmacy, Warangal, India

Rafael Henrique Holanda Pinto Laboratório de Extração (LABEX), Programa de Pós-Graduação em Ciência e Tecnologia de Alimentos, Universidade Federal do Pará, Belém, Pará, Brazil

Neha Purandare Department of Food Science and Nutrition, Lovely Professional University, Phagwara, Punjab, India

Marielba de Los Angeles Rodriguez Salazar Laboratório de Extração (LABEX), Programa de Pós-Graduação em Ciência e Tecnologia de Alimentos, Universidade Federal do Pará, Belém, Pará, Brazil

Kalpana Swain Drug Delivery Systems, Talla Padmavathi College of Pharmacy, Warangal, India

Marcus V. Tres Laboratory of Agroindustrial Processes Engineering (LAPE), Federal University of Santa Maria (UFSM), Cachoeira do Sul, RS, Brazil

Khalid Umar School of Chemical Sciences, Universiti Sains Malaysia, Penang, Malaysia

Nader Vahdat Department of Chemical Engineering, College of Engineering, Tuskegee University, Tuskegee, AL, USA

Atul Kumar Varma Coal Geology and Organic Petrology Laboratory, Department of Applied Geology, Indian Institute of Technology (Indian School of Mines), Dhanbad, Jharkhand, India

Asim Ali Yaqoob School of Chemical Sciences, Universiti Sains Malaysia, Penang, Malaysia

Mudasir Yaqoob Department of Food Science and Technology, Punjab Agricultural University, Ludhiana, Punjab, India

Department of Food Science and Nutrition, Lovely Professional University, Phagwara, Punjab, India

Tau Len-Kelly Yong Universiti Kuala Lumpur, Branch Campus Malaysian Institute of Chemical and Engineering Technology (UniKL MICET), Alor Gajah, Melaka, Malaysia

Giovani L. Zabot Laboratory of Agroindustrial Processes Engineering (LAPE), Federal University of Santa Maria (UFSM), Cachoeira do Sul, RS, Brazil

About the Editors

Dr. Inamuddin is currently working as Assistant Professor in the Chemistry Department, Faculty of Science, King Abdulaziz University, Jeddah, Saudi Arabia. He is a permanent faculty member (Assistant Professor) at the Department of Applied Chemistry, Aligarh Muslim University, Aligarh, India. He obtained Master of Science degree in Organic Chemistry from Chaudhary Charan Singh (CCS) University, Meerut, India, in 2002. He received his Master of Philosophy and Doctor of Philosophy degrees in Applied Chemistry from Aligarh Muslim University (AMU), India, in 2004 and 2007, respectively. He has extensive research experience in multidisciplinary fields of Analytical Chemistry, Materials Chemistry, and Electrochemistry and, more specifically, Renewable Energy and Environment. He has worked on different research projects as project fellow and senior research fellow funded by University Grants Commission (UGC), Government of India, and Council of Scientific and Industrial Research (CSIR), Government of India. He has received Fast Track Young Scientist Award from the Department of Science and Technology, India, to work in the area of bending actuators and artificial muscles. He has completed four major research projects sanctioned by University Grant Commission, Department of Science and Technology, Council of Scientific and Industrial Research, and Council of Science and Technology, India. He has published 171 research articles in international journals of repute and eighteen book chapters in knowledge-based book editions published by renowned international publishers. He has published 105 edited books with Springer (U.K.), Elsevier, Nova Science Publishers, Inc. (U.S.A.), CRC Press Taylor & Francis Asia Pacific, Trans Tech Publications Ltd. (Switzerland), IntechOpen Limited (U.K.), Wiley-Scrivener, (U.S.A.) and Materials Research Forum LLC (U.S.A). He is a member of various journals' editorial boards. He is also serving as Associate Editor for journals (Environmental Chemistry Letter, Applied Water Science and Euro-Mediterranean Journal for Environmental Integration, Springer-Nature), Frontiers Section Editor (Current Analytical Chemistry, Bentham Science Publishers), Editorial Board Member (Scientific Reports-Nature), Editor (Eurasian Journal of Analytical Chemistry), and Review Editor (Frontiers in Chemistry, Frontiers, U.K.) He is also guest-editing various special thematic special issues to the journals of Elsevier,

Bentham Science Publishers, and John Wiley & Sons, Inc. He has attended as well as chaired sessions in various international and national conferences. He has worked as a Postdoctoral Fellow, leading a research team at the Creative Research Initiative Center for Bio-Artificial Muscle, Hanyang University, South Korea, in the field of renewable energy, especially biofuel cells. He has also worked as a Postdoctoral Fellow at the Center of Research Excellence in Renewable Energy, King Fahd University of Petroleum and Minerals, Saudi Arabia, in the field of polymer electrolyte membrane fuel cells and computational fluid dynamics of polymer electrolyte membrane fuel cells. He is a life member of the Journal of the Indian Chemical Society. His research interest includes ion exchange materials, a sensor for heavy metal ions, biofuel cells, supercapacitors and bending actuators.

Abdullah M. Asiri is the Head of the Chemistry Department at King Abdulaziz University since October 2009, and he is the founder and the Director of the Center of Excellence for Advanced Materials Research (CEAMR) since 2010 to date. He is the Professor of Organic Photochemistry. He graduated from King Abdulaziz University (KAU) with a B.Sc. in Chemistry in 1990 and a Ph.D. from University of Wales, College of Cardiff, UK, in 1995. His research interest covers color chemistry, synthesis of novel photochromic and thermochromic systems, synthesis of novel coloring matters and dyeing of textiles, materials chemistry, nanochemistry and nanotechnology, polymers and plastics. Prof. Asiri is the principal supervisors of more than 20 M.Sc. and six Ph.D. theses. He is the main author of ten books of different chemistry disciplines. Prof. Asiri is the Editor-in-Chief of *King Abdulaziz University Journal of Science*. A major achievement of Prof. Asiri is the research of tribochromic compounds, a new class of compounds which change from slightly colored or colorless to deep colored when subjected to small pressure or when ground. This discovery was introduced to the scientific community as a new terminology published by the International Union of Pure and Applied Chemistry (IUPAC) in 2000. This discovery was awarded a patent from European Patent office and from UK patent. Prof. Asiri has involved in many committees at both KAU level and national level. He took a major role in the advanced materials committee working for King Abdulaziz City for Science and Technology (KACST) to identify the national plan for science and technology in 2007. Prof. Asiri played a major role in advancing the chemistry education and research in KAU. He has been awarded the best researchers from KAU for the past five years. He received the Young Scientist Award from the Saudi Chemical Society in 2009 and also the first prize for the distinction in science from the Saudi Chemical Society in 2012. He also received a recognition certificate from the American Chemical Society (Gulf Region Chapter) for the advancement of chemical science in the Saudi Arabia. He received a Scopus certificate as the most published scientist in Saudi Arabia in chemistry in 2008. He is also a member of the editorial board of various journals of international repute. He is the Vice-President of Saudi Chemical Society (Western Province Branch). He holds four USA patents, more than one thousand publications in international journals, several book chapters, and edited books.

Chapter 1
Supercritical Fluid Technologies: A Green Solvent Approach for Pharmaceutical Product Development

Satyanarayan Pattnaik, G. Arun, and Kalpana Swain

Contents

1.1 Introduction

Supercritical fluid (SCF) technology is in existence since the last few decades as a major tool in diverse industries including those involved with fine chemicals and pharmaceuticals (Falconer et al. 2015). The evolution of SCF technology was majorly driven by eco-friendly environmental interests that provided aptness of some SCFs for replacing hazardous industrial solvents. There have been increasing concerns about the ill-effects of the traditional solvents used in the pharmaceutical plants on the environment including the biosphere. There is a great possibility of the presence of solvent residuals in finished pharmaceutical goods which may also impose a significant health hazard. Neither the traditional organic solvents are cost-effective nor are they environment-friendly. The increasingly stringent environmental regulations, search for cost-effective alternatives and demanding needs for highly pure and superior products have fuelled significant research in the areas of green

S. Pattnaik (✉) · G. Arun · K. Swain
Drug Delivery Systems, Talla Padmavathi College of Pharmacy, Warangal, India

Inamuddin, A. M. Asiri (eds.), *Advanced Nanotechnology and Application of Supercritical Fluids*, Nanotechnology in the Life Sciences,
https://doi.org/10.1007/978-3-030-44984-1_1

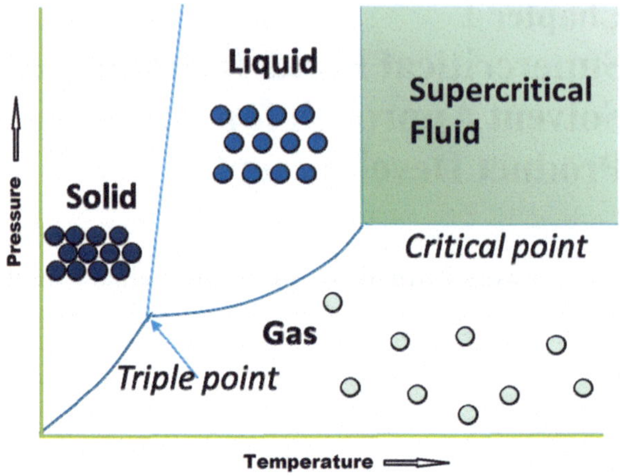

Fig. 1.1 Phase diagram showing the supercritical fluid (SCF) region

pharmaceutical technologies including supercritical fluids (Ghinet et al. 2017; Zhao et al. 2019; Soh and Lee 2019).

A fluid is called supercritical when its pressure and temperature reach above its respective critical value (T_c, critical temperature, and P_c, critical pressure). At critical points, they exhibit properties of both liquid and gas (Fig. 1.1). The density values are identical to those of liquids and flow characteristics identical to those of gases and hence are labelled supercritical fluids. The superior density of SCFs facilitates substantial solvation power. Moreover, the mass transfer phenomenon is favoured because the diffusivity of solutes is higher owing to lower viscosity of solutes in SCF.

Today, the industry has a wide choice of SCFs with varying density values and critical temperatures to meet out the processing needs. Although many SCFs render themselves as suitable for pharmaceutical processing, the most commonly used one is CO_2 because of its unique physicochemical properties including (1) low critical temperature (31.1 °C), (2) low critical pressure (7.38 M Pa), (3) non-toxic and non-flammable nature, (4) generally recognized as safe (GRAS) status and (5) inexpensiveness. Other less commonly used SCFs are water, nitrous oxide, acetone, ammonia, xenon, chlorofluro ethane, trifluromethane, etc.

1.2 Supercritical Fluid Technologies

The recent technologies involving supercritical fluids relevant to pharmaceutical industry include rapid expansion of supercritical solutions (RESS), rapid expansion from supercritical to aqueous solutions (RESAS), gas antisolvent (GAS) process, supercritical antisolvent (SAS) process, solution-enhanced dispersion by supercritical fluids (SEDS), polymerization-induced phase separation (PIPS), particles from gas-saturated solutions (PGSS), aerosol solvent extraction system (ASES), supercritical assisted atomization (SAA) and supercritical solvent impregnation (SSI)

(Girotra et al. 2013). The supercritical fluid may act as a solvent (RESS, PGSS, RESAS, etc.) or antisolvent (GAS, SAS, SEDS, ASES, SSI, etc.). The pharmaceutical industry has chosen to deploy these SF technologies to bring a green revolution in value added product development. The subsequent sections deal with the advancements in pharmaceutical product development using SCF technologies.

1.3 Pharmaceutical Product Development Using SCF Technology

In the recent past, pharmaceutical science has witnessed a lot of research dedicated to meet drug delivery and drug manufacturing demands. Development of novel technologies and suitable modifications of existing time-tested technologies are the keys to address these growing demands. Supercritical fluid (SCF) technology has been used in many fields for decades, such as the food processing industry, fine chemical industry, polymer industry, textile and fabric industry, etc., apart from pharmaceutical industry. The application of SCF technology in pharmaceutical industry is diverse spanning from bioactive extraction to recent particle engineering for novel drug delivery applications. The various application platforms in pharmaceutical industries are detailed in the following sub-sections.

1.3.1 SCF in Drug Delivery

The technologies involving supercritical fluids have evolved considerably for engineering of pharmaceutical particles and are expected, in the near future, to completely take over the traditional methods including comminution and recrystallization. This switching of technology is anticipated because of the expected superior quality of finished pharmaceutical products through deployment of SCF technology. The particle properties that influence the efficiency of many particulate drug delivery systems include particle morphology, particle size and size distribution, crystal form, etc. Based on the route of administration of the drug product and the targeting needs, the particle microstructure should be tailored. The acceptable particle size range of drug substances is different when they are administered through different routes, for example, for oral delivery, the particle size may vary in the range of 0.1–100 μm, and for parenteral delivery, the particle size should be less than 5 μm (Brazeau et al. 2011; Liu et al. 2015). In the context of drug targeting, the drug particles with size greater than 6–7 μm are preferentially targeted to the lungs. On the other hand, smaller particles with size less than 0.1 μm are collected in the bone marrows, and the particles with medium size range (0.1–7 μm) are preferably accumulated in the phagocytic organs. For effective transdermal permeation, the size of the drug particles should be preferably less than 9–10 μm. Drug particles with mass median aerodynamic diameter ranging from 1 to 5 μm are effective and achieve superior bioavailability owing to fast absorption through the alveoli. In this case, smaller particles are prone to elimination through exhalation, and bigger particles are collected near the earlier anatomical region of the respiratory tract leading to failure in drug delivery necessitating a close monitoring of optimal particle size in pulmonary drug delivery. SCF technologies adopted for particle engineering are usually RESS, PGSS and SAS processes.

Perinelli et al. (2016) used PGSS technique for production of bovine serum albumin-loaded di-block co-polymer microparticles with improved dissolution of the payload. Polyethylene glycol conjugation significantly affected the drug dissolution behaviour.

Conventional methods for drug loading into mesoporous silica micro−/nanoparticles are often associated with a major limitation of inefficient drug loading. In this context, SCF technology has emerged as a boon and helped researchers for efficient drug loading into porous carriers. Supercritical carbon dioxide (SC-CO_2) was used as a solvent for ibuprofen loading into mesoporous microparticles for sustained drug delivery (Li-Hong et al. 2013). Significantly higher quantity of ibuprofen was found loaded through the supercritical fluid technique when compared with the traditional method. Intriguingly, Gandhi et al. (2017) have used supercritical carbon dioxide for loading carbamazepine into mesoporous silica with a reportedly improved dissolution velocity.

Supercritical fluid-assisted fenofibrate loading proved efficient with loading up to 485 mg drug/g silica with significantly reduced drug crystalline state (Bouledjouidja et al. 2016). On the other hand, traditional drug loading approach resulted in inferior drug loading. The SCF technology resulted in faster and better drug loading into mesoporous silica.

Pulmonary microparticles have also been fabricated using PGSS technique (Vijayaraghavan et al. 2012, 2013). In this method, different grades of polyethylene glycols along with various lipidic carboxylic acids are used. Particle size was effectively controlled through spraying of myristic acid blend from supercritical carbon dioxide. The particle size was found optimal (2–5 μm) for deep pulmonary delivery (Vijayaraghavan et al. 2013).

The morphology of the mesopores in the carrier often dictates the drug release pattern and hence is very crucial to monitor, while synthesis of mesoporous structures for drug delivery applications and supercritical carbon dioxide has been deployed for altering the mesopore structure effectively (Hanrahan et al. 2005).

During fabrication of mesoporous structures, one of the important steps is to completely remove the template of surfactant, which is often a critical concern for those in the field of development of mesoporous composite synthesis. SCFs were also used for removal of surfactant template from mesoporous materials (Huang et al. 2005a). Methanol-modified supercritical carbon dioxide has been used for extraction of cationic surfactant templates from MCM-41-, MCM-48-, SBA-1- and SBA-3-type mesoporous materials (Huang et al. 2005b).

Sustained-release microspheres of human growth hormone (hGH) were fabricated using supercritical fluid technology for subcutaneous injection. The liquefaction of poly(lactic-co-glycolic acid)/polylactide thermoplastic polymers occurs under the influence of supercritical carbon dioxide which in turn facilitates the hGH mixing effectively. Subsequent removal of carbon dioxide yields microparticles containing hGH (Jordan et al. 2010).

Among nanoscopic vesicular drug carriers, liposomes and proliposomes have their distinct advantages for delivery of a wide variety of therapeutic agents. SCF technology has been in use for process advancement in liposome synthesis (Falconer et al. 2015; Patil and Jadhav 2014; Santo et al. 2015; Tsai and Rizvi 2017a, b; Karn et al. 2013). Melatonin-loaded liposomes were successfully prepared using RESS technology (Zhang et al. 2017). The authors reported 140 bar supercritical pressure as optimum maximizing the encapsulation efficiency (up to 82.2%).

Zhang et al. (2012) reported sirolimus liposomes fabricated using RESS technology and investigated the influence of temperature and pressure on the particle size. They found the optimal conditions for the minimal average particle size of liposome were 328 K of temperature (35 MPa) and 35 MPa of pressure (333 K).

Cyclosporin A-loaded liposomes prepared through supercritical fluid technology for the treatment of dry eye syndrome demonstrated superior efficiency compared to conventional cyclosporin A formulations (Karn et al. 2014).

A central composite design-assisted liposome formulation was developed using SEDS technology for delivery of silymarin with improved bioavailability (Yang et al. 2015). The liposomes obtained by the SEDS method exhibited better entrapment efficiency, reduced particle size and superior bioavailability (about 4.8-fold enhancement) compared to those produced by conventional methods.

Bile salt-mediated liposomes for delivery of a water-soluble but poorly permeable pharmaceutical, notoginsenoside R1, were fabricated using a novel improved supercritical reverse phase evaporation (ISCRPE) method (Fan et al. 2018). The in vitro Caco-2 studies indicated superior permeation and absorption of notoginsenoside R1 which was supported by in vivo pharmacokinetic studies with improved bioavailability.

For effective oral delivery of drug candidates, it is essential that the drug remains in a solubilized state in the gastrointestinal fluid following oral administration. The fact becomes increasingly important for poorly soluble drug substances where effective solubilization strategies have to be adopted while developing suitable formulations (Pattnaik and Pathak 2017; Mallick et al. 2007, 2008). Bicalutamide processed with polyvinylpyrrolidone by supercritical carbon dioxide revealed a four-tenfold improvement in in vitro dissolution which was supported by thermal analysis and powder X-ray diffraction (XRD) studies indicating decreased crystallinity of bicalutamide (Szafraniec et al. 2017).

To address poor and variable oral bioavailability of tacrolimus, solid dispersions were prepared with various polymers like Soluplus, PVP, HPMC and porous chitosan using supercritical fluid technology (Obaidat et al. 2017). The properties of the developed solid dispersions were found to be affected by supercritical processing parameters. In another recent investigation to improve dissolution of cefixime trihydrate, supercritical fluid-assisted solid dispersions with various polymers like polyethylene glycol 4000 and 6000 and Soluplus were explored (Obaidat et al. 2019).

The authors reported improved dissolution which may be due to the presence of co-solvent in supercritical fluid technology which converted cefixime into its amorphous form.

In an investigation carried out by Yin et al. (2015), powder XRD and FTIR (Fourier-transform infrared spectroscopy) studies indicated an interaction between itraconazole and solid dispersion excipients and confirmed the amorphous nature of itraconazole in the composite solid dispersion fabricated through SCF technology. The method reported improvement in the dissolution of itraconazole in various physiological media.

Megestrol acetate SD nanoformulations were fabricated using hydroxypropyl methylcellulose/polyvinyl pyrrolidone (HPMC/PVP) (carrier) and Ryoto sugar ester L1695/D-α-tocopheryl polyethylene glycol 1000 succinate (surfactant) using SAS technology with a fast dissolution velocity and better bioavailability (Ha et al. 2015).

A research group from China developed solid dispersion of poorly soluble co-enzyme Q10 using supercritical fluid technology for improvement of in vitro dissolution and in vivo oral bioavailability (Yang et al. 2018). The authors have reported about 80-fold improvement in dissolution and 2.43-fold and 3.0-fold improvement in area under the curve (AUC) and C_{max}, respectively.

The physicomechanical properties of solid dispersion powders should be suitable for further processing to develop a dosage form, preferably a tablet. Few studies have reported improvement in physicomechanical properties of the solid dispersion powders rendering them suitable for tablet formulation. Supercritical carbon dioxide-assisted carvedilol solid dispersions with improved tabletability and compactibility were reported by Djuris et al. (2019).

The in vitro dissolution studies of a solid dispersion (SD) formulation of bifendate revealed a significant improvement in cumulative per cent drug release from SCF processed solid dispersions (67%) when compared to a physical mixture of bifendate and silica (8%). The oral bioavailability in beagle dogs was found to be 1.6-fold higher than a commercial product (Cai et al. 2016).

Usually, in a vaccine regimen, booster dosing and multiple immunization strategies are generally adopted to induce repeated stimulation of antigen-presenting cells. Baxendale et al. (2011) have reported a single-shot polylactic acid (PLA) microparticle-based tetanus toxoid vaccine prepared using NanoMix SCF-assisted method eliminating the need for a booster dose and repeat immunization scheduling.

Inclusion complexation using a diverse derivative of cyclodextrin has attracted the attention of researchers as a platform for solubility enhancement and subsequent bioavailability improvement of drugs with dissolution rate-limited absorption properties (Mahapatra et al. 2013). A single-step, SCF-assisted technology for the preparation of olanzapine-methyl-β-cyclodextrin complexes to improve the dissolution rates of olanzapine was reported (Rudrangi et al. 2015). The complexes were prepared using four different methods like supercritical CO_2 processing, co-evaporation, lyophilization and physical blending followed by comparison of efficiencies. Complexes prepared by the SCF processing method exhibited the highest dissolution rate owing to amorphization of olanzapine in the inclusion complex.

Attempts to deliver baicalin through hydroxypropyl-β-cyclodextrin inclusion complexation using SCF technology were found efficient compared to the conventional solution mixing approach (Li et al. 2018). In another study, SCF-assisted SD formulations of glyburide showed faster dissolution from SCF-processed solid dispersion tablets (Guan et al. 2014). Furthermore, in vivo studies demonstrated a 2.01-fold improvement in the bioavailability of glyburide from the novel SCF solid dispersion tablets compared to commercial product.

Inclusion complex of apigenin-hydroxypropyl-β-cyclodextrin using SAS process was reported (Huang et al. 2016). Under the optimal processing conditions (pressure, 22.5 MPa; temperature, 50 °C; and drug concentration, 20 mg/ml), loading efficiency and encapsulation efficiency of the inclusion complex with a mean particle size of 392.13 ± 7.56 nm were 13.97% ± 0.17% and 93.22% ± 1.17%, respectively.

Flurbiprofen inclusion complex fabricated through SCF-CO_2 yielded products with enhanced dissolution behaviour (Rudrangi et al. 2016). SCF-CO_2 processed at 45 °C and 200 bar produced inclusion complexes with significant improvement in drug release characteristic (99.39 ± 2.34% at the end of 30 min) when compared to raw flurbiprofen (1.11 ± 0.09% at the end of 60 min). Miconazole/hydroxypropyl-gamma-cyclodextrin/L-tartaric acid inclusion complex produced by supercritical carbon dioxide processing demonstrated significantly higher bioavailability in pigs (Barillaro et al. 2005). Yet, in another study to improve bioavailability of itraconazole through supercritical carbon dioxide-assisted inclusion complexation with beta-cyclodextrin revealed improved drug levels in circulation (Hassan et al. 2007).

1.3.2 Sterilization of Products

Supercritical methods have established their presence for disinfection and reduction of microbial load in pharmaceutical products (Sikin and Rizvi 2011). The conventional methods adopted to achieve these goals are associated with many drawbacks which, in many cases, affect adversely the quality of the product. In this scenario, SC-CO_2 has been found efficient in achieving product sterility without any adverse effects (White et al. 2006). The method reported by White et al. (2006) is relatively gentle, as the morphology, ultrastructure and protein profiles of inactivated microbes are maintained, making it an ideal choice for many pharmaceuticals and biologicals, viz. animal tissue-based products, protein and peptide therapeutics, etc.

Investigation using two types of bacterial spores to assess suitability of supercritical carbon dioxide (with water or hydrogen peroxide) for sterilization purposes was reported (Hemmer et al. 2007). The sterility conditions were achieved within 25 min at 105 °C and 300 atm without any additive. The method suggested possible quick elimination of resistant bacterial spores with supercritical carbon dioxide (with added hydrogen peroxide) under low temperature compared to conventional method.

Shieh et al. (2009) reported a supercritical carbon dioxide-based sterilization method effective for metal surfaces and heat-sensitive electronic devices. The authors reported effective inactivation of bacterial spores in 45 min using optimized modifier (hydrogen peroxide, tert-butyl hydroperoxide, formic acid and Triton X-100) concentrations. The sterilization method reported by Checinska et al. (2011) used supercritical carbon dioxide with 3.3% water and 0.1% hydrogen peroxide to achieve effective control of sterility within half an hour.

1.3.3 Supercritical Fluid Chromatography (SFC) and Chiral Separation

The chirality of optically active substances is of paramount importance in an industrial and therapeutic perspective. The enantiomers usually follow different metabolism pathways in the body and often elicit different therapeutic activities. One of the enantiomers may have desired therapeutic activity, while other may cause toxicity. Recently, a quantum of work has been devoted to exploit supercritical fluid technology for achieving resolution of racemic mixtures of pharmaceutical agents (Kalíková et al. 2014; Lemasson et al. 2016; Zawatzky et al. 2016; Ghinet et al. 2017; Wolrab et al. 2017; Xin et al. 2017; Michaels et al. 2019; Zhao et al. 2019).

Though traditionally non-chlorinated cellulose- and amylose-derivatized stationary phases have been used in SFC to affect resolution of a number of chiral compounds, some tough compounds are not well resolved on these traditional stationary phases owing to the absence of chiral selectivity. In this scenario, chlorinated chiral stationary phases (e.g. Lux Cellulose-4, Lux Cellulose-2, Lux Amylose-2, Chiralpak IC, etc.) may be of significance (Wu et al. 2016).

A method of perfluoro-1-methylheptane sulfonate (1 m-PFOS) enantioseparation by supercritical fluid chromatography was reported by Zhao et al. (2019). The resolution and retention time were 0.88 and 130 min, respectively, under optimal conditions [column, Chiralpak QN-AX; mobile phase, CO_2/2-propanol at 70:30; flow rate, 1 ml/min; temperature, 32 °C; back pressure regulator (BPR) pressure, 1800 psi].

A small-scale chiral separation method was developed by Zehani et al. (2017). The researchers reported preparation of 10 mg each of the eight enantiomers using a Chiralpak® AD-H column with varying ethyl alcohol concentration as a co-solvent.

Simultaneous estimation of risperidone and its chiral metabolites in rat plasma was reported adopting a sensitive SCF-tandem mass spectrometry method (Prasad et al. 2017). The method resolved risperidone and its chiral metabolites quickly in 6-min runtime.

1.3.4 Supercritical Fluid Extraction (SFE)

Bioactive materials including volatile and essential oils are usually thermolabile and hence are prone to degradation at elevated temperature. The traditional extraction method like steam distillation is not suitable for such extraction of bioactives. Supercritical fluid extraction methods have drawn much attention for extraction of active materials from biological sources (Wrona et al. 2017; Yen et al. 2015).

Zoccali et al. (2017) have developed a system of supercritical fluid-based extraction/chromatography for natural bioactives with high efficiency. The yield following initial extraction was reported as high as 37.4% to 65.4% at 80 °C.

Polyphenol extracts from black tea leftover obtained by SCE were found to be cleaner and richer in polyphenols as compared to those obtained by traditional extraction process (Mushtaq et al. 2017).

The operating pressure in SCE has a significant impact on the efficiency of the extraction procedure and, hence, the quality of the product. Majdoub et al. (2019) have assessed the impact of pressure on the nature of extractable substances from wild carrots. The authors reported that increased operating pressure (when varied from 100 to 300 bar) resulted in higher concentration of geranyl acetate and β-bisabolene in the extracts.

Recently, the influence of ultrasound on the efficiency of SCF-assisted bioactive extraction was studied by Santos-Zea et al. (2019). Multiplate transducers, in the study, found to intensify the SCF extraction of saponins and antioxidants from agave bagasse. The study has thrown light on the effective extraction method of bioactives which may be translated to suitable application in pharmaceutical industries.

1.4 Occupational Safety

Though SCF technology has created momentum in the industry as a green environment-friendly platform for pharmaceutical and other industries, the potential safety aspects of the technology can never be ruled out (Lucas et al. 2003). Handling of SC-CO_2 is associated with several bio-hazards including potential threat to humans, hence requiring due attention. The types of hazards may include mechanical, chemical, biological and thermodynamic (Soares and Coelho 2012). The personnel involved during the SCF processing are prone to mechanical hazards due to the excessively high operating pressure in the plant. Hence, it is advised to enclose the processing unit with special enclosing barriers for the protection of the personnel. Furthermore, in case of carbon dioxide leakage in the closed compartment may lead to asphyxiation, as it can replace the oxygen in the surroundings. Hence, it is recommended to install carbon dioxide sensors in the plant site and other areas especially underground cellars, where CO_2 that is heavier than air may accumulate. Accidental exposure of personnel to high-pressure environment may cause eardrum and lung damage (Soares and Coelho 2012). However, comprehensive and cautious training of the personnel in proper consultation with the equipment designer may substantially improve the hazard incidences.

1.5 Future Perspectives

SCF technology at industrial scale is still in a primitive state with limited fundamental studies to understand the phase behaviour of the multi-component blend of diverse compounds. So far, there are no pharmaceutical products manufactured by the SCF technology available in the market. Hence, there is a strong need for industrialization of the process especially in pharmaceutical sectors. There is a strong need for proper optimization of process variables so as to render the process scalable. The vast opportunity of future application of the SCF technology in the improvement of the existing drug products and new drug development by appropriate particle engineering is quite visible.

Undoubtedly, supercritical fluid technology, as a clean and green technology, is about to set a green revolution in the chemical processing industries creating avenues for development of better products with improved therapeutic efficacy and stability in an eco-friendly approach.

References

Barillaro V, Evrard B, Delattre L, Piell G (2005) Oral bioavailability in pigs of a miconazole/hydroxypropyl-gamma-cyclodextrin/L-tartaric acid inclusion complex produced by supercritical carbon dioxide processing. AAPS J 7(1):E149–E155

Baxendale A, van Hooff P, Durrant LG, Spendlove I, Howdle SM, Woods HM, Whitaker MJ, Davies OR, Naylor A, Lewis AL, Illum L (2011) Single shot tetanus vaccine manufactured by a supercritical fluid encapsulation technology. Int J Pharm 413(1–2):147–154. https://doi.org/10.1016/j.ijpharm.2011.04.053

Bouledjouidja A, Masmoudi Y, Van Speybroeck M, Schueller L, Badens E (2016) Impregnation of fenofibrate on mesoporous silica using supercritical carbon dioxide. Int J Pharm 499(1–2):1–9. https://doi.org/10.1016/j.ijpharm.2015.12.049

Brazeau G, Sauberan SL, Gatlin L, Wisniecki P, Shah J (2011) Effect of particle size of parenteral suspensions on in vitro muscle damage. Pharm Dev Technol 16(6):591–598. https://doi.org/10.3109/10837450.2010.542161

Cai C, Liu M, Li Y, Guo B, Chang H, Zhang X, Yang X, Zhang T (2016) A silica-supported solid dispersion of bifendate using supercritical carbon dioxide method with enhanced dissolution rate and oral bioavailability. Drug Dev Ind Pharm 42(3):412–417. https://doi.org/10.3109/03639045.2015.1071833

Checinska A, Fruth IA, Green TL, Crawford RL, Paszczynski AJ (2011) Sterilization of biological pathogens using supercritical fluid carbon dioxide containing water and hydrogen peroxide. J Microbiol Methods 87(1):70–75. https://doi.org/10.1016/j.mimet.2011.07.008

Djuris J, Milovanovic S, Medarevic D, Dobricic V, Dapčević A, Ibric S (2019) Selection of the suitable polymer for supercritical fluid assisted preparation of carvedilol solid dispersions. Int J Pharm 554:190–200. https://doi.org/10.1016/j.ijpharm.2018.11.015

Falconer JR, Svirskis D, Adil AA, Wu Z (2015) Supercritical fluid technologies to fabricate proliposomes. J Pharm Pharm Sci 18(5):747–764

Fan Q, Zhang Y, Hou X, Li Z, Zhang K, Shao Q, Feng N (2018) Improved oral bioavailability of notoginsenoside R1 with sodium glycocholate-mediated liposomes: preparation by supercritical fluid technology and evaluation in vitro and in vivo. Int J Pharm 552(1–2):360–370. https://doi.org/10.1016/j.ijpharm.2018.10.005

Gandhi AV, Thipsay P, Kirthivasan B (2017) Squillante E adsorption onto mesoporous silica using supercritical fluid technology improves dissolution rate of carbamazepine-a poorly soluble compound. AAPS PharmSciTech 18(8):3140–3150. https://doi.org/10.1208/s12249-017-0784-3

Ghinet A, Zehani Y, Lipka E (2017) Supercritical fluid chromatography approach for a sustainable manufacture of new stereoisomeric anticancer agent. J Pharm Biomed Anal 145:845–853. https://doi.org/10.1016/j.jpba.2017.08.006

Girotra P, Singh SK, Nagpal K (2013) Supercritical fluid technology: a promising approach in pharmaceutical research. Pharm Dev Technol 18(1):22–38

Guan J, Han J, Zhang D, Chu C, Liu H, Sun J, He Z, Zhang T (2014) Increased dissolution rate and oral bioavailability of hydrophobic drug glyburide tablets produced using supercritical CO_2 silica dispersion technology. Eur J Pharm Biopharm 86(3):376–382. https://doi.org/10.1016/j.ejpb.2013.10.008

Ha ES, Kim JS, Baek IH, Yoo JW, Jung Y, Moon HR, Kim MS (2015) Development of megestrol acetate solid dispersion nanoparticles for enhanced oral delivery by using a supercritical antisolvent process. Drug Des Devel Ther 9:4269–4277. https://doi.org/10.2147/DDDT.S90706

Hanrahan JP, Copley MP, Ziegler KJ, Spalding TR, Morris MA, Steytler DC, Heenan RK, Schweins R, Holmes JD (2005) Pore size engineering in mesoporous silicas using supercritical CO2. Langmuir 21(9):4163–4167

Hassan HA, Al-Marzouqi AH, Jobe B, Hamza AA, Ramadan GA (2007) Enhancement of dissolution amount and in vivo bioavailability of itraconazole by complexation with beta-cyclodextrin using supercritical carbon dioxide. J Pharm Biomed Anal 45(2):243–250

Hemmer JD, Drews MJ, LaBerge M, Matthews MA (2007) Sterilization of bacterial spores by using supercritical carbon dioxide and hydrogen peroxide. J Biomed Mater Res B Appl Biomater 80(2):511–518

Huang L, Poh C, Ng SC, Hidajat K, Kawi S (2005a) Preparation of supported mesoporous thin films concerning template removal by supercritical fluid extraction. Langmuir 21(4):1171–1174

Huang L, Kawi S, Poh C, Hidajat K, Ng SC (2005b) Extraction of cationic surfactant templates from mesoporous materials by CH(3)OH-modified CO(2) supercritical fluid. Talanta 66(4):943–951. https://doi.org/10.1016/j.talanta.2004.12.057

Huang Y, Zu Y, Zhao X, Wu M, Feng Z, Deng Y, Zu C, Wang L (2016) Preparation of inclusion complex of apigenin-hydroxypropyl-β-cyclodextrin by using supercritical antisolvent process for dissolution and bioavailability enhancement. Int J Pharm 511(2):921–930. https://doi.org/10.1016/j.ijpharm.2016.08.007

Jordan F, Naylor A, Kelly CA, Howdle SM, Lewis A, Illum L (2010 Jan 25) Sustained release hGH microsphere formulation produced by a novel supercritical fluid technology: in vivo studies. J Control Release 141(2):153–160. https://doi.org/10.1016/j.jconrel.2009.09.013

Kalíková K, Slechtová T, Vozka J, Tesařová E (2014) Supercritical fluid chromatography as a tool for enantioselective separation; a review. Anal Chim Acta 821:1–33. https://doi.org/10.1016/j.aca.2014.02.036

Karn PR, Cho W, Hwang SJ (2013) Liposomal drug products and recent advances in the synthesis of supercritical fluid-mediated liposomes. Nanomedicine (Lond) 8(9):1529–1548. https://doi.org/10.2217/nnm.13.131

Karn PR, Kim HD, Kang H, Sun BK, Jin SE, Hwang SJ (2014) Supercritical fluid-mediated liposomes containing cyclosporin A for the treatment of dry eye syndrome in a rabbit model:

comparative study with the conventional cyclosporin A emulsion. Int J Nanomedicine 9:3791–3800. https://doi.org/10.2147/IJN.S65601

Lemasson E, Bertin S, West C (2016) Use and practice of achiral and chiral supercritical fluid chromatography in pharmaceutical analysis and purification. J Sep Sci 39(1):212–233. https://doi.org/10.1002/jssc.201501062

Li Y, He ZD, Zheng QE, Hu C, Lai WF (2018) Hydroxypropyl-β-cyclodextrin for delivery of baicalin via inclusion complexation by supercritical fluid encapsulation. Molecules 23(5). pii: E1169. https://doi.org/10.3390/molecules23051169

Li-Hong W, Xin C, Hui X, Li-Li Z, Jing H, Mei-Juan Z, Jie L, Yi L, Jin-Wen L, Wei Z, Gang C (2013) A novel strategy to design sustained-release poorly water-soluble drug mesoporous silica microparticles based on supercritical fluid technique. Int J Pharm 454(1):135–142. https://doi.org/10.1016/j.ijpharm.2013.07.027

Liu G, Jiang Y, Wang X (2015) Tailoring particle microstructures via supercritical CO_2 processes for particular drug delivery. Curr Pharm Des 21(19):2543–2562

Lucas S, Alonso E, Sanz JA, Cocero MJ (2003) Safety study in a supercritical extraction plant. Chem Eng Technol 26(4):449–461

Mahapatra AK, Murthy PN, Patra RK, Pattnaik S (2013) Solubility enhancement of modafinil by complexation with β-cyclodextrin and hydroxypropyl β-cyclodextrin: a response surface modeling approach. Drug Deliv Lett 3:210–219

Majdoub S, El Mokni R, Muradalievich AA, Piras A, Porcedda S, Hammami S (2019) Effect of pressure variation on the efficiency of supercritical fluid extraction of wild carrot (Daucus carota subsp. maritimus) extracts. J Chromatogr B Analyt Technol Biomed Life Sci 1125:121713. https://doi.org/10.1016/j.jchromb.2019.121713

Mallick S, Pattnaik S, Swain K, De PK (2007) Current perspectives of solubilization: potential for improved bioavailability. Drug Dev Ind Pharm 33(8):865–873

Mallick S, Pattnaik S, Swain K, De PK, Saha A, Ghoshal G, Mondal A (2008) Formation of physically stable amorphous phase of ibuprofen by solid state milling with kaolin. Eur J Pharm Biopharm 68(2):346–351

Michaels P, Neef J, Galyan K, Ginsburg-Moraff C, Zhou X, Dunstan D, Poirier J, Reilly J (2019) Enabling chiral separations in discovery chemistry with open-access chiral supercritical fluid chromatography. Chirality 31(8):575–582. https://doi.org/10.1002/chir.23081

Mushtaq M, Sultana B, Akram S, Anwar F, Adnan A, Rizvi SSH (2017) Enzyme-assisted supercritical fluid extraction: an alternative and green technology for non-extractable polyphenols. Anal Bioanal Chem 409(14):3645–3655. https://doi.org/10.1007/s00216-017-0309-7

Obaidat RM, Tashtoush BM, Awad AA, Al Bustami RT (2017) Using supercritical fluid technology (SFT) in preparation of tacrolimus solid dispersions. AAPS PharmSciTech 18(2):481–493. https://doi.org/10.1208/s12249-016-0492-4

Obaidat RM, Khanfar M, Ghanma R (2019) A comparative solubility enhancement study of cefixime trihydrate using different dispersion techniques. AAPS PharmSciTech 20(5):194. https://doi.org/10.1208/s12249-019-1395-y

Patil YP, Jadhav S (2014 Jan) Novel methods for liposome preparation. Chem Phys Lipids 177:8–18. https://doi.org/10.1016/j.chemphyslip.2013.10.011

Pattnaik S, Pathak K (2017) Mesoporous silica molecular sieve based nanocarriers: transpiring drug dissolution research. Curr Pharm Des 23(3):467–480. https://doi.org/10.2174/1381612822666161026162005

Perinelli DR, Cespi M, Bonacucina G, Naylor A, Whitaker M, Lam JK, Howdle SM, Casettari L, Palmieri GF (2016) PEGylated biodegradable polyesters for PGSS microparticles formulation: processability, physical and release properties. Curr Drug Deliv 13(5):673–681

Prasad TR, Joseph S, Kole P, Kumar A, Subramanian M, Rajagopalan S, Kr P (2017) Enantioselective supercritical fluid chromatography-tandem mass spectrometry method for simultaneous estimation of risperidone and its 9-hydroxyl metabolites in rat plasma. Bioanalysis 9(22):1739–1750. https://doi.org/10.4155/bio-2017-0168

Rudrangi SR, Trivedi V, Mitchell JC, Wicks SR, Alexander BD (2015) Preparation of olanzapine and methyl-β-cyclodextrin complexes using a single-step, organic solvent-free supercritical fluid process: an approach to enhance the solubility and dissolution properties. Int J Pharm 494(1):408–416. https://doi.org/10.1016/j.ijpharm.2015.08.062

Rudrangi SR, Kaialy W, Ghori MU, Trivedi V, Snowden MJ, Alexander BD (2016) Solid-state flurbiprofen and methyl-β-cyclodextrin inclusion complexes prepared using a single-step, organic solvent-free supercritical fluid process. Eur J Pharm Biopharm 104:164–170. https://doi.org/10.1016/j.ejpb.2016.04.024

Santo IE, Campardelli R, Albuquerque EC, Vieira De Melo SAB, Reverchon E, Porta GD (2015) Liposomes size engineering by combination of ethanol injection and supercritical processing. J Pharm Sci 104(11):3842–3850. https://doi.org/10.1002/jps.24595

Santos-Zea L, Gutiérrez-Uribe JA, Benedito J (2019) Effect of ultrasound intensification on the supercritical fluid extraction of phytochemicals from *Agave salmiana* bagasse. J Supercrit Fluids 144:98–107

Shieh E, Paszczynski A, Wai CM, Lang Q, Crawford RL (2009) Sterilization of Bacillus pumilus spores using supercritical fluid carbon dioxide containing various modifier solutions. J Microbiol Methods 76(3):247–252. https://doi.org/10.1016/j.mimet.2008.11.005

Sikin AM, Rizvi SS (2011) Recent patents on the sterilization of food and biomaterials by supercritical fluids. Recent Pat Food Nutr Agric 3(3):212–225

Soares VB, Coelho GLV (2012) Safety study of an experimental apparatus for extraction with supercritical CO2. Braz J Chem Eng 29(3):677–682

Soh SH, Lee LY (2019) Microencapsulation and nanoencapsulation using supercritical fluid (SCF) techniques. Pharmaceutics 11(1). pii: E21. https://doi.org/10.3390/pharmaceutics11010021

Szafraniec J, Antosik A, Knapik-Kowalczuk J, Kurek M, Syrek K, Chmiel K, Paluch M, Jachowicz R (2017) Planetary ball milling and supercritical fluid technology as a way to enhance dissolution of bicalutamide. Int J Pharm 533(2):470–479. https://doi.org/10.1016/j.ijpharm.2017.03.078

Tsai WC, Rizvi SSH (2017a) Microencapsulation and characterization of liposomal vesicles using a supercritical fluid process coupled with vacuum-driven cargo loading. Food Res Int 96:94–102. https://doi.org/10.1016/j.foodres.2017.03.027

Tsai WC, Rizvi SSH (2017b) Simultaneous microencapsulation of hydrophilic and lipophilic bioactives in liposomes produced by an ecofriendly supercritical fluid process. Food Res Int 99(Pt 1):256–262. https://doi.org/10.1016/j.foodres.2017.05.029

Vijayaraghavan M, Stolnik S, Howdle SM, Illum L (2012) Suitability of polymer materials for production of pulmonary microparticles using a PGSS supercritical fluid technique: thermodynamic behaviour of fatty acids, PEGs and PEG-fatty acids. Int J Pharm 438(1–2):225–231. https://doi.org/10.1016/j.ijpharm.2012.08.044

Vijayaraghavan M, Stolnik S, Howdle SM, Illum L (2013) Suitability of polymer materials for production of pulmonary microparticles using a PGSS supercritical fluid technique: preparation of microparticles using PEG, fatty acids and physical or chemicals blends of PEG and fatty acids. Int J Pharm 441(1–2):580–588. https://doi.org/10.1016/j.ijpharm.2012.10.040

White A, Burns D, Christensen TW (2006) Effective terminal sterilization using supercritical carbon dioxide. J Biotechnol 123:504–515

Wolrab D, Frühauf P, Gerner C, Kohout M, Lindner W (2017) Consequences of transition from liquid chromatography to supercritical fluid chromatography on the overall performance of a chiral zwitterionic ion-exchanger. J Chromatogr A 1517:165–175. https://doi.org/10.1016/j.chroma.2017.08.022

Wrona O, Rafińska K, Możeński C, Buszewski B (2017) Supercritical fluid extraction of bioactive compounds from plant materials. J AOAC Int 100(6):1624–1635. https://doi.org/10.5740/jaoacint.17-0232

Wu DR, Yip SH, Li P, Sun D, Mathur A (2016) From analytical methods to large scale chiral supercritical fluid chromatography using chlorinated chiral stationary phases. J Chromatogr A 1432:122–131. https://doi.org/10.1016/j.chroma.2015.12.072

Xin H, Dai Z, Cai J, Ke Y, Shi H, Fu Q, Jin Y, Liang X (2017) Rapid purification of diastereoisomers from Piper kadsura using supercritical fluid chromatography with chiral stationary phases. J Chromatogr A 1509:141–146. https://doi.org/10.1016/j.chroma.2017.06.020
Yang G, Zhao Y, Zhang Y, Dang B, Liu Y, Feng N (2015) Enhanced oral bioavailability of silymarin using liposomes containing a bile salt: preparation by supercritical fluid technology and evaluation in vitro and in vivo. International Journal of Nanomedicine 6633
Yang R, Li Y, Li J, Liu C, Du P, Zhang T (2018) Application of scCO2 technology for preparing CoQ10 solid dispersion and SFC-MS/MS for analyzing in vivo bioavailability. Drug Dev Ind Pharm 44(2):289–295. https://doi.org/10.1080/03639045.2017.1391833
Yen HW, Yang SC, Chen CH, Jesisca, Chang JS (2015) Supercritical fluid extraction of valuable compounds from microalgal biomass. Bioresour Technol 184:291–296. https://doi.org/10.1016/j.biortech.2014.10.030
Yin X, Daintree LS, Ding S, Ledger DM, Wang B, Zhao W, Qi J, Wu W, Han J (2015) Itraconazole solid dispersion prepared by a supercritical fluid technique: preparation, in vitro characterization, and bioavailability in beagle dogs. Drug Des Devel Ther 9:2801–2810. https://doi.org/10.2147/DDDT.S81253
Zawatzky K, Biba M, Regalado EL, Welch CJ (2016) MISER chiral supercritical fluid chromatography for high throughput analysis of enantiopurity. J Chromatogr A 1429:374–379. https://doi.org/10.1016/j.chroma.2015.12.057
Zehani Y, Lemaire L, Millet R, Lipka E (2017) Small scale separation of isoxazole structurally related analogues by chiral supercritical fluid chromatography. J Chromatogr A 1505:106–113. https://doi.org/10.1016/j.chroma.2017.05.028
Zhang W, Sun Y, Li Y, Shen R, Ni H, Hu D (2012) Preparation and influencing factors of sirolimus liposome by supercritical fluid. Artif Cells Blood Substit Immobil Biotechnol 40(1–2):62–65. https://doi.org/10.3109/10731199.2011.585618
Zhang Q, Ou C, Ye S, Song X, Luo S (2017) Construction of nanoscale liposomes loaded with melatonin via supercritical fluid technology. J Microencapsul 34(7):687–698. https://doi.org/10.1080/02652048.2017.1376001
Zhao L, Chen F, Guo F, Liu W, Liu K (2019) Enantioseparation of chiral perfluorooctane sulfonate (PFOS) by supercritical fluid chromatography (SFC): effects of the chromatographic conditions and separation mechanism. Chirality 31:870. https://doi.org/10.1002/chir.23120
Zoccali M, Giuffrida D, Dugo P, Mondello L (2017) Direct online extraction and determination by supercritical fluid extraction with chromatography and mass spectrometry of targeted carotenoids from red habanero peppers (Capsicum chinense Jacq.). J Sep Sci 40(19):3905–3913. https://doi.org/10.1002/jssc.201700669

Chapter 2
Supercritical Green Solvent for Amazonian Natural Resources

Fernanda Wariss Figueiredo Bezerra, Priscila do Nascimento Bezerra, Vânia Maria Borges Cunha, Marielba de Los Angeles Rodriguez Salazar, Jhonatas Rodrigues Barbosa, Marcilene Paiva da Silva, Mozaniel Santana de Oliveira, Wanessa Almeida da Costa, Rafael Henrique Holanda Pinto, Jorddy Neves da Cruz, and Raul Nunes de Carvalho Junior

Contents

F. W. F. Bezerra (✉) · P. do Nascimento Bezerra · V. M. B. Cunha
M. de Los Angeles Rodriguez Salazar · J. R. Barbosa · M. P. da Silva
M. S. de Oliveira · R. H. H. Pinto
Laboratório de Extração (LABEX), Programa de Pós-Graduação em Ciência e Tecnologia de Alimentos, Universidade Federal do Pará, Belém, Pará, Brazil

W. A. da Costa
Programa de Pós-Graduação em Engenharia de Recursos Naturais da Amazônia, Universidade Federal do Pará, Belém, Pará, Brazil

J. N. da Cruz
Laboratório Adolpho Ducke, Museu Paraense Emílio Goeldi, Belém, Pará, Brazil

R. N. de Carvalho Junior
Laboratório de Extração (LABEX), Programa de Pós-Graduação em Ciência e Tecnologia de Alimentos, Universidade Federal do Pará, Belém, Pará, Brazil

Programa de Pós-Graduação em Engenharia de Recursos Naturais da Amazônia, Universidade Federal do Pará, Belém, Pará, Brazil
e-mail: raulncj@ufpa.br

Inamuddin, A. M. Asiri (eds.), *Advanced Nanotechnology and Application of Supercritical Fluids*, Nanotechnology in the Life Sciences,
https://doi.org/10.1007/978-3-030-44984-1_2

2.1 Introduction

Amazon is a natural complex that comprises about 40% of South America and has the largest rainforest and river basin (Amazon basin) in the world, common to Brazil (63%), Peru (10%), Colombia (7%), Bolivia (6%), Venezuela (6%), Guyana (3%), Suriname (2%), and Ecuador (1.5%) (Picq 2016). The region presents many sources of plants and fruits that have high economic importance to this locality, for they are applied in technological and industrial processes. Due to this, the extraction sector has been expanding in recent years, causing an enlargement of the local and international markets (da Costa et al. 2017).

Some of these raw materials have bioactive constituents, such as carotenoids, phenolic compounds, unsaturated fatty acids, and vitamins, that are associated with the prevention of various diseases such as cancer and diabetes (dos Santos et al. 2015; Virgolin et al. 2017). In order to obtain such bioactive compounds, many extraction methods have been applied such as Soxhlet and other techniques that use organic solvents; however, the search for new mechanisms eco-friendly, such as pressurized solvent extraction, has gained great importance due to environmental concerns (da Costa et al. 2019).

One of these eco-friendly methods is the supercritical fluid extraction, which can use many solvents, such as ethanol (EtOH), water (H_2O), and carbon dioxide (CO_2). The most used solvent is CO_2 because it is non-flammable and non-toxic, can be easily taken out of the extracts, and is also relatively cheap (de Oliveira et al. 2018). Given the above, the objective of this work was to present the main solvents used in pressurized solvent extraction, highlighting supercritical fluid extraction, as well as the main bioactive compounds obtained in such processes, and their consequent biological applications.

2.2 Green Methods

Bioactive compounds from plant matrices are usually extracted using organic solvents. However, a growing awareness of environmental issues has been very relevant for extraction processes such as the reduction of energy consumption, less generation of harmful waste, efficient conversion of reagents into products, lower generation of by-products, and higher quality and safety of end products. In this context, high-pressure technologies are a relatively new tool to meet these demands as they develop processes that result in products with specific characteristics.

Among these tools, the extraction with supercritical fluids offers many advantages due to their excellent physicochemical properties (Herrero and Ibañez 2018; Knez et al. 2019). And also the possibility of using non-toxic and non-flammable solvents, meeting strict environmental and health requirements for obtaining natural products of food or medical character (De Melo et al. 2014; El Asbahani et al. 2015).

A pure component can be considered a supercritical fluid when its temperature and pressure are above its critical point; under these conditions, there is no change in the component properties (Brunner 2005). In this way, the extraction with a supercritical fluid provides fast and efficient processes, requires moderate temperatures, and eliminates subsequent removal steps of noxious organic solvents. In addition, it offers greater selectivity compared to conventional extraction technologies. Supercritical fluid extraction makes it possible to obtain various fractions of oils or extracts, mainly in terms of mass yield and chemical composition (Pinto et al. 2018; de Oliveira et al. 2019a; Yousefi et al. 2019).

The supercritical fluid extraction process basically consists of the extraction of soluble compounds present in the raw material by the fluid and the separation of these compounds after the depressurization. To carry out an effective and satisfactory extraction, with the maximum possible yield, some factors must be considered, such as the careful control of the operating conditions and the process step optimization (De Melo et al. 2014; da Silva et al. 2016b). Therefore, the supercritical fluid extraction process selectivity, that is, the extraction of the compounds of interest present in the plant matrices, depends on the solvent type used in the process and the following factors:

1. *Density:* one of the main supercritical fluid characteristics is the possibility of modifying the fluid density by changing its pressure and temperature. As the density is related to solubility, by changing the extraction pressure, the fluid solvating force can be modified (da Silva et al. 2016b).
2. *Pressure and temperature:* the control of pressure allows a selective extraction. Also, a temperature increase at constant pressure causes a decrease in the supercritical fluid density and, therefore, in its solvation power. Thus, small changes in pressure or temperature result in large changes in density (Bhusnure et al. 2015).
3. *Solubility:* it is highly dependent on the fluid pressure, temperature, and density. The solubility of most components in supercritical fluid increases with increasing fluid density, which can be achieved by increasing the extraction pressure (Cunha et al. 2019; Silva et al. 2019).
4. *Cosolvents:* carbon dioxide is the main fluid used in supercritical fluid extraction, and for its low polarity, addition of cosolvents to SC-CO_2 can increase the extraction efficiency by increasing polar solute solubility. The most commonly used "green" cosolvents are ethanol (EtOH) and water (H_2O) (Salazar et al. 2018).
5. *Solvent flow rate:* an increase in the solvent flow rate leads to a shorter extraction time, causing an increase in the extraction capacity. This is due to the increase in the solvent surface velocity on the particles, which therefore increases the rate of convective and diffusive extraction (Pires et al. 2019).
6. *Partical size, shape, and porosity of the solid material:* reduction of particle size increases the extraction efficiency. This occurs when the particle size decreases, the mass transfer area increases, and thus, the extraction rate increases as well. The porosity of the solid material is a parameter in the contact of the matrix surface and the supercritical fluid so that a good transfer rate of mass and heat in the process occurs (Salazar et al. 2019).
7. *Solid material moisture:* the water in the plant matrix can negatively influence the extraction of several substances of low polarity, such as lipids, reducing the process yield. For this reason, plant material drying is recommended for this type of extraction (Salazar et al. 2019).

In Brazil there are several research groups studying supercritical technology, but there are still few industrial applications. Being the biggest obstacle to the supercritical technology use in the industry is the high investment in the extraction unit installation. Table 2.1 shows studies carried out with Amazonian natural resources using supercritical technology in a laboratory-scale extraction method.

Table 2.1 Examples of studies carried out with Amazonian natural resources using supercritical technology as extraction method

Species	Process conditions	Extraction solvents	Compound of interest	Reference
Leaves and stems				
Jambú (*Spilanthes acmella*)	50 °C/250 bar; flow rate, 3.5×10^{-3} kg/min; and extraction time (min), 30 static/180 dynamic	CO_2	Alkylamides (spilanthol) and phenolic compounds	Dias et al. (2012)
Caju (*Anacardium occidentale* L.)	35–55 °C/100–300 bar; flow rate, 3.6×10^{-6} kg/s; and extraction time, 120 min	CO_2 + 5% EtOH	Essential oils, flavonoids, and antioxidant compounds	Leitão et al. (2013)
Copaíba (*Copaifera* sp.)	60 °C/200 bar; flow rate, 8.3×10^{-5} kg/s; and extraction time (min), 30 static/120 dynamic	CO_2	Antioxidant compounds	Botelho et al. (2015)
Pepper-rosmarin (*Lippia sidoides* Cham.)	60 °C/400 bar; flow rate, 2.5 g/min; and extraction time, 360 min	CO_2 + EtOH/H_2O	Thymol and trans-caryophyllene	Garmus et al. (2015)
Juçara (*Euterpe edulis* Mart.)	60 °C/200 bar; flow rate, 2.0×10^{-4} kg/s; and extraction time (min), 7 static/39 dynamic	CO_2 + 10% EtOH/H_2O	Anthocyanins	del Pilar Garcia-Mendoza et al. (2017)
Cipó-pucá (*Cissus sicyoides* L.)	40 °C/400 bar; flow rate, 4.5 g/min; and extraction time (min), 30 static/180 dynamic	CO_2 + 10% EtOH	Phenolic compounds	Salazar et al. (2018)
Jaborandi-manso or "beetle" (*Piper divaricatum*)	35 °C/300 bar; flow rate, 8.85×10^{-5} kg/s; and extraction time (min), 30 static/180 dynamic	CO_2	Terpenoids and phenylpropanoids (methyl eugenol)	de Oliveira et al. (2019a)
Guayusa (*Ilex guayusa* Loes.)	45–75 °C/150–250 bar; flow rate, 0.5 kg/h; and extraction time, 180 min	CO_2 + 7% EtOH	Caffeine, squalene, and α-amyrin	del Pilar Garcia-Mendoza et al. (2017)
Seeds and almonds				
Guaraná (*Paullinia cupana*)	40–70 °C/400 bar; flow rate, 5.7 g/min; and extraction time, 210 min	CO_2 + H_2O	Caffeine	Saldaña et al. (2002)
Brazil nut (*Bertholletia excelsa* H.B.K)	50–60 °C/300 bar; flow rate, 7.04 g/min; and extraction time, 120 min	CO_2	Fatty acids (oleic and linoleic acids)	Santos et al. (2013)
Cupuassu (*Theobroma grandiflorum* Shum.)	50 °C/300–350 bar; flow rate, 11.50×10^{-5} kg/s; and extraction time (min), 15 static/140 dynamic	CO_2	Phenolic compounds, tocopherols, and fatty acids	Cavalcanti et al. (2016)
Cumbaru (*Dipteryx alata* Vogel)	40–50 °C/150–350 bar; flow rate, 1.91×10^{-4} kg/s; and extraction time, 120 min	CO_2	Fatty acids (oleic and linoleic acids)	Dos Santos et al. (2016)

Ucuúba (*Virola surinamensis*)	40–80 °C/350 bar; flow rate, 7.9×10^{-5} kg/s; and extraction time (min), 30 static/180 dynamic	CO_2	Fatty acids (lauric and myristic acid), steroids, terpenes, coumarins, and phenolic compounds	Cordeiro et al. (2018)
Annatto (*Bixa orellana* L.)	40–60 °C/100–200 bar; flow rate, 5.0 g/min; and extraction time, 250–325 min	CO_2	Diterpene (geranylgeraniol) and tocotrienols	Vardanega et al. (2019)
Sapucaia (*Lecythis pisonis* Camb.)	60 °C/300 bar and flow rate, 15 l/min	CO_2	Fatty acids (oleic, linoleic, and palmitic acids)	dos Santos et al. (2019)
Babassu (*Orbignya phalerata*)	70–80 °C/250–350 bar; flow rate, 1.6×10^{-4} kg/s; and extraction time (min), 20 static/120 dynamic	CO_2	Fatty acids (lauric, oleic, and myristic acids) and phytosterol (β-sitosterol)	de Oliveira et al. (2019b)
Fruits				
Pupunha (*Guilielma speciosa*)	45–50 °C/250–300 bar and flow rate, 3×10^{-4} kg/s	CO_2	Fatty acids (oleic, stearic, and palmitic acids)	Araújo et al. (2000)
Jucá (*Libidibia ferrea*)	50 °C/250 bar; flow rate, 3.50×10^{-3} kg/min; and extraction time (min), 30 static/180 dynamic	CO_2	Fatty acids, terpenoids, lupenone, and gamma-sitosterol	Dias et al. (2013)
Tucumã (*Astrocaryum aculeatum* Meyer)	40 °C/300 bar; flow rate, 15 g/min; and extraction time, 180 min	CO_2	Fatty acids (oleic, linoleic, and palmitic acids) and carotenes	Costa et al. (2016)
Pequi (*Caryocar brasiliense*)	60 °C/400 bar and flow rate, 2.93×10^{-4} kg/s	CO_2	Fatty acids (oleic and palmitic acids)	Johner et al. (2018)
Bacaba (*Oenocarpus bacaba*)	60 °C/420 bar; flow rate, 8.85×10^{-5} kg/s; and extraction time (min), 30 static/180 dynamic	CO_2	Fatty acids (oleic acids)	Pinto et al. (2018)
Bacaba-de-leque (*Oenocarpus distichus* Mart.)	60 °C/270 bar; flow rate, 5.31 g/min; and extraction time (min), 30 static/180 dynamic	CO_2	Fatty acids (oleic, linoleic, and palmitic acids) and antioxidant compounds	Cunha et al. (2019)
Açaí (*Euterpe oleracea* Mart.)	70 °C/490 bar; flow rate, 8.85×10^{-5} kg/s; and extraction time (min), 30 static/180 dynamic	CO_2	Fatty acids (oleic, linoleic, palmitic, and palmitoleic acids)	Silva et al. (2019)
Murici (*Byrsonima crassifolia*)	50–60 °C/150–420 bar; flow rate, 8.85×10^{-5} kg/s; and extraction time (min), 30 static/60–180 dynamic	CO_2 + 10% EtOH	Lutein, fatty acids (oleic, linoleic, and palmitic acids), and phenolic compounds	Pires et al. (2019)

2.3 Greener Supercritical (Co)Solvents

During the last decades, there has been a greater concern about the adverse effects caused by reagents and solvents used in conventional chemical processes. Therefore, the need to develop and apply ecological natural solvents has now become the discussion focus in various areas of science and technologies of extraction and chemical separation. Solvents considered green can be defined as substances capable of reducing environmental, health, and safety problems, thus causing less impact on the life cycle (Capeletto et al. 2016).

For this reason, there is a continuing interest in modern extraction technologies that apply green solvents in order to minimize the use of organic solvents, as they cause environmental damage and undesirable effects on food and other end products' quality and safety due to their toxicity. In order to be applied, green solvents must have low toxicity properties, be easily biodegradable in the environment without harmful effects, have natural origin, be produced from renewable sources, have low vapor pressure, and not require a step evaporation at the end of the process (Pena-Pereira and Tobiszewski 2017; Herrero and Ibañez 2018).

In high-pressure extraction processes such as supercritical fluid extraction, carbon dioxide (CO_2), water (H_2O), and ethanol (EtOH) are the most commonly used "green" solvents as they are ecologically applicable and have suitable transport properties for the recovery of compounds of interest. They are applied at temperatures and pressures above their critical point. When a substance is above its critical pressure (P_c) and approaching its critical temperature (T_c), there is a decrease in the interface between the liquid and gaseous phases, thus converging to a single phase—*the supercritical fluid*. Solvents, with temperature and pressure above their critical point, are characterized by their physical properties between gas and liquid, that is, their viscosity resembles the gas viscosity and its density is similar to liquid density. These characteristics allow a rapid mass transfer in the separation processes (Machida et al. 2011; Knez 2016; del Pilar Sánchez-Camargo et al. 2019).

Carbon dioxide is the most used solvent in supercritical processes due to the advantages related to its properties such as being non-toxic, non-flammable, low cost, and ecologically easy to recycle; having low critical point, $T_c = 304.15$ K and $P_c = 73.7$ bar, thus being able to reach the supercritical state quickly, and high purity; and providing products exempt from solvents, since it is a gaseous solvent at ambient temperature and pressure. Furthermore, in the supercritical state, small variations of pressure and temperature cause significant changes in its density and therefore its properties. CO_2 also offers advantages over organic solvents: high diffusivity, low viscosity, and low surface tension (York et al. 2004; del Pilar Sánchez-Camargo et al. 2019).

One of the main applications of CO_2 in the supercritical state is in low-polarity compound extraction, which is also its limitation. However, the supercritical solvent polarity can be altered with a cosolvent addition such as water and ethanol, which increases the solubility of specific compounds and may also create specific interactions with the solute or active sites of the solid matrix (Bhusnure et al. 2015; Shimizu and Abbott 2016). There are also other applications for SC-CO_2, such as supercritical antisolvent (SAS), in the processes of rapid expansion of supercritical solutions (RESS), supercritical fluid extraction of emulsions (SFEE), particles from gas-saturated suspensions (PGSS), and polymer foam drying and forming (Soh and Lee 2019).

Ethanol is widely used as cosolvent in natural product extraction with SC-CO_2, since it is a non-toxic, renewable solvent and has high CO_2 miscibility and temperature and pressure critical points at 516.25 K and 63.79 bar, respectively (Danh et al. 2010; Song et al. 2010). Studies report (Danh et al. 2010; Sánchez-Camargo et al. 2012; Solana et al. 2014; Mezzomo and Ferreira 2016; Salazar et al. 2018; Pires et al. 2019) that the addition of ethanol as cosolvent at low concentrations (1–15% w:w) increases the yield of high-polarity compounds due to the increased solubility of these compounds in SC-CO_2 caused by hydrogen bonding formation, dipole-dipole interactions, and other interactions with organic acids, availability, and stability for ethanol hydrogenation, as well as the raw material pore expansion (Fan et al. 2010; Sánchez-Camargo et al. 2012; Mezzomo and Ferreira 2016).

Another green solvent is supercritical or subcritical water that has applications in industry processes. As well as CO_2, water also has the advantages of being non-toxic, non-flammable, environmentally clean, abundant, easily accessible, and low cost. It becomes supercritical under conditions of temperature and pressure above its critical point, $T_c = 647.15$ K and $P_c = 221$ bar, from 373.15 K up to its critical temperature, which is called subcritical, almost critical, or pressurized water. In supercritical conditions, water has opposite properties in relation to its liquid state at room temperature, such as low dielectric constant and low polarity. This behavior is very similar to what occurs in an organic solvent, which makes the supercritical water to act as a non-polar solvent with high solubility, being an excellent alternative for use in low-polarity organic compound extraction, and capable to replace organic solvents (Chemat et al. 2019; Zhou et al. 2019).

Another important characteristic of supercritical water is that its density can be continuously changed from higher values, such as liquid density, to lower values, as a gas, with no phase transition, varying only the pressure and the temperature. As the temperature increases, there is a decrease in the permittivity and viscosity and an increase in the diffusion rate. In addition, the increase in hydrogen and hydroxyl ion content owing to increased water dissociation makes the water a strong base and acid simultaneously. These reactions are possible in neutral conditions, which can reduce the waste generation. The most important effect that can be observed with the water property improvement certainly is the weakening of the hydrogen bonds, which leads to a lower dielectric constant (Bubalo et al. 2015; Chemat et al. 2019; Zhou et al. 2019).

Table 2.1 in *topic 2* shows several extraction technologies that apply SC-CO_2 as solvent and added cosolvents, such as EtOH and H_2O, to obtain compounds of interest from Amazonian raw materials. Figure 2.1 is a *P-T* diagram schematic representation to demonstrate the critical properties of CO_2, H_2O, and EtOH.

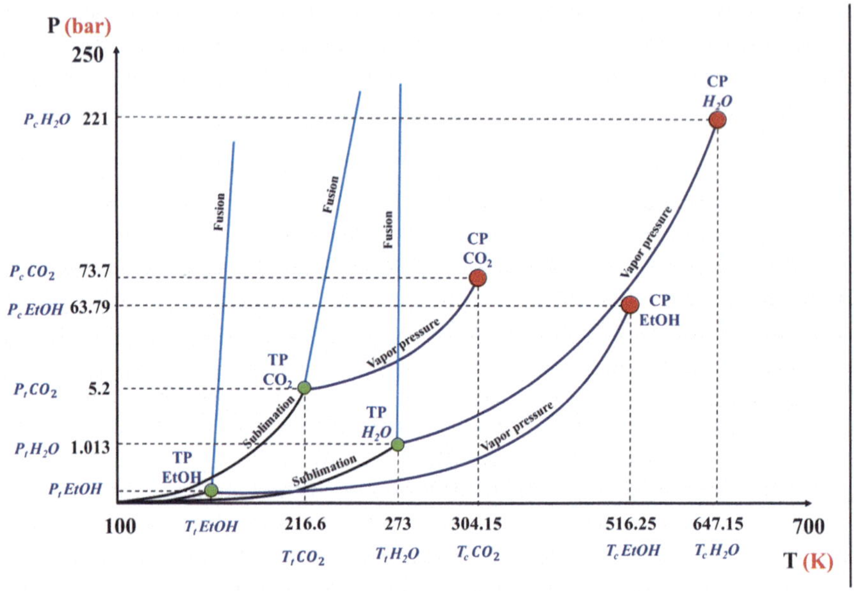

Fig. 2.1 Solid-liquid-gas-supercritical fluid phase diagram. TP = triple point. CP = critical point. P_c = critical pressure. T_c = critical temperature. T_t = triple point temperature. P_t = triple point pressure

2.4 Biological Compounds Extracted

The supercritical technology development for biologically active compound extraction from different plant sources, especially those from the Amazon region, is based on extensive research carried out in the last decades. Many classes of biologically active compounds have been extracted by the SC-CO_2 technology (Ribas et al. 2014; Queiroz and Cajaiba 2016; Pascoal et al. 2017), supercritical and subcritical H_2O (Ravber et al. 2015; Chan et al. 2018; Knez et al. 2018; Munir et al. 2018) around the world. Recent investigations (De Melo et al. 2014; Ibáñez et al. 2016; Silva et al. 2016; Stubbs 2016; Yoon and Lee 2018) have produced significant advances in the understanding of mechanisms, routes, and process parameters, relevant during the extraction of metabolites from plant sources. The understanding of the supercritical process is the key point in the development of new technologies with the use of bioactive compounds, thus defining the directions for industrial application and development (Batista et al. 2016a).

Supercritical technology has been used to extract different forms of biologically active metabolites. The SC-CO_2 technology is probably the most known and studied. This technology was used to extract α-bisabolol from *Eremanthus erythroppapus* (Santos et al. 2017); osthole, isogeijerin, hexadecane, and squalene from *Citrus aurantium* L. var. *amara* (Trabelsi et al. 2016); gallic acid, catechin, and epicatechin from *Copaifera langsdorffii* (Botelho et al. 2015; Batista et al. 2016b); and germacrene D, pipercallosidine, 14-oxy-α-muuroleno, bicyclogermacrene, and (E)-caryophyllene from *Piper klotzschianum* (Lima et al. 2019). These secondary metabolites are important groups of bioactive compounds of some plants from the Amazon region, with importance and relevance in pharmaceutical and food application scenario.

Understanding the extraction process and its influence on the molecules' properties is not yet adequately elucidated. However, recent works (Natolino and da Porto 2019; Santos et al. 2019) suggest that the supercritical fluid extraction does not affect the molecules' chemical structure significantly, which makes the process ideal for metabolites of high added value extraction, since it has several advantages over conventional processes, such as producing extracts with relevant concentrations for industrial applications. Next, some studies on Amazonian plants that have relevant industrial applications will be presented and discussed some works with Amazonian plants that demonstrate relevance in the setting of industrial applications.

Açaí (*Euterpe oleracea* Mart.), one of the most popular fruits in the Amazon, has several nutritional attributes. Recent works (Yamaguchi et al. 2015; Barbosa et al. 2016; Pala et al. 2018; Lee 2019; Machado et al. 2019) suggest that this plant has sufficient chemical potential for the generation of high added value inputs, applicable to the chemical, pharmaceutical, and food industries. According to Batista et al. (2016b), the bioactive compound extraction from açaí using SC-CO_2 is influenced by the solvent temperature and density. The analysis revealed a range of phenolic compounds, mainly anthocyanins in the extraction residual cake, besides monounsaturated fatty acids in the obtained oil. Silva et al. (2019) corroborate the results described above when evaluating the influence of the parameters in the supercritical fluid extraction process of açaí oil. It was observed that 70 °C temperature and 490 bar pressure provide improved oil yield. The fatty acid profile revealed that the açaí oil is rich in oleic acid. In addition, the theoretical results predicted from the Peng–Robinson equations (Peng and Robinson 1976) under the same operating conditions showed that overall yields increased as pressure and temperature increased.

Another genus of Amazonian plant with substantial potential for industrial applications is the *Oenocarpus*. This genus has several genotypes of bacaba-de-leque (*Oenocarpus distichus* Mart.) distributed throughout the Amazon region (Aguiar and Souza 2018). Recent works (Carvalho et al. 2016; Barros et al. 2017; de Sousa et al. 2018) show that this genus is rich in phenolic compounds with antioxidant activity, such as anthocyanins, 3,4-dihydroxybenzoic acid, epicatechin, and ferulic acid. Cunha et al. (2019) explored the extraction of *Oenocarpus distichus* oil by SC-CO_2. Maximum yields were reached between 50–60 °C and 270–350 bar. The results of chemical composition revealed the presence of oleic, palmitic, and linoleic acids, as well as triglycerides and phenolic compounds such as anthocyanins, in the composition of the extraction residual cake, independent of the extraction conditions.

Murici (*Byrsonima crassifolia*) is an Amazonian plant that produces a fruit rich in bioactive compounds of industrial interest. The fruit is consumed *in natura* or processed for sweets and juices formulation and for cosmetics and pharmaceutical applications, due to the presence of important metabolites such as carotenoids, phenolic compounds, and unsaturated fatty acids (da Silva et al. 2016a; Uekane et al. 2017). Pires et al. (2019), using supercritical extraction, with cosolvent (CO_2 + EtOH), found that the increase in the solvent flow rate and the decrease of the dynamic extraction period significantly increased the extraction yield. The extracts presented antioxidant activity and phenolic compounds, as well as lutein.

As observed, Amazonian plants are an important source of biologically active compounds of high added value. These compounds are an important part of cell biology and represent the biochemical evolution that plants have undergone (Rodriguez et al. 2019). These plants have peculiar classes of compounds, which demonstrates the species evolutionary capacity. Thus, their chemical biodiversity represents a wide range of possibilities, including research and applications (Segonzac and Monaghan 2019). The supercritical technology allied with all this biodiversity contributes to the advancement in the knowledge of new bioactive molecules. The process also contributes to the sequential exploration of target molecule groups with potential industrial applications that can lead to a new technological level in the Amazon (Knez et al. 2019).

2.5 Biological Applications

Natural products such as essential oils, extracts, and isolated molecules have the ability to inhibit the growth and survival of microorganisms such as bacteria, fungi, and protozoa (DeCorte 2016; Khan 2018). With the emergence of pharmacotherapy-resistant microorganisms, researchers around the world have sought alternatives for new drug discovery (Newman and Cragg 2009; Bergström and Lindmark 2019). Bioactive compounds of natural origin, mainly vegetal, are also used in several researches in the execution of in vivo test protocols and enzymatic assays (Gu et al. 2013; de Oliveira et al. 2019a). In this way, extracts obtained from Amazonian plants have been used as agents with medicinal, cosmetic, food, and agrochemical activity potential. In this section, we highlight the applicability of bioactive compounds extracted by "green processes" from these matrices.

Capeletto et al. (2016) used SC-CO_2 to obtain *Campomanesia xanthocarpa* Berg extract. The obtained extract presented a great amount of terpenoids, mainly β-caryophyllene. The extract biological activity was evaluated against the *Pseudomonas aeruginosa* gram-negative bacteria. The minimum inhibitory concentration (MIC) and minimum bactericidal concentration (MBC) were, respectively, 5.0 mg/ml and 10.0 mg/ml. While Cordeiro et al. (2018) studied the antimicrobial effects of fixed oils extracted from *Virola surinamensis*, the authors demonstrate that the oils can act as fungi and bacteria inhibitors. This oils' mechanism of action is diverse, and their activity cannot be correlated to a single molecule that integrates the oil or extract. Thus, several mechanisms of action have been proposed in the literature. For example, essential oils which have large amounts of phenolic compounds are able to diffuse into the cell membrane, preventing functional membrane activity and causing cellular fluid leakage. Essential oil compounds may interfere with the adenosine triphosphate (ATP) production process in the membrane and are also capable of altering the mitochondria and the cell membrane permeability, making the structure more permeable to liquids and ions (Saad et al. 2013; Nazzaro et al. 2013; Chouhan et al. 2017).

Botelho et al. (2015) evaluated the *Copaifera* sp. neuroprotective activity, used in traditional Brazilian medicine for inflammation and skin wound treatment, as well as presenting antiseptic and expectorant potential. The leaf extracts were obtained by SC-CO_2 fluid extraction. The authors concluded, from the results of the histological and immunohistochemical analysis, that the extract obtained at 60 °C and 200 bar have anti-inflammatory effects on experimental cerebral ischemia in rats.

de Oliveira et al. (2019a) evaluated the chemical composition of *Piper divaricatum* essential oils' neuroprotective activity, obtained with SC-CO_2, and verified that the different fractions obtained with supercritical fluid had acetylcholinesterase inhibitory activity, indicating that this plant matrix may be an important source of potentially active drugs for improvements in synaptic cleft cholinergic manifestations. Furthermore, essential oils and fixed oils obtained with SC-CO_2 from Amazonian plant matrices have presented activities with potential use in agriculture, since they can be promoters of phytotoxic activities, mainly improving the productive chain, as they can replace some synthetic herbicides harmful to the humans' and animals' health. In the work by Batista et al. (2016b), the açaí fixed oils obtained with SC-CO_2 presented two types of allelopathic effects: one stimulatory for *Senna obtusifolia* seed germination and another phytotoxic for *Mimosa pudica* seeds; in relation to the radicle and hypocotyl stretches, the oils had inhibitory effects for both receptor species.

2.6 Considerations

It can be concluded that the use of supercritical technology is an alternative to conventional extraction methods that use toxic organic solvents, allowing the selective extraction of more pure and standardized compounds by the simple variation of process parameters such as temperature, pressure, density, and solvent flow. It is an important tool to obtain chemically active compounds from Amazonian raw materials for applications in sustainable management areas, improving the quality of food, pharmaceutical, cosmetic, and agrochemical products that apply these compounds, without providing risks to the consumer and to the environment.

Acknowledgments Fernanda Wariss Figueiredo Bezerra thanks the Coordenação de Aperfeiçoamento de Pessoal de Nível Superior—Brazil (CAPES)—Finance Code 001, for the scholarship.

References

Aguiar JPL, Souza FCA (2018) Bacaba-de-leque (*Oenocarpus distichus*): a new wet tropics nutritional source. African J Agric Res 13:803–805. https://doi.org/10.5897/ajar2018.13027

Araújo ME, Machado NT, França LF, Meireles MAA (2000) Supercritical extraction of pupunha (*Guilielma speciosa*) oil in a fixed bed using carbon dioxide. Brazilian J Chem Eng 17:297–306. https://doi.org/10.1590/S0104-66322000000300005

Barbosa PO, Pala D, Silva CT, de Souza MO, do Amaral JF, Vieira RAL, Folly GAF, Volp ACP, de Freitas RN (2016) Açai (*Euterpe oleracea* Mart.) pulp dietary intake improves cellular antioxidant enzymes and biomarkers of serum in healthy women. Nutrition 32:674–680. https://doi.org/10.1016/j.nut.2015.12.030

Barros RGC, Andrade JKS, Denadai M, Nunes ML, Narain N (2017) Evaluation of bioactive compounds potential and antioxidant activity in some Brazilian exotic fruit residues. Food Res Int 102:84–92. https://doi.org/10.1016/j.foodres.2017.09.082

Batista ÂG, Ferrari AS, da Cunha DC, da Silva JK, Cazarin CBB, Correa LC, Prado MA, Carvalho-Silva LB, Esteves EA, Maróstica Júnior MR (2016a) Polyphenols, antioxidants, and antimutagenic effects of Copaifera langsdorffii fruit. Food Chem 197:1153–1159. https://doi.org/10.1016/j.foodchem.2015.11.093

Batista CCR, Oliveira MS, Araújo ME, Rodrigues AMC, Botelho JRS, Filho APSS, Machado NT, Carvalho RN Jr (2016b) Supercritical CO_2 extraction of açaí (*Euterpe oleracea*) berry oil: global yield, fatty acids, allelopathic activities, and determination of phenolic and anthocyanins total compounds in the residual pulp. J Supercrit Fluids 107:364–369. https://doi.org/10.1016/j.supflu.2015.10.006

Bergström F, Lindmark B (2019) Accelerated drug discovery by rapid candidate drug identification. Drug Discov Today 24:1237–1241. https://doi.org/10.1016/J.DRUDIS.2019.03.026

Bhusnure OG, Gholve SB, Giram PS, Borsure VS, Jadhav PP, Satpute VV, Sangshetti JN (2015) Importance of supercritical fluid extraction techniques in pharmaceutical industry: a review. Indo Am J Pharm Res 5:3785–3801

Botelho JRS, Santos AG, Araújo ME, Braga MEM, Gomes-Leal W, Carvalho Junior RN, Meireles MAA, Oliveira MS (2015) Copaíba (*Copaifera* sp.) leaf extracts obtained by CO_2 supercritical fluid extraction: isotherms of global yield, kinetics data, antioxidant activity and neuroprotective effects. J Supercrit Fluids 98:167–171. https://doi.org/10.1016/j.supflu.2014.12.006

Brunner G (2005) Supercritical fluids: technology and application to food processing. J Food Eng 67:21–33. https://doi.org/10.1016/j.jfoodeng.2004.05.060

Bubalo MC, Vidovi S, Redovnikovic IR, Jokic S (2015) Green solvents for green technologies. J Chem Technol Biotechnol 90:1631–1639. https://doi.org/10.1002/jctb.4668

Capeletto C, Conterato G, Scapinello J, Rodrigues FS, Copini MS, Kuhn F, Tres MV, Dal Magro J, Oliveira JV (2016) Chemical composition, antioxidant and antimicrobial activity of guavirova (*Campomanesia xanthocarpa* Berg) seed extracts obtained by supercritical CO_2 and compressed n-butane. J Supercrit Fluids 110:32–38. https://doi.org/10.1016/j.supflu.2015.12.009

Carvalho AV, da Silveira TF, de Sousa SHB, de Moraes MR, Godoy HT (2016) Phenolic composition and antioxidant capacity of bacaba-de-leque (*Oenocarpus distichus* Mart.) genotypes. J Food Compos Anal 54:1–9. https://doi.org/10.1016/j.jfca.2016.09.013

Cavalcanti RN, Albuquerque CLC, Meireles MAA (2016) Supercritical CO_2 extraction of cupuassu butter from defatted seed residue: experimental data, mathematical modeling and cost of manufacturing. Food Bioprod Process 97:48–62. https://doi.org/10.1016/j.fbp.2015.10.004

Chan YH, Quitain AT, Yusup S, Uemura Y, Sasaki M, Kida T (2018) Optimization of hydrothermal liquefaction of palm kernel shell and consideration of supercritical carbon dioxide mediation effect. J Supercrit Fluids 133:640–646. https://doi.org/10.1016/j.supflu.2017.06.007

Chemat F, Abert-Vian M, Fabiano-Tixier AS, Strube J, Uhlenbrock L, Gunjevic V, Cravotto G (2019) Green extraction of natural products. Origins, current status, and future challenges. Trends Anal Chem 118:248–263. https://doi.org/10.1016/j.trac.2019.05.037

Chouhan S, Sharma K, Guleria S, Chouhan S, Sharma K, Guleria S (2017) Antimicrobial activity of some essential oils—present status and future perspectives. Medicines 4:1–21. https://doi.org/10.3390/medicines4030058

Cordeiro RM, Silva APS, Pinto RHH, da Costa WA, da Silva SHM, Pinheiro WBS, Arruda MSP, Carvalho Junior RN (2018) Supercritical CO_2 extraction of ucuúba (*Virola surinamensis*) seed oil: global yield, kinetic data, fatty acid profile, and antimicrobial activities. Chem Eng Commun 206:1–12. https://doi.org/10.1080/00986445.2018.1474741

Costa BET, dos Santos OV, Corrêa NCF, de França LF (2016) Comparative study on the quality of oil extracted from two tucumã varieties using supercritical carbon dioxide. Food Sci Technol 36:322–328. https://doi.org/10.1590/1678-457x.0094

Cunha VMB, da Silva MP, de Sousa SHB, do Nascimento Bezerra P, Menezes EGO, da Silva NJN, da Silva Banna DAD, Araújo ME, de Carvalho Junior RN (2019) Bacaba-de-leque (*Oenocarpus distichus* Mart.) oil extraction using supercritical CO_2 and bioactive compounds determination in the residual pulp. J Supercrit Fluids 144:81–90. https://doi.org/10.1016/j.supflu.2018.10.010

da Costa WA, de Oliveira MS, da Silva MP, Cunha VMB, Pinto RHH, Bezerra FWF, de Carvalho Junior RN (2017) Açaí (*Euterpe oleracea*) and Bacaba (*Oenocarpus bacaba*) as functional food. In: Superfood and functional food-an overview of their processing and utilization. IntechOpen, pp 155–172

da Costa WA, Bezerra FWF, de Oliveira MS, de Aguiar Andrade EH, dos Santos APM, Cunha VMB, dos Santos DCS, da Silva Banna DAD, Teixeira E, de Carvalho Junior RN (2019) Supercritical CO_2 extraction and transesterification of the residual oil from industrial palm kernel cake with supercritical methanol. J Supercrit Fluids 147:179–187. https://doi.org/10.1016/j.supflu.2018.10.012

da Silva PMC, Neves LC, Bastos VJ, Lima CGB, de Araújo KGM, Roberto SR (2016a) Harvesting period of Murici (*Byrsonima crassifolia* Kunth) fruit in relation to physical and chemical parameters evaluated during fruit development. Sci Hortic (Amsterdam) 200:66–72. https://doi.org/10.1016/j.scienta.2015.12.041

da Silva RPFF, Rocha-Santos TAP, Duarte AC (2016b) Supercritical fluid extraction of bioactive compounds. TrAC Trends Anal Chem 76:40–51. https://doi.org/10.1016/j.trac.2015.11.013

Danh LT, Truong P, Mammucari R, Foster N (2010) Extraction of vetiver essential oil by ethanol-modified supercritical carbon dioxide. Chem Eng J 165:26–34. https://doi.org/10.1016/j.cej.2010.08.048

De Melo MMR, Silvestre AJD, Silva CM (2014) Supercritical fluid extraction of vegetable matrices: applications, trends and future perspectives of a convincing green technology. J Supercrit Fluids 92:115–176. https://doi.org/10.1016/j.supflu.2014.04.007

de Oliveira MS, da Costa WA, Bezerra FWF, Araújo ME, Ferreira GC, de Carvalho Junior RN (2018) Phytochemical profile and biological activities of *Momordica charantia* L. (Cucurbitaceae): a review. African J Biotechnol 17:829–846. https://doi.org/10.5897/ajb2017.16374

de Oliveira MS, da Cruz JN, Gomes Silva S, da Costa WA, de Sousa SHB, Bezerra FWF, Teixeira E, da Silva NJN, de Aguiar Andrade EH, de Jesus Chaves Neto AM, de Carvalho RN (2019a) Phytochemical profile, antioxidant activity, inhibition of acetylcholinesterase and interaction mechanism of the major components of the *Piper divaricatum* essential oil obtained by supercritical CO_2. J Supercrit Fluids 145:74–84. https://doi.org/10.1016/j.supflu.2018.12.003

de Oliveira NA, Mazzali MR, Fukumasu H, Gonçalves CB, de Oliveira AL (2019b) Composition and physical properties of babassu seed (*Orbignya phalerata*) oil obtained by supercritical CO_2 extraction. J Supercrit Fluids 150:21–29. https://doi.org/10.1016/j.supflu.2019.04.009

de Sousa SHB, de Andrade MR, Chisté RC, Carvalho AV (2018) Phenolic compounds are highly correlated to the antioxidant capacity of genotypes of *Oenocarpus distichus* Mart. fruits. Food Res Int 108:405–412. https://doi.org/10.1016/j.foodres.2018.03.056

DeCorte BL (2016) Underexplored opportunities for natural products in drug discovery. J Med Chem 59:9295–9304. https://doi.org/10.1021/acs.jmedchem.6b00473

del Pilar Garcia-Mendoza M, Espinosa-Pardo FA, Baseggio AM, Barbero GF, Maróstica Junior MR, Rostagno MA, Martínez J (2017) Extraction of phenolic compounds and anthocyanins from juçara (*Euterpe edulis* Mart.) residues using pressurized liquids and supercritical fluids. J Supercrit Fluids 119:9–16. https://doi.org/10.1016/j.supflu.2016.08.014

del Pilar Sánchez-Camargo A, Bueno M, Parada-Alfonso F, Cifuentes A, Ibáñez E (2019) Hansen solubility parameters for selection of green extraction solvents. Trends Anal Chem 118:227–237. https://doi.org/10.1016/j.trac.2019.05.046

Dias AMA, Santos P, Seabra IJ, Júnior RNC, Braga MEM, De Sousa HC (2012) Spilanthol from *Spilanthes acmella* flowers, leaves and stems obtained by selective supercritical carbon dioxide extraction. J Supercrit Fluids 61:62–70. https://doi.org/10.1016/j.supflu.2011.09.020

Dias AMA, Rey-Rico A, Oliveira RA, Marceneiro S, Alvarez-Lorenzo C, Concheiro A, Júnior RNC, Braga MEM, De Sousa HC (2013) Wound dressings loaded with an anti-inflammatory jucá (*Libidibia ferrea*) extract using supercritical carbon dioxide technology. J Supercrit Fluids 74:34–45. https://doi.org/10.1016/j.supflu.2012.12.007

dos Santos M, Mamede R, Rufino M, de Brito E, Alves R (2015) Amazonian native palm fruits as sources of antioxidant bioactive compounds. Antioxidants 4:591–602. https://doi.org/10.3390/antiox4030591

Dos Santos P, De Aguiar AC, Viganó J, Boeing JS, Visentainer JV, Martínez J (2016) Supercritical CO_2 extraction of cumbaru oil (*Dipteryx alata* Vogel) assisted by ultrasound: global yield, kinetics and fatty acid composition. J Supercrit Fluids 107:75–83. https://doi.org/10.1016/j.supflu.2015.08.018

dos Santos OV, Carvalho RN Jr, da Costa CEF, da Silva Lannes SC (2019) Chemical, chromatographic-functional, thermogravimetric-differential and spectroscopic parameters of the sapucaia oil obtained by different extraction methods. Ind Crop Prod 132:487–496

El Asbahani A, Miladi K, Badri W, Sala M, Addi EHA, Casabianca H, El Mousadik A, Hartmann D, Jilale A, Renaud FNR, Elaissari A (2015) Essential oils: from extraction to encapsulation. Int J Pharm 483:220–243. https://doi.org/10.1016/j.ijpharm.2014.12.069

Fan J, Hou Y, Wu W, Zhang J, Ren S, Chen X (2010) Levulinic acid solubility in supercritical carbon dioxide with and without ethanol as cosolvent at different temperatures. J Chem Eng Data 55:2316–2321. https://doi.org/10.1021/je900727r

Garmus TT, Paviani LC, Queiroga CL, Cabral FA (2015) Extraction of phenolic compounds from pepper-rosmarin (Lippia sidoides Cham.) leaves by sequential extraction in fixed bed extractor using supercritical CO_2, ethanol and water as solvents. J Supercrit Fluids 99:68–75. https://doi.org/10.1016/j.supflu.2015.01.016

Gu J, Gui Y, Chen L, Yuan G, Lu H-Z, Xu X (2013) Use of natural products as chemical library for drug discovery and network pharmacology. PLoS One 8:e62839. https://doi.org/10.1371/journal.pone.0062839

Herrero M, Ibañez E (2018) Green extraction processes, biorefineries and sustainability: recovery of high added-value products from natural sources. J Supercrit Fluids 134:252–259. https://doi.org/10.1016/j.supflu.2017.12.002

Ibáñez E, Mendiola JA, Castro-Puyana M (2016) Supercritical fluid extraction. *In* Caballero B, Finglas P, Toldrá F. Encycl Food Heal 227–233. Academic Press, Oxford. https://doi.org/10.1016/B978-0-12-384947-2.00675-9

Johner JCF, Hatami T, Meireles MAA (2018) Developing a supercritical fluid extraction method assisted by cold pressing for extraction of pequi (*Caryocar brasiliense*). J Supercrit Fluids 137:34–39. https://doi.org/10.1016/j.supflu.2018.03.005

Khan RA (2018) Natural products chemistry: the emerging trends and prospective goals. Saudi Pharm J 26:739–753. https://doi.org/10.1016/J.JSPS.2018.02.015

Knez Ž (2016) Chapter 20 - Food processing using supercritical fluids. In: Emerging and traditional technologies for safe, healthy and quality food. Washington State University, USA, pp 413–442

Knez Ž, Hrnčič MK, Čolnik M, Škerget M (2018) Chemicals and value added compounds from biomass using sub- and supercritical water. J Supercrit Fluids 133:591–602. https://doi.org/10.1016/j.supflu.2017.08.011

Knez Ž, Pantić M, Cör D, Novak Z, Hrnčič MK (2019) Are supercritical fluids solvents for the future? Chem Eng Process Process Intensif 141:107532

Lee J (2019) Anthocyanins of açai products in the United States. NFS J 14–15:14–21. https://doi.org/10.1016/j.nfs.2019.05.001

Leitão NCMCS, Prado GHC, Veggi PC, Meireles MAA, Pereira CG (2013) *Anacardium occidentale* L. leaves extraction via SFE: global yields, extraction kinetics, mathematical modeling and economic evaluation. J Supercrit Fluids 78:114–123. https://doi.org/10.1016/j.supflu.2013.03.024

Lima RN, Ribeiro AS, Cardozo-Filho L, Vedoy D, Alves PB (2019) Extraction from leaves of *Piper klotzschianum* using supercritical carbon dioxide and co-solvents. J Supercrit Fluids 147:205–212. https://doi.org/10.1016/j.supflu.2018.11.006

Machado AK, Cadoná FC, Assmann CE, Andreazza AC, Duarte MMMF, Branco CS, Zhou X, de Souza DV, Ribeiro EE, da Cruz IBM (2019) Açaí (*Euterpe oleracea* Mart.) has anti-inflammatory potential through NLRP3-inflammasome modulation. J Funct Foods 56:364–371. https://doi.org/10.1016/j.jff.2019.03.034

Machida H, Takesue M, Smith RL (2011) Green chemical processes with supercritical fluids: properties, materials, separations and energy. J Supercrit Fluids 60:2–15. https://doi.org/10.1016/j.supflu.2011.04.016

Mezzomo N, Ferreira SRS (2016) Carotenoids functionality, sources, and processing by supercritical technology: a review. J Chem 2016:16. https://doi.org/10.1155/2016/3164312

Munir MT, Kheirkhah H, Baroutian S, Quek SY, Young BR (2018) Subcritical water extraction of bioactive compounds from waste onion skin. J Clean Prod 183:487–494. https://doi.org/10.1016/j.jclepro.2018.02.166

Natolino A, da Porto C (2019) Supercritical carbon dioxide extraction of pomegranate (*Punica granatum* L.) seed oil: kinetic modelling and solubility evaluation. J Supercrit Fluids 151:30–39. https://doi.org/10.1016/j.supflu.2019.05.002

Nazzaro F, Fratianni F, De Martino L, Coppola R, De Feo V, Nazzaro F, Fratianni F, De Martino L, Coppola R, De Feo V (2013) Effect of essential oils on pathogenic bacteria. Pharmaceuticals 6:1451–1474. https://doi.org/10.3390/ph6121451

Newman D, Cragg G (2009) Natural products in medicinal chemistry. Bioorg Med Chem 17:2120–2120. https://doi.org/10.1038/mp.2011.182.doi

Pala D, Barbosa PO, Silva CT, de Souza MO, Freitas FR, Volp ACP, Maranhão RC, Freitas RN (2018) Açai (*Euterpe oleracea* Mart.) dietary intake affects plasma lipids, apolipoproteins, cholesteryl ester transfer to high-density lipoprotein and redox metabolism: a prospective study in women. Clin Nutr 37:618–623. https://doi.org/10.1016/j.clnu.2017.02.001

Pascoal DRC, Cabral-albuquerque ECM, Velozo ES, Sousa HC, Vieira de Melo SAB, Braga MEM (2017) Copaiba oil-loaded commercial wound dressings using supercritical CO_2: a potential alternative topical antileishmanial treatment. J Supercrit Fluids 129:106–115. https://doi.org/10.1016/j.supflu.2017.02.012

Pena-Pereira F, Tobiszewski M (2017) The application of green solvents in separation processes. Elsevier, Amsterdam

Peng D, Robinson DB (1976) A new two-constant equation of state. Ind Eng Chem Fundam 15:59–64

Picq M (2016) Rethinking IR from the Amazon. Rev Bras Política Int 59:e003. https://doi.org/10.1590/0034-7329201600203

Pinto RHH, Sena C, Santos OV, Da Costa WA, Rodrigues AMC, Carvalho Junior RN (2018) Extraction of bacaba (*Oenocarpus bacaba*) oil with supercritical CO_2: global yield isotherms, fatty acid composition, functional quality, oxidative stability, spectroscopic profile and antioxidant activity. Grasas Aceites 69:246. https://doi.org/10.3989/gya.0883171

Pires FCS, Silva APS, Salazar MAR, Costa WA, Costa HSC, Lopes AS, Rogez H, Carvalho Junior RN (2019) Determination of process parameters and bioactive properties of the murici pulp (*Byrsonima crassifolia*) extracts obtained by supercritical extraction. J Supercrit Fluids 146:128–135. https://doi.org/10.1016/j.supflu.2019.01.014

Queiroz A, Cajaiba J (2016) A sustainable process for (−)-α-bisabolol extraction from *Eremanthus erythropappus* using supercritical CO_2 and ethanol as co-solvent. J Supercrit Fluids 110:39–46. https://doi.org/10.1016/j.supflu.2015.12.011

Ravber M, Knez Ž, Škerget M (2015) Simultaneous extraction of oil- and water-soluble phase from sunflower seeds with subcritical water. Food Chem 166:316–323. https://doi.org/10.1016/j.foodchem.2014.06.025

Ribas MC, Mantovani D, Awadallak JA, Canevesi RL, Tazinafo NM, Cardozo Filho L, Palú F, da Silva EA (2014) Study of candeia oil extraction using pressurized fluids and purification by adsorption process. J Supercrit Fluids 92:177–182. https://doi.org/10.1016/j.supflu.2014.04.009

Rodriguez PA, Rothballer M, Chowdhury SP, Nussbaumer T, Gutjahr C, Falter-Braun P (2019) Systems biology of plant-microbiome interactions. Mol Plant 12:804–821. https://doi.org/10.1016/j.molp.2019.05.006

Saad NY, Muller CD, Lobstein A (2013) Major bioactivities and mechanism of action of essential oils and their components. Flavour Fragr J 28:269–279. https://doi.org/10.1002/ffj.3165

Salazar MAR, Costa JV, Urbina GRO, Cunha VMB, Silva MP, Bezerra PN, Pinheiro WBS, Gomes-Leal W, Lopes AS, Carvalho Junior RN (2018) Chemical composition, antioxidant activity, neuroprotective and anti-inflammatory effects of cipó-pucá (*Cissus sicyoides* L.) extracts obtained from supercritical extraction. J Supercrit Fluids 138:36–45. https://doi.org/10.1016/j.supflu.2018.03.022

Salazar MDLAR, Urbina GRO, Bezerra PN, Cunha VMB, da Silva MP, Pires FCS, Silva AOS, Sousa SHB, de Carvalho Jr RN (2019) Antioxidant and biological activity of *Cissus sicyoides* and *Rosmarinus officinalis* extracts. In: Shalaby EA (ed) Antioxidants. IntechOpen, p 20

Saldaña MDA, Zetzl C, Mohamed RS, Brunner G (2002) Extraction of methylxanthines from guaraná seeds, maté leaves, and cocoa beans using supercritical carbon dioxide and ethanol. J Agric Food Chem 50:4820–4826. https://doi.org/10.1021/jf020128v

Sánchez-Camargo AP, Meireles MÂA, Ferreira ALK, Saito E, Cabral FA (2012) Extraction of ω-3 fatty acids and astaxanthin from Brazilian redspotted shrimp waste using supercritical CO_2 + ethanol mixtures. J Supercrit Fluids 61:71–77. https://doi.org/10.1016/j.supflu.2011.09.017

Santos OV, Corrêa NCF, Carvalho RN, Costa CEF, Lannes SCS (2013) Yield, nutritional quality, and thermal-oxidative stability of Brazil nut oil (*Bertolletia excelsa* H.B.K) obtained by supercritical extraction. J Food Eng 117:499–504. https://doi.org/10.1016/j.jfoodeng.2013.01.013

Santos KA, Frohlich PC, Hoscheid J, Tiuman TS, Gonçalves JE, Cardozo-Filho L, da Silva EA (2017) Candeia (*Eremanthus erythroppapus*) oil extraction using supercritical CO_2 with ethanol and ethyl acetate cosolvents. J Supercrit Fluids 128:323–330. https://doi.org/10.1016/j.supflu.2017.03.029

Santos KA, Klein EJ, da Silva C, da Silva EA, Cardozo-Filho L (2019) Extraction of vetiver (*Chrysopogon zizanioides*) root oil by supercritical CO_2, pressurized-liquid, and ultrasound-assisted methods and modeling of supercritical extraction kinetics. J Supercrit Fluids 150:30–39. https://doi.org/10.1016/j.supflu.2019.04.005

Segonzac C, Monaghan J (2019) Modulation of plant innate immune signaling by small peptides. Curr Opin Plant Biol 51:22–28. https://doi.org/10.1016/j.pbi.2019.03.007

Shimizu S, Abbott S (2016) How entrainers enhance solubility in supercritical carbon dioxide. J Phys Chem B 120:3713–3723. https://doi.org/10.1021/acs.jpcb.6b01380

Silva RPFF, Rocha-Santos TAP, Duarte AC (2016) Supercritical fluid extraction of bioactive compounds. Trends Anal Chem 76:40–51. https://doi.org/10.1016/j.trac.2015.11.013

Silva MP, Cunha VMB, Sousa SHB, Menezes EGO, Bezerra PN, De Farias Neto JT, Filho GNR, Araújo ME, De Carvalho RN (2019) Supercritical CO_2 extraction of lyophilized açaí (*Euterpe oleracea* Mart.) pulp oil from three municipalities in the state of Pará, Brazil. J CO2 Util 31:226–234. https://doi.org/10.1016/j.jcou.2019.03.019

Soh SH, Lee L (2019) Microencapsulation and nanoencapsulation using supercritical fluid (SCF) techniques. Pharmaceutics 11:1–17. https://doi.org/10.3390/pharmaceutics11010021

Solana M, Boschiero I, Dall'Acqua S, Bertucco A (2014) Extraction of bioactive enriched fractions from Eruca sativa leaves by supercritical CO_2 technology using different co-solvents. J Supercrit Fluids 94:245–251. https://doi.org/10.1016/j.supflu.2014.08.022

Song Q, Zhu J, Wan J, Cao X (2010) Measurement and modeling of epigallocatechin gallate solubility in supercritical carbon dioxide fluid with ethanol cosolvent. J Chem Eng Data 55:3946–3951. https://doi.org/10.1021/je901025f

Stubbs JM (2016) Molecular simulations of supercritical fluid systems. J Supercrit Fluids 108:104–122. https://doi.org/10.1016/j.supflu.2015.10.027

Trabelsi D, Aydi A, Zibetti AW, Della Porta G, Scognamiglio M, Cricchio V, Langa E, Abderrabba M, Mainar AM (2016) Supercritical extraction from *Citrus aurantium amara* peels using CO_2 with ethanol as co-solvent. J Supercrit Fluids 117:33–39. https://doi.org/10.1016/j.supflu.2016.07.003

Uekane TM, Nicolotti L, Griglione A, Bizzo HR, Rubiolo P, Bicchi C, Rocha-Leão MHM, Rezende CM (2017) Studies on the volatile fraction composition of three native Amazonian-Brazilian fruits: Murici (*Byrsonima crassifolia* L., Malpighiaceae), bacuri (*Platonia insignis* M., Clusiaceae), and sapodilla (*Manilkara sapota* L., Sapotaceae). Food Chem 219:13–22. https://doi.org/10.1016/j.foodchem.2016.09.098

Vardanega R, Nogueira GC, Nascimento CDO, Faria-Machado AF, Meireles MAA (2019) Selective extraction of bioactive compounds from annatto seeds by sequential supercritical CO_2 process. J Supercrit Fluids 150:122–127. https://doi.org/10.1016/j.supflu.2019.01.013

Virgolin LB, Seixas FRF, Janzantti NS (2017) Composition, content of bioactive compounds, and antioxidant activity of fruit pulps from the Brazilian Amazon biome. Pesqui Agropecu Bras 52:933–941. https://doi.org/10.1590/S0100-204X2017001000013

Yamaguchi KKL, Pereira LFR, Lamarão CV, Lima ES, Da Veiga-Junior VF (2015) Amazon acai: chemistry and biological activities: a review. Food Chem 179:137–151. https://doi.org/10.1016/j.foodchem.2015.01.055

Yoon TJ, Lee Y-W (2018) Current theoretical opinions and perspectives on the fundamental description of supercritical fluids. J Supercrit Fluids 134:21–27. https://doi.org/10.1016/j.supflu.2017.11.022

York P, Kompella UB, Shekunov BY (2004) Supercritical fluid technology for drug product development, Drugs and the pharmaceutical sciences. CRC Press, New York

Yousefi M, Rahimi-Nasrabadi M, Pourmortazavi SM, Wysokowski M, Jesionowski T, Ehrlich H, Mirsadeghi S (2019) Supercritical fluid extraction of essential oils. Trends Anal Chem 118:182–193. https://doi.org/10.1016/j.trac.2019.05.038

Zhou F, Hearne Z, Li C-J (2019) Water – the greenest solvent overall. Curr Opin Green Sustain Chem 18:118–123. https://doi.org/10.1016/j.cogsc.2019.05.004

Chapter 3
Non-catalytic and Catalytic Supercritical Water Oxidation of Phenol in the Wastewaters of Petroleum and Other Industries

Shahryar Jafarinejad and Nader Vahdat

Contents

Abbreviations

AOPs	Advanced oxidation processes
API	American Petroleum Institute
CSCWO	Catalytic supercritical water oxidation
CWAO	Catalytic wet air oxidation
DDT	Dichlorodiphenyltrichloroethane

S. Jafarinejad (✉) · N. Vahdat
Department of Chemical Engineering, College of Engineering, Tuskegee University, Tuskegee, AL, USA
e-mail: sjafarinejad@tuskegee.edu

Inamuddin, A. M. Asiri (eds.), *Advanced Nanotechnology and Application of Supercritical Fluids*, Nanotechnology in the Life Sciences,
https://doi.org/10.1007/978-3-030-44984-1_3

LHHW	Langmuir-Hinshelwood-Hougen-Watson
MODEC	Modell Environmental Corp.
MW	Microwave
PCBs	Polychlorinated biphenyls
PFD	Process flow diagram
SCWO	Supercritical water oxidation
US	Ultrasound
VUV	Vacuum ultraviolet
WAO	Wet air oxidation
ZVI	Zerovalent iron

3.1 Introduction

The petroleum industry processes and activities can generate large amounts of aqueous waste containing different contaminants that must be treated before discharging. Discharging of these wastewaters without treatment not only contaminates/destructs the environment and endangers water resources and human health but also decreases the reuse capability of oil and water (Jafarinejad 2014a, b, 2015a, b, c, d, e, 2017a, d, e, f, 2019; Jafarinejad et al. 2017, 2018, 2019; Yu et al. 2017; Zhong et al. 2003). In other words, compliance with the regulations and/or standards, the environmental protection, and water reuse issues are driving forces to treat wastewater before discharge (Ghorbanian et al. 2014; Jafarinejad 2017e, 2019).

Primary treatment (e.g., American Petroleum Institute (API) separators), secondary treatment (e.g., suspended- and/or attached-growth biological processes), and tertiary treatment are various steps and/or methods that have been used to manage/treat petroleum industry wastewater (IPIECA 2010; European Commission and Joint Research Center 2013; Jafarinejad 2017a, c, d, 2019). Compliance with discharge limits/standards and/or water reuse issues are tertiary treatment purposes (U.S. EPA 1995; Goldblatt et al. 2014; Jafarinejad 2017a). Sand and membrane filtration processes, ion exchange, chemical oxidation, advanced oxidation processes (AOPs), etc. are different technologies that can be used as tertiary treatment (Jafarinejad 2017a).

AOPs are green physicochemical processes that produce the reactive hydroxyl radicals ($HO^\bullet$) for destruction of organic matters in a medium. In other words, AOPs can totally mineralize recalcitrant organic pollutants and convert them into inorganic substances (CO_2 and H_2O) or can partially mineralize these compounds and convert them into more biodegradable substances (Covinich et al. 2014; Jafarinejad 2017b). According to the number of phases involved and light usage in these processes, the AOPs are classified into homogeneous/heterogeneous and non-photochemical/photochemical processes, respectively (Mota et al. 2008; Litter and Quici 2010; Santos et al. 2011; Loures et al. 2013; Deng and Zhao 2015; Jafarinejad 2015e, 2017a, b). Different AOPs are demonstrated in Fig. 3.1 (Jafarinejad 2017b). In recent years, AOPs have extensively been used for the treatment of industrial and municipal aqueous wastes.

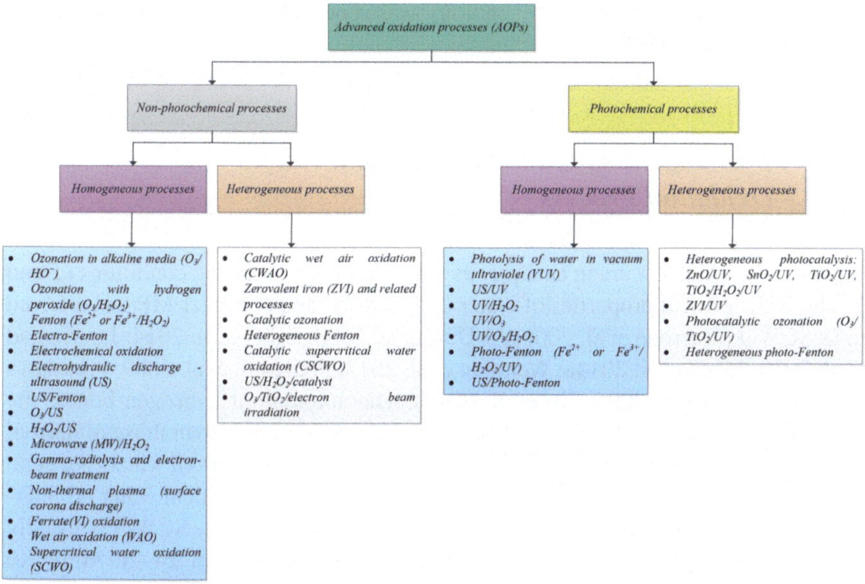

Fig. 3.1 Advanced oxidation processes (AOPs) (Modified after Jafarinejad 2017b)

Phenol is one of the main toxic pollutants in wastewater streams from the petroleum (De Silva et al. 2017) and other industries. It is usually very difficult to decompose phenol because of its resistance to biodegradation (Pruden and Le 1976; Yanan 2012). Direct biological treatment is not feasible for wastewaters containing more than 200 mg/L phenol (Singh et al. 2004; Yanan 2012). Thus, it is essential to discover an efficient treatment technique to degrade phenol in wastewater to comply with standards. In this chapter, historical background, fundamentals, classification, reactor designs, industrial applications, and commercialization of supercritical water oxidation (SCWO) technology are discussed. In addition, reaction pathways and kinetics for phenol oxidation in supercritical water and non-catalytic and catalytic SCWO of phenol in petroleum industry wastewater and other wastewaters are reviewed.

3.2 Supercritical Water Oxidation Process

3.2.1 Historical Background

SCWO is an attractive green process for the destruction of organic matters in wastewaters which was proposed and patented by Modell (1982). Since then, it has been applied efficiently and extensively over various toxic and hazardous industrial wastes. Several researchers have reviewed these applications (Marrone et al. 2004; Xu et al. 2012, Guolin et al. 2012; Yu et al. 2017; Zhang et al. 2017). Note that the first commercialization of this technology was reported in 1994 (McBrayer 1995; Zhang and Savage 1998).

3.2.2 *Fundamentals of the Supercritical Water Oxidation Process*

The process which occurs in an aqueous medium under operating conditions (T and P) above the critical properties of water (T_c = 374 °C and P_c = 22.1 MPa) is referred to as SCWO (Marrone et al. 2004; Al-Duri et al. 2008; Sabet et al. 2014; Jafarinejad et al. 2010; Jafarinejad 2014a; Wenbing et al. 2013; Fourcault et al. 2009; Paraskeva and Diamadopoulos 2006, Xu et al. 2012). The amounts of hydrogen bonds, the dielectric constant, and viscosity of supercritical water are less than those of normal water, whereas its diffusion coefficient is higher than that of normal water (Xu et al. 2012). In reality, these properties and conditions make supercritical water an interesting and desired reaction medium to destruct organic matters. SCWO is usually conducted under temperatures of 400–650 °C and pressures of 24–28 MPa (Xu et al. 2012; Sabet et al. 2014; Jafarinejad 2014a). According to Tester et al. (1993), complete degradation of organic compounds with reactor residence times of 1 min or less can be achievable at temperatures of 550–650 °C and pressure of 250 bar for commercial systems (Tester et al. 1993; Marrone et al. 2004).

Treatment of organic wastes is the main application of SCWO process (Yu et al. 2017; Jafarinejad 2014a). A single phase is formed by completely dissolving organic compounds and oxidants (such as air, or pure O_2, or H_2O_2) into supercritical water, which is a desired reaction medium with outstanding transport properties and accelerated the reaction rate (Conelly 1996; Al-Duri et al. 2008; Xu et al. 2012). This ideal reaction medium of the SCWO process can result in rapid and complete destruction of organic wastes into environmentally acceptable matters such as CO_2, N_2, H_2O, mineral acids, and inorganic salts without undesirable products such as NO_x and SO_x (Marrone et al. 2004; Son et al. 2008; Xu et al. 2012; Zhang et al. 2017). Therefore, this process is a green, non-pollution, environmental benign technology for treating toxic and hazardous organic wastes. Higher than 99% removal efficiency of organic wastes has been reported to attain with reactor residence times less than 1 min (Tester and Cline 1999; Wenbing et al. 2013; Yu et al. 2017). Due to its short residence time, a small SCWO plant can attain large-scale treatment capacity (Gidner and Stenmark 2001; Xu et al. 2012).

A common SCWO system may include feed preparation and pressurization unit, preheater (optional), reactor, alternate solid elimination unit, heat exchanger, pressure letdown, gas/liquid separator, and solid separation unit (optional) (Marrone et al. 2004). The schematic process flow diagram (PFD) of a SCWO system (MODAR type) is demonstrated in Fig. 3.2. The feed mixture consisting of the organic matters is pressurized and preheated before entering the SCWO reactor. The pressurized oxidant is also injected and mixed with the feed. In SCWO reactor, a single phase is formed by completely dissolving organic compounds and oxidants into supercritical

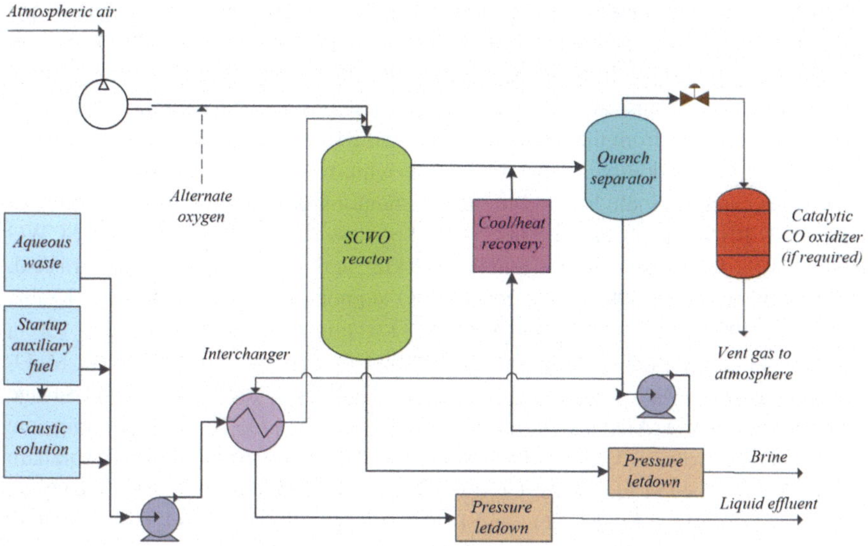

Fig. 3.2 The typical process flow diagram (PFD) of a SCWO system (MODAR type) (Modified after Barner et al. 1992, National Research Council 1993; Jafarinejad 2017a)

water. This single phase is a desired reaction medium so that the reaction takes place with no interfacial mass transfer resistances (Barner et al. 1992; Marrone et al. 2004). Inorganic salts are precipitated and removed in the form of a brine solution from the bottom of the SCWO reactor after pressure letdown. Gaseous products primarily CO_2 and H_2O and small amount of dissolved salts are emitted from the top of the reactor. The pollutants such as NO_x and SO_x are not formed, and the process has the privilege of in situ acid gas neutralization (Barner et al. 1992). Cooling gaseous product stream and recovering some of the system heat can be done using heat transfer unit. Gas/liquid separation is performed using quench separator vessel by gravity. A fixed bed catalytic CO oxidizer may be used to treat the gaseous stream before discharging to the environment, if needed (Barner et al. 1992; National Research Council 1993).

3.2.3 Classification of the Process

Based on the catalyst usage, this technology can be categorized into non-catalytic supercritical water oxidation (SCWO) and catalytic supercritical water oxidation (CSCWO) (Guolin et al. 2012; Jafarinejad 2015e, 2017a, b; De Silva et al. 2017).

Selection of the suitable catalyst and effective application of the catalyst not only can decrease the reactor temperature, the energy use of the system, the residence time of the reactor, and the oxidant concentration but also can enhance destruction rate of recalcitrant organic pollutants, leading to reducing the cost (Guolin et al. 2012; Jafarinejad 2017b; De Silva et al. 2017). Also, catalysts can be useful for better selectivity of competing reaction pathways (Ding et al. 1996; Youngprasert et al. 2010; Jafarinejad 2017b). Homogeneous and heterogeneous catalysts are used in the CSCWO technology (Jafarinejad, 2017b). It should be noted that the homogeneous catalysts can increase operational costs because an additional separation step is necessary for the catalyst ion removal or recovery from the treated effluent owing to their toxicity. For the heterogeneous catalysts, no additional separation unit is required (Guolin et al. 2012).

Various catalysts including noble metals, metal oxides, and transition metal salts (Han et al. 2015) have been used in CSCWO process. The catalysts used in CSCWO include $H_4Si_{12}O_{40}$ (Arslan-Alaton and Ferry 2002); activated carbon (Matsumura et al. 2002; Nunoura et al. 2003a); CuO/ZnO/CoO supported on porous cement (Krajnc and Levec 1997); MnO_2/CeO_2 and V_2O_5/Al_2O_3 (Ding et al. 1995a); Carulite 150, a commercial catalyst (Zhang and Savage 1998); TiO_2 (Yu and Savage 2000a); MnO_2 (Yu and Savage 1999; Youngprasert et al. 2010); CuO and CuO/Al_2O_3 (Yu and Savage 2000b); MnO_2/CuO (Angeles-Herna'ndez et al. 2008); Mn_2O_3/Ti-Al oxide composite catalyst (Dong et al. 2015); Ni/Al_2O_3 (Civan et al. 2015); Pt/γ-Al_2O_3 (Aki and Abraham 1998); NaOH (Lee et al. 2002); $CuSO_4$ (Han et al. 2015; Qi et al. 2002); $CuNO_3$, $FeCl_3$, $(Fe)_2(SO4)_3$, V_2O_5, $MnSO_4$, and $FeSO_4$ (Qi et al. 2002); $ZnSO_4$ (Kazemi et al. 2015); TiO_2-supported MnO_2 catalyst (De Silva et al. 2017); etc.

3.2.4 Reactors

Batch and continuous reactors can be utilized for SCWO process; however, continuous reactors are preferred for industrial applications (Zhang et al. 2017). The tubular reactor (Barner et al. 1992; Bermejo and Cocero 2006) and/or reversible flow tubular reactor (Whiting 1996, 1997; Marrone et al. 2004), reverse flow tank reactor with brine pool (MODAR reactor) (Hong et al. 1989; Marrone et al. 2004), transpiring wall reactor (Daman 1996, 1998; Marrone et al. 2004), reactors with adsorption/reaction on fluidized solid phase (Ross et al. 1995; Ross and Jayaweera 1998; Marrone et al. 2004), fixed bed flow reactor (Zhang and Savage 1998; Guolin et al. 2012), centrifuge reactor (Reid and Halff 1995; Marrone et al. 2004), lined reactor (Hazelbeck et al. 1995; Gloyna and Li 1995), double-wall reactor (Fauvel et al. 2003; Zhang et al. 2017), cool wall reactor (Bermejo et al. 2009; Zhang et al. 2017), etc. have been reported for the application in SCWO process.

3.2.5 Applications of the Supercritical Water Oxidation Process

SCWO process is utilized to manage and treat municipal and industrial wastes. SCWO process has been studied to treat wastewater from textile industry; wastewater from terephthalic acids; wastes from food industry; alcohol distillery wastewater; excess sludge from municipal wastewater treatment plant; sewage sludge; pond sludge (Brunner 2009; Xu et al. 2012); wastewater from acrylic acid production units (Gong et al., 2009; Xu et al. 2012); wastewater from dye industry (Gong et al. 2008; Xu et al. 2012); polychlorinated biphenyls (PCBs) (Kawasaki et al. 2006; Marulanda and Bolanos 2010; Wei et al. 2007; Xu et al. 2012); chlorinated wastes (Schroer et al. 2006; Xu et al. 2012); oily wastes (Portela et al. 2001a, b; Xu et al. 2012); oil/gas produced water (De Silva et al. 2017); wastes from pharmaceutical and biopharmaceutical industry; wastes such as pesticide dichlorodiphenyltrichloroethane (DDT), bacteria, dioxins, phenol, chlorophenol, chlorobenzene, chlorocarbon γ-hexachlorocyclohexane, hexachlorocyclohexane, chemical weapons stockpiles, smokes, pyrotechnics, VX [O-ethyl S-(2-diisopropylamino)ethyl methylphospho-nohioate], GB [sarin, isopropylmethylphosphonofluoridate] (Wagner et al. 2001; Xu et al. 2012), and solid rocket propellants; wastes from military industry; etc. (Xu et al. 2012).

3.2.6 Commercialization of the Supercritical Water Oxidation Process

The world's first commercialization of SCWO technology to treat industrial wastewater was reported in 1994 (McBrayer 1995; Svensson 1995; Ding et al. 1996; Zhang and Savage 1998). According to Marrone et al. (2004) and Xu et al. (2012), General Atomics; MODAR; Modell Environmental Corp. (MODEC); Oxidyne Corp.; Hanwha Chemical; Foster-Wheeler Development Corp.; EcoWaste Technologies; Abitibi-Price, Inc.; SRI International; Hydro-Processing L.L.C.; KemShredder, Ltd.; Chematur Engineering AB; Supercritical Fluids International; etc. are companies that have attempted to commercialize SCWO systems (Xu et al. 2012; Marrone et al. 2004). There are commercial-scale supercritical water oxidation plants in the United States, the United Kingdom, Japan, China, Sweden, etc. (Marrone et al. 2004; Xu et al. 2012; Yu et al. 2017).

Corrosion, salt deposition and plugging, high energy need, and high operating costs are the main obstacles that inhibit the industrialization of SCWO process and even inactivate some commercial-scale plants. These issues should be resolved before any large-scale commercialization is possible (Marrone et al. 2004; Bermejo and Cocero 2006; Vadillo et al. 2011; Xu et al. 2012, Abelleira et al. 2013; Zhang et al. 2017).

3.3 Application of Supercritical Water Oxidation Technology for Phenol Removal from Wastewater

Phenol is usually utilized as a raw material to produce herbicides, drugs, paints, cosmetics, lubricants, resins, etc. It is necessary to remove phenol from the waste streams of various industries such as petroleum (refineries and petrochemical plants), coke operations, coal processing, fiberglass production, textiles, pharmaceuticals, plastics, wood production, paint, pulp and paper, etc. (Mohammadi et al. 2015). Mohammadi et al. (2015) reviewed traditional and emerging processes used for phenol removal from various wastewaters (Mohammadi et al. 2015). In addition, Ding et al. (1996) reviewed catalytic oxidation in supercritical water (Ding et al. 1996). Furthermore, Guolin et al. (2012) reviewed heterogeneous CSCWO of recalcitrant organic contaminants such as phenolics, carboxylic acids, and N-containing compounds in industrial wastewaters (Guolin et al. 2012). In this section, reaction pathways and kinetics to oxidize phenol in supercritical water and non-catalytic and catalytic SCWO of phenol in petroleum industry wastewater and other wastewaters are discussed.

3.3.1 *Reaction Pathways and Kinetics*

Several studies have reported that catalytic and/or non-catalytic SCWO of phenol follow a complex pathway including parallel and consecutive reactions (Thornton and Savage 1992a; Ding et al. 1995a; Gopalan and Savage 1995; Guolin et al. 2012). According to Guolin et al. (2012), mechanisms of SCWO may usually be similar to those for CSCWO (Guolin et al. 2012). Several researchers have investigated mechanisms for the oxidation of phenol (Devlin and Harris 1984; Thornton and Savage 1992a; Ding et al. 1995a; Krajnc and Levec 1997; Guolin et al. 2012). Reaction pathways for phenol oxidation in supercritical water are demonstrated in Fig. 3.3. The following general reaction describes the complete oxidation of phenol (Fourcault et al. 2009):

$$C_6H_6O + 7O_2 \rightarrow 3H_2O + 6CO_2 \tag{3.1}$$

Several researchers have investigated the kinetics of phenol destruction, and many of these researches (Thornton and Savage 1992b; Li et al. 1994; Gopalan and Savage 1995; Krajnc and Levec 1996; Koo et al. 1997; Oshima et al. 1998; Portela et al. 2001b; Pérez et al. 2004) have expressed the following form for the rate of phenol oxidation (r_{Phenol}):

$$-r_{\mathrm{Phenol}} = -\frac{d[\mathrm{Phenol}]}{dt} = A\exp\left(\frac{-E_{\mathrm{a}}}{RT}\right)[\mathrm{Phenol}]^a [\mathrm{O_2}]^b [\mathrm{H_2O}]^c \tag{3.2}$$

where A, E_{a}, R, T, t, and [species] denote the pre-exponential factor of Arrhenius law or the frequency factor, activation energy, gas constant, absolute temperature, time, and compound concentrations in mole per liter, respectively. The phenol, O_2, and

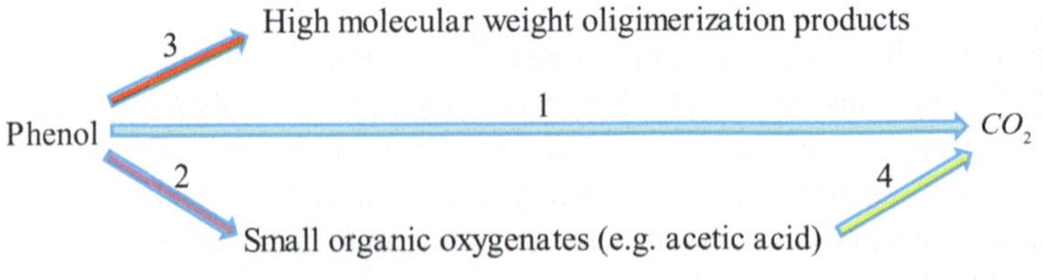

Fig. 3.3 Reaction pathways for phenol oxidation in supercritical water (Modified after Ding et al. 1995a; Guolin et al. 2012)

H_2O reaction orders are shown by a, b, and c, respectively. Arrhenius form has been used to express the reaction rate coefficient (Portela et al. 2001b; Pérez et al. 2004).

On the other hand, some researchers (Gopalan and Savage 1994; Zhang and Savage 1998; Yu and Savage 2000b, Portela et al. 2001b, Fourcault et al. 2009; Guolin et al. 2012) have assumed that water does not take part considerably in the phenol oxidation reaction. They used the following global power law rate equation and showed that it fits experimental data accurately (Guolin et al. 2012):

$$-r_{\text{Phenol}} = -\frac{d[\text{Phenol}]}{dt} = A\exp\left(\frac{-E_{\text{a}}}{RT}\right)[\text{Phenol}]^{a}[\text{O}_2]^{b} \quad (3.3)$$

Mechanism-based rate equations such as dual-site Langmuir-Hinshelwood-Hougen-Watson (LHHW) model and Mars-van Krevelen model have also been applied for CSCWO of phenol. The LHHW rate equation is (Krajnc and Levec 1997; Yu and Savage 2000b; Guolin et al. 2012):

$$-r_{\text{Phenol}} = \frac{kK_{\text{Phenol}}K_{\text{O}_2}^{0.5}[\text{Phenol}][\text{O}_2]^{0.5}}{\left(1+K_{\text{Phenol}}[\text{Phenol}]\right)\left(1+K_{\text{O}_2}^{0.5}[\text{O}_2]^{0.5}\right)} \quad (3.4)$$

where k denotes the rate constant, that Arrhenius form may be used to express the reaction rate constant, and the adsorption equilibrium constants for phenol and O_2 on the catalyst surface are shown by K_{Phenol} and K_{O2}, respectively.

The Mars-van Krevelen rate equation is:

$$-r_{\text{Phenol}} = \frac{1}{\dfrac{1}{k_1[\text{Phenol}]} + \dfrac{\beta}{k_2[\text{O}_2]^{m}}} \quad (3.5)$$

where k_1 and k_2 are rate constants, β denotes the number of O_2 atoms taking part in the surface reaction step, and m shows the oxygen reaction order for the surface reoxidation reaction (Yu and Savage 2000b; Guolin et al. 2012).

3.3.2 *Literature Review of Non-catalytic and Catalytic Supercritical Water Oxidation of Phenol in the Wastewaters from Petroleum and Other Industries*

Refineries can generate wastewaters containing phenol concentration levels of 20–200 mg/L (World Bank Group 1998). In petrochemical plants, process wastewaters may contain phenol levels of up to 200 mg/L (MIGA 2004). In the other literature (Jafarinejad 2017a), phenol concentration levels in wastewaters from refineries and petrochemicals have been reported to be 0.2–200 mg/L (Jafarinejad 2017a). In addition, according to Mohammadi et al. (2015), concentrations of phenols in wastewaters from refineries and petrochemical plants can be in the ranges of 6–500 and 2.8–1220 mg/L, respectively (Mohammadi et al. 2015). On the other hand, the maximum effluent level for phenol from the petroleum refineries and petrochemical plants must be 0.5 mg/L (World Bank Group 1998). Phenol levels in wastewaters from other industries such as coke operations and coal processing are 28–3900 mg/L and 9–6800 mg/L, respectively (Mohammadi et al. 2015). Therefore, an efficient treatment technique for phenol removal from industrial aqueous wastes is required.

Several studies have been reported to remove phenol from wastewater using supercritical water; but, in most cases, the sources of wastewater are not known in the papers. A summary of studies on non-catalytic and catalytic SCWO of phenol in wastewaters from petroleum and other industries is presented in Table 3.1. Phenol conversion in supercritical water generally depends on the initial phenol concentration, temperature, pressure, type and amount of catalysts (components, manufacturing process, and morphology), residence time in the reactor, reactor configuration, oxidant excess, etc. (Ding et al. 1995a; Ding et al. 1996; Pérez et al. 2004).

De Silva et al. (2017) studied SCWO of phenol in oil/gas produced water using continuous packed bed reactor with TiO_2-supported MnO_2 catalyst at operating conditions of 382 °C and 24.1 MPa. With no O_2 supplied to the reactor, phenol conversion was reported to be 9.3% at 382 °C, 24.1 MPa, and 5 s of residence time with 6% (w/w) Mn/TiO_2 catalyst, whereas it was 5.9% without catalyst. At O_2 ratio of 1 (the moles of O_2 supplied to the reactor/the stoichiometric moles of O_2 needed for complete phenol conversion), the phenol conversion from non-catalytic homogeneous SCWO was 46% in comparison with a conversion of 66% with 6% (w/w) Mn/TiO_2 catalyst. Also, 70% phenol conversion was reported at 12% (w/w) Mn in the catalyst with O_2 ratio of 1. They concluded that both Mn loading and amount of O_2 had positive influences on phenol oxidation reaction (De Silva et al. 2017).

According to Table 3.1, a variety of catalysts including CuO, ZnO, MnO_2, MnO_2/CeO, V_2O_5, MnO_2/CeO_2, V_2O_5/Al_2O_3, Cr_2O_3/Al_2O_3, CuO/Al_2O_3, TiO_2, and TiO_2-supported MnO_2 catalyst; transition-metal oxide catalyst; Carulite 150, a commercial catalyst; and carbonaceous materials (e.g., bamboo charcoal, coke, graphite, etc.) have been used in the SCWO of phenol. Under the same operating conditions, the phenol conversion and carbon dioxide molar yield from the CSCWO can be much higher than those from conventional non-catalytic SCWO of phenol (Yu and Savage 1999). The phenol conversion and the selectivity of phenol conversion to carbon dioxide are not only enhanced by the catalysts, but also the yield of organic byproducts is reduced (De Silva et al. 2017).

In the case of utilizing carbonaceous material such as activated carbon for SCWO of phenol, the carbonaceous material can also be oxidized, although at a slower rate than phenol. The application and addition of activated carbon not only can increase the oxygen utilization efficiency for phenol destruction, the reaction rate, and the gas yield but also can reduce the tarry product yield (Matsumura et al. 2002).

Lee et al. (2002) investigated the effects of the addition of NaOH on the SCWO of phenol. NaOH neutralizes acids during SCWO process which results in

Table 3.1 Summary of studies on non-catalytic and catalytic SCWO of phenol in petroleum industry wastewater and other wastewaters

Non-catalytic or catalytic	Reactor type	Reaction conditions	Results	Reference
Non-catalytic	Plug flow reactor	300–420 °C, 188–278 atm	0–100% phenol conversions	Thornton and Savage (1990)
Non-catalytic	Batch and flow reactors	Temperatures of 300, 380, and 420 °C; pressures of 218, 233, and 278 atm	4–100% phenol conversions in batch reactor	Thornton et al. (1991)
Non-catalytic	Batch and flow reactors	Temperatures of 300, 380, and 420 °C; pressures of 218 and 278 atm	17.2–100% phenol conversions in batch reactor and 4.3–99.4% phenol conversions in flow reactor	Thornton and Savage (1992a)
Non-catalytic	Flow reactor	300–420 °C, 188–278 atm	Less than 1% to greater than 99% phenol conversions	Thornton and Savage (1992b)
Non-catalytic and catalytic: CuO, ZnO	Flow reactor	380–390 °C, 23–23.5 MPa	12–60% phenol conversions for *non-catalytic SCWO and greater than 95%* phenol conversion for *catalytic SCWO*	Krajnc and Levec (1994)
Non-catalytic	Flow reactor	420–480 °C, 250 atm	6.6–48.73% phenol conversions	Gopalan and Savage (1995)
Catalytic: MnO_2/CeO or V_2O_5 catalysts	Packed bed reactor	390–410 °C, 450 °C	85–99.98% phenol conversions	Ding et al. (1995a)
Catalytic: MnO_2/CeO_2, V_2O_5/Al_2O_3, and Cr_2O_3/Al_2O_3	Flow reactor	390–460 °C, 23.8 MPa	Greater than 90% to approximately 100% phenol conversions	Ding et al. (1995b)
Non-catalytic	Tubular reactor	380–440 °C, 190–270 atm	11–99% phenol conversions	Koo et al. (1997)
Catalytic: transition-metal oxide catalyst	Fixed bed reactor	400–440 °C, 230–250 bar	34–93% phenol conversions	Krajnc and Levec (1997)
Catalytic: Carulite 150	Fixed bed flow reactor	380–430 °C, 250 atm	70–95% phenol conversions; 100% calculated phenol conversion	Zhang and Savage (1998)
Non-catalytic	Flow reactor	370–430 °C, 22.9–33.7 MPa	9.4–34.2% phenol conversions	Oshima et al. (1998)
Catalytic: MnO_2	Flow reactor	380–420 °C, 21.9–30 MPa	40 to near 80% phenol conversions	Yu and Savage (1999)

(continued)

Table 3.1 (continued)

Non-catalytic or catalytic	Reactor type	Reaction conditions	Results	Reference
Catalytic: MnO_2	Tubular flow reactor	425 °C, 22.7–27.2 MPa	Over 90% phenol conversion	Oshima et al. (1999)
Non-catalytic and catalytic: TiO_2	Tubular flow reactor	380–420 °C, 21.9–30 MPa	3–97% phenol conversions	Yu and Savage (2000a)
Catalytic: CuO/Al_2O_3	Packed bed flow reactor	380–450 °C, 219–300 atm	27–100% phenol conversions	Yu and Savage (2000b)
Non-catalytic	Tubular reactor	623–723 K, 25 MPa	52.8–59.2% phenol conversions	Matsumura et al. (2000)
Non-catalytic	Batch and flow reactors	400–500 °C, 25 MPa	13–82.9% phenol conversions in flow reactor	Portela et al. (2001b)
Catalytic: activated carbon	Packed bed flow reactor	400 °C, 25 MPa	Greater than 40% phenol conversions	Nunoura et al. (2002)
Catalytic: activated carbon	Packed bed reactor	400 °C, 25 MPa	Approximately 40–67.5% phenol conversions	Matsumura et al. (2002)
Catalytic: NaOH	Plug flow reactor	713 K, 26 MPa	7% phenol conversion at a residence time of 0.51 s without adding NaOH; approximately 60% phenol conversion with adding NaOH	Lee et al. (2002)
Catalytic: carbonaceous materials (e.g., coconut shell activated carbon, bamboo charcoal, coke, graphite, and carbon fiber)	Packed bed flow	400 °C, 25 MPa	All of the studied carbonaceous materials demonstrated a catalytic effect to increase the oxidative decomposition of phenol which activated carbon was the best	Nunoura et al. (2003b)
Catalytic: activated carbon	Packed bed flow reactor	400 °C, 25 MPa	Greater than 40% phenol conversions	Nunoura et al. (2003a)
Catalytic: MnO_2	Tubular flow reactor	425 °C, 24.7 MPa	5–84% phenol destructions	Tomita and Oshima (2004)
Non-catalytic	Tubular reactor	666–778 K, 25 MPa	94–99.98% phenol destructions	Pérez et al. (2004)
Non-catalytic	Continuous flow tubular reactor	400–500 °C, 25 MPa	A mathematical model used which the conversion ratios gained in numerical and experimental cases had errors between 3% and 10%	Fourcault et al. (2009)
Catalytic: TiO_2-supported MnO_2 catalysts	Continuous packed bed reactor	382 °C, 24.1 MPa	70% phenol conversion at 12% (w/w) Mn in the catalyst with O_2 ratio of 1	De Silva et al. (2017)

inhibition of reactor corrosion. In addition, it accelerates the destruction of phenol as a catalyst and reduces the generation of dimers and/or intermediates (Lee et al. 2002).

To develop a successful CSCWO process for phenol, catalysts should be more stable due to the harsh medium formed by the SCWO process, water adsorption, sintering, and dissolution of catalyst ingredients (Ding et al. 1996). Ding et al. (1996) and De Silva et al. (2017) mentioned that correlation of catalytic activity with physicochemical properties including redox potential, acid-base characteristics, oxygen adsorption, and surface structure can help in the selection of catalysts (Ding et al. 1996; De Silva et al. 2017). In addition to activity and stability of catalysts, simplicity of preparation, cost, nontoxicity, environmental friendly, etc. should be considered in the catalyst selection (Jafarinejad, 2017b).

Yu and Savage (2000a) studied catalyst stability and activity during CSCWO of phenol over TiO_2 and reported that this catalyst was stable and active in SCWO of phenol for over 120-h operation (Yu and Savage 2000a).

3.4 Conclusions

Phenol conversion in supercritical water can generally be affected by the initial phenol concentration, temperature, pressure, type and amount of catalysts (components, manufacturing process, and morphology), residence time in the reactor, reactor configuration, amount of oxidant, etc.

Variety of catalysts including CuO, ZnO, MnO_2, MnO_2/CeO, V_2O_5, MnO_2/CeO_2, V_2O_5/Al_2O_3, Cr_2O_3/Al_2O_3, CuO/Al_2O_3, TiO_2, and TiO_2-supported MnO_2 catalyst; transition-metal oxide catalyst; Carulite 150, a commercial catalyst; and carbonaceous materials (e.g., bamboo charcoal, coke, graphite, etc.) have been used in the SCWO of phenol. Under the same operating conditions, the phenol conversion and carbon dioxide molar yield from the CSCWO can be much higher than those from conventional non-catalytic SCWO of phenol. The phenol conversion and the selectivity of phenol conversion to carbon dioxide are not only enhanced by the catalysts, but also the yield of organic byproducts is reduced.

References

Abelleira J, Sánchez-Oneto J, Portela JR, Martínez de la Ossa EJ (2013) Kinetics of supercritical water oxidation of isopropanol as an auxiliary fuel and co-fuel. Fuel 111:574–583. https://doi.org/10.1016/j.fuel.2013.03.033

Aki SNVK, Abraham MA (1998) An economic evaluation of catalytic supercritical water oxidation: comparison with alternative waste treatment technologies. Environ Prog Sustain Energy 17(4):246–255. https://doi.org/10.1002/ep.670170415

Al-Duri B, Pinto L, Ashraf-Ball NH, Santos RCD (2008) Thermal abatement of nitrogen-containing hydrocarbons by non-catalytic supercritical water oxidation (SCWO). J Mater Sci 43:1421–1428. https://doi.org/10.1007/s10853-007-2285-3

Angeles-Hernandez MJ, Leeke GA, Santos RC (2008) Catalytic supercritical water oxidation for the destruction of quinoline over MnO_2/CuO mixed catalyst. Ind Eng Chem Res 48:1208–1214. https://doi.org/10.1021/ie8006402

Arslan-Alaton I, Ferry JL (2002) $H_4SiW_{12}O_{40}$-catalyzed oxidation of nitrobenzene in supercritical water: kinetic and mechanistic aspects. Appl Catal B Environ 38:283–293. https://doi.org/10.1016/S0926-3373(02)00059-0

Barner HE, Huang CY, Johnson T, Jacobs G, Martch MA (1992) Supercritical water oxidation: an emerging technology (paper presented at ACHEMA '91). J Hazard Mater 31(1):1–17. https://doi.org/10.1016/0304-3894(92)87035-E

Bermejo MD, Cocero MJ (2006) Supercritical water oxidation: a technical review. AICHE J 52:3933–3951. https://doi.org/10.1002/aic.10993

Bermejo MD, Rincon D, Martin A, Cocero MJ (2009) Experimental performance and modeling of a new cooled-wall reactor for the supercritical water oxidation. Ind Eng Chem Res 48:6262–6272. https://doi.org/10.1021/ie900054e

Brunner G (2009) Near and supercritical water. Part II. Oxidative processes. J Supercrit Fluids 47:382–390. https://doi.org/10.1016/j.supflu.2008.09.001

Civan F, Ozaltun DH, Kıpcak E, Akgün M (2015) The treatment of landfill leachate over Ni/Al_2O_3 by supercritical water oxidation. J Supercrit Fluids 100:7–14. https://doi.org/10.1016/j.supflu.2015.02.018

Conelly JF (1996) Solubility of hydrocarbons in water near the critical solution temperatures. J Chem Eng Data 11:13–16. https://doi.org/10.1021/je60028a003

Covinich LG, Bengoechea DI, Fenoglio RJ, Area MC (2014) Advanced oxidation processes for wastewater treatment in the pulp and paper industry: a review. Am J Environ Eng 4(3):56–70. https://doi.org/10.5923/j.ajee.20140403.03

Daman EL (1996) Process and apparatus for supercritical water oxidation. U.S. Patent #5571423

Daman EL (1998) Process and apparatus for supercritical water oxidation. U.S. Patent #5723045

De Silva CL, Garlapalli RK, Trembly JP (2017) Removal of phenol from oil/gas produced water using supercritical water treatment with TiO_2 supported MnO_2 catalyst. J Environ Chem Eng 5:488–493. https://doi.org/10.1016/j.jece.2016.12.015

Deng Y, Zhao R (2015) Advanced oxidation processes (AOPs) in wastewater treatment. Curr Pollution Rep 1:167–176. https://doi.org/10.1007/s40726-015-0015-z

Devlin HR, Harris IJ (1984) Mechanism of the oxidation of aqueous phenol with dissolved oxygen. Ind Eng Chem Fundam 23:387–392. https://doi.org/10.1021/i100016a002

Ding ZY, Aki SN, Abraham MA (1995a) Catalytic supercritical water oxidation: phenol conversion and product selectivity. Environ Sci Technol 29:2748–2753. https://doi.org/10.1021/es00011a008

Ding ZY, Aki SNVK, Abraham MA (1995b) In: Hutchenson KW, Foster NR (eds) Innovations in supercritical fluids. American Chemical Society, Washington, DC, pp 232–245

Ding ZY, Frisch MA, Li L, Gloyna EF (1996) Catalytic oxidation in supercritical water. Ind Eng Chem Res 35:3257–3279. https://doi.org/10.1021/ie960022n

Dong X, Gan Z, Lu X, Jin W, Yu Y, Zhang M (2015) Study on catalytic and non-catalytic supercritical water oxidation of p-nitrophenol wastewater. Chem Eng J 277:30–39. https://doi.org/10.1016/j.cej.2015.04.134

European Commission, Joint Research Center (2013) Best available techniques (BAT) Reference Document for the refining of mineral oil and gas, Industrial Emissions Directive 2010/75/EU (Integrated Pollution Prevention and Control), Joint Research center, Institute for Prospective Technological Studies Sustainable Production and Consumption Unit European IPPC Bureau.

Fauvel E, Joussot-Dubien C, Pomier E, Guichardon P, Charbit G, Charbit F, Sarrade S (2003) Modeling of a porous reactor for supercritical water oxidation by a residence time distribution study. Ind Eng Chem Res 42:2122–2130. https://doi.org/10.1021/ie0207616

Fourcault A, Garcia-Jarana B, Sanchez-Oneto J, Mariasa F, Portela JR (2009) Supercritical water oxidation of phenol with air. Experimental results and modeling. Chem Eng J 152:227–233. https://doi.org/10.1016/j.cej.2009.04.029

Ghorbanian M, Moussavi G, Farzadkia M (2014) Investigating the performance of an up-flow anoxic fixed-bed bioreactor and a sequencing anoxic batch reactor for the biodegradation of hydrocarbons in petroleum-contaminated saline water. Int Biodeterior Biodegrad 90:106–114. https://doi.org/10.1016/j.ibiod.2014.02.009

Gidner A, Stenmark L (2001) Supercritical water oxidation of sewage sludge - state of the art. In: IBC's conference on sewage sludge and disposal options

Gloyna EF, Li L (1995) Supercritical water oxidation research and development update. Environ Prog 14(3):182–192. https://doi.org/10.1002/ep.670140318

Goldblatt ME, Gucciardi JM, Huban CM, Vasconcellos SR, Liao WP (2014) New polyelectrolyte emulsion breaker improves oily wastewater cleanup at lower usage rates, Technical paper, GE water and power, Water & Process Technologies, General Electric Company, tp382.doc 06, 1-6

Gong WJ, Li F, Xi DL (2008) Oxidation of industrial dyeing wastewater by supercritical water oxidation in transpiring-wall reactor. Water Environ Res 80:186–192

Gong WJ, Li F, Xi DL (2009) Supercritical water oxidation of acrylic acid production wastewater in transpiring wall reactor. Environ Eng Sci 6:131–136. https://doi.org/10.1089/ees.2007.0279

Gopalan S, Savage PE (1994) Reaction mechanism for phenol oxidation in supercritical water. J Phys Chem 98:646–652. https://doi.org/10.1021/j100099a031

Gopalan S, Savage PE (1995) A reaction network model for phenol oxidation in supercritical water. AICHE J 41:1864–1873. https://doi.org/10.1002/aic.690410805

Guolin J, Mingming L, Tingting C (2012) Heterogeneous catalytic supercritical water oxidation of refractory organic pollutants in industrial wastewaters. J Adv Oxid Technol 15(1):134–145. https://doi.org/10.1515/jaots-2012-0115

Han D, Zhang H, Fang L, Lin C (2015) Continuous monitoring of total organic carbon based on supercritical water oxidation improved by $CuSO_4$ catalyst. J Anal Bioanal Tech S13:002. https://doi.org/10.4172/2155-9872.S13-002

Hazelbeck DA, Downey KW, Elliott JP, Spritzer MH (1995) Design of corrosion resistant HTO systems for DoD hazardous wastes, presented at the first international workshop on supercritical water oxidation, Jacksonville, FL, February 6-9, 1995

Hong GT, Killilea WR, Thomason TB (1989) Method for solids separation in a wet oxidation type process. U.S. Patent #4822497

IPIECA (2010) Petroleum refining water/wastewater use and management, IPIECA operations best practice series. United Kingdom, London

Jafarinejad S (2014a) Supercritical water oxidation (SCWO) in oily wastewater treatment, national e-conference on advances in basic sciences and engineering (AEBSCONF), Iran

Jafarinejad S (2014b) Electrochemical oxidation process in oily wastewater treatment, national e-conference on advances in basic sciences and engineering (AEBSCONF), Iran

Jafarinejad S (2015a) Ozonation advanced oxidation process and place of its use in oily sludge and wastewater treatment, 1st international conference on environmental engineering (eiconf), Tehran, Iran

Jafarinejad S (2015b) Heterogeneous photocatalysis oxidation process and use of it for oily wastewater treatment, 1st international conference on environmental engineering (eiconf), Tehran, Iran

Jafarinejad S (2015c) Recent advances in nanofiltration process and use of it for oily wastewater treatment, 1st international conference on environmental engineering (eiconf), Tehran, Iran

Jafarinejad S (2015d) Investigation of advanced technologies for wastewater treatment from petroleum refinery processes, 2nd E-conference on recent research in science and technology, Kerman, Iran, Summer 2015

Jafarinejad S (2015e) Recent advances in determination of herbicide paraquat in environmental waters and its removal from aqueous solutions: a review. Inter Res J App Basic Sci 9(10):1758–1774

Jafarinejad S (2017a) Petroleum waste treatment and pollution control, 1st edn. Elsevier/Butterworth-Heinemann, Cambridge, MA

Jafarinejad S (2017b) Cost-effective catalytic materials for AOP treatment units. In: Gil A, Galeano L, Vicente M (eds) Applications of advanced oxidation processes (AOPs) in drinking water treatment, The handbook of environmental chemistry, vol 67. Springer, Cham. https://doi.org/10.1007/698_2017_77

Jafarinejad S (2017c) Cost estimation and economical evaluation of three configurations of activated sludge process for a wastewater treatment plant (WWTP) using simulation. Appl Water Sci 7:2513–2521. https://doi.org/10.1007/s13201-016-0446-8

Jafarinejad S (2017d) A comprehensive study on the application of reverse osmosis (RO) technology for the petroleum industry wastewater treatment. J Water Environ Nanotechnol 2(4):243–264. https://doi.org/10.22090/jwent.2017.04.003

Jafarinejad S (2017e) Activated sludge combined with powdered activated carbon (PACT process) for the petroleum industry wastewater treatment: a review. Chem Int 3(4):268–277

Jafarinejad S (2017f) Recent developments in the application of sequencing batch reactor (SBR) technology for the petroleum industry wastewater treatment. Chem Int 3(3):241–250

Jafarinejad S (2019) Simulation for the performance and economic evaluation of conventional activated sludge process replacing by sequencing batch reactor technology in a petroleum refinery wastewater treatment plant. ChemEng 3(2):45. https://doi.org/10.3390/chemengineering3020045

Jafarinejad S, Abolghasemi H, Golzary A, Moosavian MA, Maragheh MG (2010) Fractional factorial design for the optimization of hydrothermal synthesis of lanthanum oxide under supercritical water condition. J Supercrit Fluids 52:292–297. https://doi.org/10.1016/j.supflu.2010.01.013

Jafarinejad S, Faraji M, Jafari P, Mokhtari-Aliabad J (2017) Removal of lead ions from aqueous solutions using novel modified magnetic nanoparticles: optimization, isotherm, and kinetics studies. Desalination Water Treatment 92:267–274. https://doi.org/10.5004/dwt.2017.21562

Jafarinejad S, Faraji M, Norouzi Z, Mokhtari-Aliabad J (2018) Application of sulfur-modified magnetic nanoparticles for cadmium removal from aqueous solutions. J Water Environ Nanotechnol 3(1):58–69. https://doi.org/10.22090/jwent.2018.01.006

Jafarinejad S, Park H, Mayton H, Walker SL, Jiang SC (2019) Concentrating ammonium in wastewater by forward osmosis using a surface modified nanofiltration membrane. Environ Sci Water Res Technol 5:246–255. https://doi.org/10.1039/C8EW00690C

Kawasaki SI, Oe T, Anjon N, Nakamori T, Suzuki A, Arai K (2006) Practical supercritical water reactor for destruction of high concentration polychlorinated biphenyls (PCB) and dioxin waste streams. Process Saf Environ Prot 84:317–324. https://doi.org/10.1205/psep.05187

Kazemi N, Tavakoli O, Seif S, Nahangi M (2015) High-strength distillery wastewater treatment using catalytic sub- and supercritical water. J Supercrit Fluids 97:74–80. https://doi.org/10.1016/j.supflu.2014.10.025

Koo M, Lee WK, Lee CH (1997) New reactor system for supercritical water oxidation and its application on phenol destruction. Chem Eng Sci 52:1201–1214. https://doi.org/10.1016/S0009-2509(96)00477-0

Krajnc M, Levec J (1994) Catalytic oxidation of toxic organics in supercritical water. J Appl Catal B 3:L101–L107. https://doi.org/10.1016/0926-3373(93)E0038-D

Krajnc M, Levec J (1996) On the kinetics of phenol oxidation in supercritical water. AICHE J 42:1977–1984. https://doi.org/10.1002/aic.690420718

Krajnc M, Levec J (1997) Oxidation of phenol over a transition-metal oxide catalyst in supercritical water. Ind Eng Chem Res 36:3439–3445. https://doi.org/10.1021/ie9701130

Lee G, Nunoura T, Matsumura Y, Yamamoto K (2002) Comparison of the effects of the addition of NaOH on the decomposition of 2-chlorophenol and phenol in supercritical water and under supercritical water oxidation conditions. J Super Fluids 24:239–250. https://doi.org/10.1016/S0896-8446(02)00150-X

Li L, Chen P, Gloyna EF (1994) Chemical oxidation: technology for the nineties. In: Eckenfelder W, Bowers A, Roth J (eds) Proceedings of the third international symposium on the chemical oxidation: technology for the nineties. Vanderbilt University, Nashville, TN

Litter MI, Quici N (2010) Photochemical advanced oxidation processes for water and wastewater treatment. Rec Pat Eng 4:217–241. https://doi.org/10.2174/187221210794578574

Loures CCA, Alcantara MAK, Filho HJI, Teixeira ACSC, Silva FT, Paiva TCB, Samanamud GRL (2013) Advanced oxidative degradation processes: fundamentals and applications. Inter Rev Chem Eng 5(2):102–120

Marrone PA, Hodes M, Smith KA, Tester JW (2004) Salt precipitation and scale control in supercritical water oxidation-part B: commercial/full-scale applications. J Supercrit Fluids 29:289–312. https://doi.org/10.1016/S0896-8446(03)00092-5

Marulanda V, Bolanos G (2010) Supercritical water oxidation of a heavily PCB-contaminated mineral transformer oil: laboratory-scale data and economic assessment. J Supercrit Fluids 54:258–265. https://doi.org/10.1016/j.supflu.2010.04.008

Matsumura Y, Nunoura T, Urase T, Yamamoto K (2000) Supercritical water oxidation of high concentrations of phenol. J Hazard Mater 73(3:245–254. https://doi.org/10.1016/S0304-3894(99)00187-9

Matsumura Y, Urase T, Yamamoto K, Nunoura T (2002) Carbon catalyzed supercritical water oxidation of phenol. J Supercrit Fluids 22(2):149–156. https://doi.org/10.1016/S0896-8446(01)00112-7

McBrayer R (1995) Design and operation of the first commercial supercritical water oxidation facility. Presented at the first international workshop on supercritical water oxidation, Mohiuddin J, workshop chairman, Jacksonville, FL, 1995

Modell M (1982) Processing methods for the oxidation of organics in supercritical water. U.S. Patent 4,338,199

Mohammadi S, Kargari A, Sanaeepur H, Abbassian K, Najafi A, Mofarrah E (2015) Phenol removal from industrial wastewaters: a short review. Desalin Water Treat 53. https://doi.org/10.1080/19443994.2014.883327

Mota ALN, Albuquerque LF, Beltrame LTC, Chiavone-Filho O, Machulek A Jr, Nascimento CAO (2008) Advanced oxidation processes and their application in the petroleum industry: a review. Brazil J Petrol Gas 2(3):122–142

Multilateral Investment Guarantee Agency (MIGA) (2004) Environmental guidelines for petrochemicals manufacturing, pp 461e467. [Online] https://www.miga.org/documents/Petrochemicals.pdf

National Research Council (1993) Alternative technology for the destruction of chemical agents and munitions, 7 Processes at medium and high temperatures. National Academy Press, Washington, DC

Nunoura T, Lee GH, Matsumura Y, Yamamoto K (2002) Modeling of supercritical water oxidation of phenol catalyzed by activated carbon. Chem Eng Sci 57:3061–3071. https://doi.org/10.1016/S0009-2509(02)00187-2

Nunoura T, Lee G, Matsumura Y, Yamamoto K (2003a) Reaction engineering model for supercritical water oxidation of phenol catalyzed by activated carbon. Ind Eng Chem Res 42:3522–3531. https://doi.org/10.1021/ie0206047

Nunoura T, Lee GH, Matsumura Y, Yamamoto K (2003b) Effect of carbonaceous materials on the oxidation of phenol in supercritical water: a preliminary study. Ind Eng Chem Res 42:3718–3720. https://doi.org/10.1021/ie020847z

Oshima Y, Hori K, Toda M, Chommanad T, Koda S (1998) Phenol oxidation kinetics in supercritical water. J Supercrit Fluids 13:241–246. https://doi.org/10.1016/S0896-8446(98)00057-6

Oshima Y, Tomita K, Koda S (1999) Kinetics of the catalytic oxidation of phenol over manganese oxide in supercritical water. Ind Eng Chem Res 38:4183–4188. https://doi.org/10.1021/ie9902939

Paraskeva P, Diamadopoulos E (2006) Technologies for olive mill wastewater (OMW) treatment: a review. J Chem Technol Biotechnol 81:1475–1485. https://doi.org/10.1002/jctb.1553

Pérez IV, Rogak S, Branion R (2004) Supercritical water oxidation of phenol and 2,4-dinitrophenol. J Supercrit Fluids 30(1):71–87. https://doi.org/10.1016/S0896-8446(03)00166-9

Portela JR, López J, Nebot E, Martínez de la Ossa E (2001a) Elimination of cutting oil wastes by promoted hydrothermal oxidation. J Hazard Mater 88:95–106. https://doi.org/10.1016/S0304-3894(01)00295-3

Portela JR, Nebot E, Martínez de la Ossa E (2001b) Kinetic comparison between subcritical and supercritical water oxidation of phenol. Chem Eng J 81:287–299. https://doi.org/10.1016/S1385-8947(00)00226-6

Pruden BB, Le H (1976) Wet air oxidation of soluble components in waste water. Can J Chem Eng 54:319–321. https://doi.org/10.1002/cjce.5450540413

Qi XH, Zhuan YY, Yuan YC, Gu WX (2002) Decomposition of aniline in supercritical water. J Hazard Mater B 90:51–62. https://doi.org/10.1016/S0304-3894(01)00330-2

Reid AF, Halff AH (1995) Method for separation and removal of impurities from liquids. U.S. Patent #5425883

Ross DS, Jayaweera I (1998) Environmentally acceptable waste disposal by hydrothermal decomposition of labile compounds with nitrite. U.S. Patent #5709800

Ross DS, Jayaweera I, Nguyen L, Hum GP, Haag WR (1995) Environmentally acceptable waste disposal by conversion of hydrothermally labile compounds. U.S. Patent #5409617

Sabet JK, Jafarinejad S, Golzary A (2014) Supercritical water oxidation for the recovery of dysprosium ion from aqueous solutions. Int Res J Appl Basic Sci 8(8):1079–1083

Santos MSF, Alves A, Madeira LM (2011) Paraquat removal from water by oxidation with Fenton's reagent. Chem Eng J 175:279–290. https://doi.org/10.1016/j.cej.2011.09.106

Schroer C, Konys J, Novotny J, Hausselt J (2006) Aspects of the corrosion kinetics for binary nickel–chromium alloys under conditions found in supercritical water oxidation plants processing chlorinated substances. Corrosion 65:444–459. https://doi.org/10.5006/1.3278281

Singh A, Pant KK, Nigam KDP (2004) Catalytic wet oxidation of phenol in a trickle bed reactor. Chem Eng J 103:51–57. https://doi.org/10.1016/j.cej.2004.06.006

Son S, Lee J, Lee C (2008) Corrosion phenomena of alloys by subcritical and supercritical water oxidation of 2-chlorophenol. J Supercrit Fluids 44:370–378. https://doi.org/10.1016/j.supflu.2007.09.027

Svensson P (1995) Look no stack: supercritical water destroys organic wastes. Chem Technol Eur 2:16

Tester W, Cline JA (1999) Hydrolysis and oxidation in subcritical and supercritical water: connecting process engineering science to molecular interaction. Corrosion 5:1088–1100

Tester JW, Holgate HR, Armellini FJ, Webley PA, Killilea WR, Hong GT, Barner HE (1993) Supercritical water oxidation technology: process development and fundamental research. In: Tedder WD, Pohland FG (eds) Emerging technologies in hazardous waste management III, ACS symposium series, vol 518. American Chemical Society, Washington, DC, p 35

Thornton TD, Savage PE (1990) Phenol oxidation in supercritical water. J Supercrit Fluids 3:240–248. https://doi.org/10.1016/0896-8446(90)90029-L

Thornton TD, Savage PE (1992a) Phenol oxidation pathways in supercritical water. Ind Eng Chem Res 31:2451–2456. https://doi.org/10.1021/ie00011a007

Thornton TD, Savage PE (1992b) Kinetics of phenol oxidation in supercritical water. AICHE J 38:321. https://doi.org/10.1002/aic.690380302

Thornton TD, LaDue DE, Savage PE (1991) Phenol oxidation in supercritical water: formation of dibenzofuran, and related compounds. Environ Sci Technol 25:1507–1510. https://doi.org/10.1021/es00020a023

Tomita K, Oshima Y (2004) Stability of manganese oxide in catalytic supercritical water oxidation of phenol. Ind Eng Chem Res 43:7740–7743. https://doi.org/10.1021/ie040154d

United States Environmental Protection Agency (U.S. EPA) (1995) Profile of the petroleum refining industry, EPA Office of compliance Sector Notebook Project, EPA/310-R-95-013, September 1995, Office of Compliance, Office of Enforcement and Compliance Assurance. U.S. Environmental Protection Agency, Washington, DC

Vadillo V, García-Jarana MB, Sánchez-Oneto J, Portela JR, Martínez de la Ossa EJ (2011) Supercritical water oxidation of flammable industrial wastewaters: economic perspectives of an industrial plant. J Chem Technol Biotechnol 86:1049–1057

Wagner GW, Procell LR, O'Connor RJ, Munavalli S, Carnes CL, Kapoor PN, Klabunde KJ (2001) Reactions of VX, GB GD, and HD with nanosize Al2O3. Formation of aluminophosphonates. J Am Chem Soc 123:1636–1644. https://doi.org/10.1021/ja003518b

Wei CH, Yan B, Hu CS (2007) PCBs treatment by sub-supercritical water catalytic oxidation, thermolysis and reduction. Prog Chem 19:1275–1281

Wenbing M, Hongpeng L, Xuemei M (2013) Study on supercritical water oxidation of oily wastewater with ethanol. Res J Appl Sci Eng Technol 6(6):1007–1011

Whiting P (1996) Reversible flow supercritical reactor and method for operating same. U.S. Patent #5560823

Whiting P (1997) Reversible flow supercritical reactor and method for operating same. U.S. Patent #5667698

World Bank Group (1998) Petroleum refining, project guidelines: industry sector guidelines. In: Pollution prevention and abatement handbook, pp 377–381. http://www.ifc.org/wps/wcm/connect/b99a2e804886589db69ef66a6515bb18/petroref_PPAH.pdf?MOD¼AJPERES

Xu D, Wang S, Tang X, Gong Y, Guo Y, Wang Y, Zhang J (2012) Design of the first pilot scale plant of China for supercritical water oxidation of sewage sludge. Chem Eng Res Des 90:288–297. https://doi.org/10.1016/j.cherd.2011.06.013

Yanan J (2012) Hydrothermal oxidation of phenol with H_2O_2. Adv Mater Res 535-537:2475–2478. https://doi.org/10.4028/www.scientific.net/AMR.535-537.2475

Youngprasert B, Poochinda K, Ngamprasertsith S (2010) Treatment of acetonitrile by catalytic supercritical water oxidation in compact-sized reactor. J Water Resource Protect 2:222–226. https://doi.org/10.4236/jwarp.2010.23025

Yu J, Savage PE (1999) Catalytic oxidation of phenol over MnO_2 in supercritical water. Ind Eng Chem Res 38:3793–3801. https://doi.org/10.1021/ie990277b

Yu J, Savage PE (2000a) Kinetics of catalytic supercritical water oxidation of phenol over TiO_2. Environ Sci Technol 34:3191–3198. https://doi.org/10.1021/es9914588

Yu J, Savage PE (2000b) Phenol oxidation over CuO/Al_2O_3 in supercritical water. Appl Catal B Environ 28(34):275–288. https://doi.org/10.1016/S0926-3373(00)00184-3

Yu L, Han M, He F (2017) A review of treating oily wastewater. Arab J Chem 10:S1913–S1922. https://doi.org/10.1016/j.arabjc.2013.07.020

Zhang X, Savage PE (1998) Fast catalytic oxidation of phenol in supercritical water. Catal Today 40:333–342. https://doi.org/10.1016/S0920-5861(98)00062-5

Zhang S, Zhang Z, Zhao R, Gu J, Liu J, Örmeci B, Zhang J (2017) A review of challenges and recent progress in supercritical water oxidation of wastewater. J Chem Eng Commun 204(2). https://doi.org/10.1080/00986445.2016.1262359

Zhong J, Sun X, Wang C (2003) Treatment of oily wastewater produced from refinery processes using flocculation and ceramic membrane filtration. Sep Purif Technol 32:93–98. https://doi.org/10.1016/S1383-5866(03)00067-4

Chapter 4
Production of Platform Chemicals Using Supercritical Fluid Technology

Tau Len-Kelly Yong, Khairul Faizal Pa'ee, Norfahana Abd-Talib, and Nurabiyiah Mohamad

Contents

Abbreviations

5-HMF	5-Hydroxymethylfurfural
CO_2	Carbon dioxide
THF	Tetrahydrofuran

4.1 Introduction

The exponential population growth pushes towards expanding the utilization of fossil fuels especially in the energy and chemical production (Mohan et al. 2015). However, this impacted the economic, environmental, political, and life quality of the population. It increases the emissons of greenhouse gases, further aggravating

T. L.-K. Yong (✉) · K. F. Pa'ee · N. Abd-Talib · N. Mohamad
Universiti Kuala Lumpur, Branch Campus Malaysian Institute of Chemical and Engineering Technology (UniKL MICET), Alor Gajah, Melaka, Malaysia
e-mail: kytlen@unikl.edu.my; khairulfaizal@unikl.edu.my; norfahana@unikl.edu.my; nurabiyiah.mohamad29@s.unikl.edu.my

Inamuddin, A. M. Asiri (eds.), *Advanced Nanotechnology and Application of Supercritical Fluids*, Nanotechnology in the Life Sciences,
https://doi.org/10.1007/978-3-030-44984-1_4

global warming (Naik et al. 2010). Thus, the urgency to seek other processes with the prerequisite of sustainable feedstock is crucial. This has been the motivation behind the studies on organic wastes and agricultural crops as renewable feedstock. One of the potential solutions could be the utilization of lignocellulosic and protein biomass as alternative resources (Jana and Ulla 2011). Biomass can be defined as substances that are derived from animal and plant resources and utilized as energy or as materials in a substantial amount. Biomass is also called as "phytomass" and is often translated as bio-resource or bio-derived resource (Yokoyama and Matsumura 2008). The resources of biomass are abundant. It consists of energy crops (barley, corn, wheat, sugar crops, grasses), forest (tree, wood, logging residues, shrubs), or aquatic biomass (water hyacinth, algae, water weed) (Huber et al. 2006). The C5 and C6 sugars of lignocellulosic biomass can be exploited to produce fuel chemical intermediates via platform chemical pathways especially 5-HMF, furfural, and levulinic acid.

Petroleum refineries use platform chemicals as the building blocks to produce complex molecules, for example, polymers. In biorefineries, sugar monomers from biomass can be converted to bio-based polymers, fuels, and chemicals or other complex molecules through various methods (Boisen et al. 2009; Werpy et al. 2004). Agriculture residues including oil palm biomass, corn stover, pine wood, corncobs, wheat straw, bagasse, sugarcane bagasse, oat hulls, rice straw, and maple wood have been used to produce platform chemicals (Mohamad and Yong 2015; Weiss et al. 2010; Marzialetti et al. 2008; Qing et al. 2017; de Carvalho et al. 2017; Zeitsch 2000; Asghari and Yoshida 2010; Kaar et al. 1998). However, resource limitation is often cited as the challenge to produce bio-based chemicals from agricultural crops. The versatility of the biomass resources demonstrates its locality potential where the choice is largely determined by the geographical availability and policies in place for its implementation. For example, China is the largest producer and consumer of bio-based chemicals from corncobs, as it is a primary crop in the country (Mamman et al. 2008). On the other hand, the availability of these crops in certain countries, for example, Indonesia, Malaysia, and Africa, is minimal as corncobs are considered secondary crops and its harvested area is not as large as other plants, such as oil palm biomass.

Supercritical fluids are defined as fluids with properties (temperature and pressure) higher than its critical point. Supercritical fluid is widely used as a reaction medium as it possesses unique properties. The reduction in dielectric constant as the temperature increases, with significant reduction near the critical point, will affect its polarity, thus increasing its solubility and stability which gives them superior properties for advanced applications (Kang et al. 2013). The water phase diagram is shown in Fig. 4.1.

Figure 4.2 shows the water properties that change with temperature. In subcritical temperatures and sufficiently high pressure, water acts as an acid/base catalyst precursor due to its high ionic product, which is found to be as much as three times higher than under ambient conditions. Moreover, with the sufficient properties of its dielectric constant, water at this region is an excellent reaction medium for ionic reactions including the degradation of biomass to produce liquid and gaseous

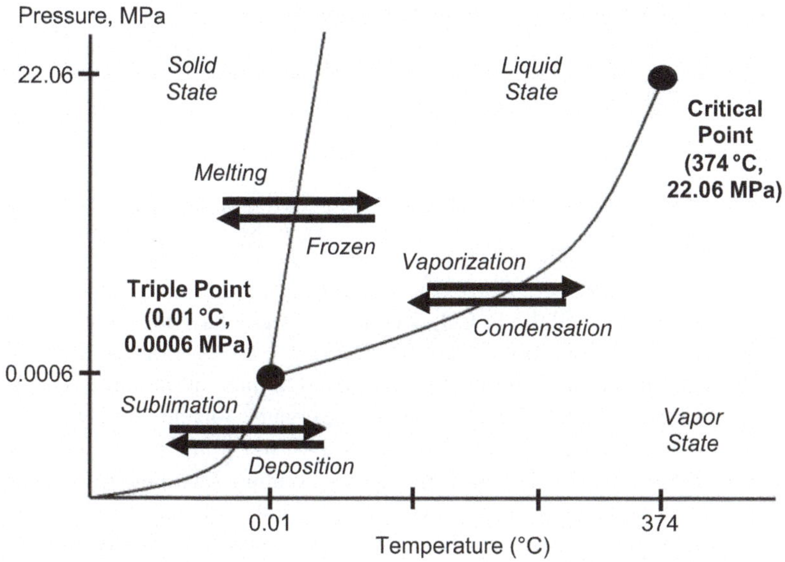

Fig. 4.1 Water phase diagram

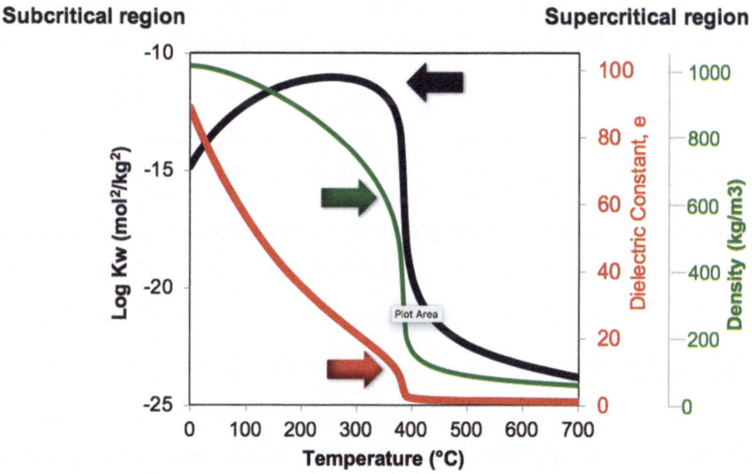

Fig. 4.2 Density, dielectric constant (ε), and ionic product of water (K_w) as a function of temperature at 25 MPa (JSME steam tables, 1983)

compounds (Kruse and Dinjus 2007). However, water at its critical state has a lower dielectric constant with the number and persistence of hydrogen bonds which is further reduced (Savage 1999). Therefore, water at this condition behaved similarly to many organic solvents with high solubility at near-critical condition and complete miscibility at its critical condition.

Table 4.1 Critical conditions of the common supercritical fluids

Substance	Molecular weight	Critical temperature (°C)	Critical pressure (bar)	Critical density (g/cm³)
Water	18.02	373.98	220.55	0.322
Carbon dioxide	44.01	31.040	73.82	0.468
Methanol	32.04	239.43	80.96	0.272
Ethanol	46.07	243.10	63.84	0.276

Among all supercritical fluids, CO_2, alcohol, and water have the highest potential as reaction media for platform chemical production. Comparatively, water has a higher critical pressure and temperature compared to most of organic solvents and CO_2 as shown in Table 4.1. Therefore, by using organic solvents or CO_2, the reaction will occur in a medium with a significantly lower temperature and pressure (Gairola and Smirnova 2012). Supercritical CO_2 combined with hydrothermal process is particularly interesting especially in the production of platform chemicals, as it can be used at high temperatures, has easy recovery processes, is environmentally friendly, and possesses acidifying properties, advantageous for biomass extraction (Gairola and Smirnova 2012).

Supercritical liquids exhibit a wide range of chemical and physical properties, without going through a phase change which is important to dissolve organic compounds. Its high diffusivities and low surface tension allow its diffusion into the biomass recalcitrant structure, properties that are very beneficial in the gasification and liquefaction of biomass (Yong and Matsumura 2012, 2013). However, the high critical temperature and pressure of water is a major concern in the production industry. On the other hand, the critical temperature and pressure of alcohols such ethanol and methanol are considerably lower and therefore offer milder conditions for the reaction (Minami and Saka 2005). The significant decrease of the dielectric constant at the critical point can dissolve various non-polar organic substances.

4.2 Furfural

Furfural (furan-2-carbaldehyde, furaldehyde, 2-furancarboxaldehyde, 2-furfuraldehyde, fural, 2-furanaldehyde, or furfurol) is one of the top 30 value-added biomass-based chemicals by the US Department of Energy (Werpy et al. 2004; Cai et al. 2014; Yan et al. 2014). It has a heteroaromatic furan ring with an aldehyde functional group (Dias et al. 2010). Its high potential lies in its ability to compete with oil derivatives to produce various derived chemicals including furfuryl alcohol, furan, and tetrahydrofuran. Furfural is widely used in various fields. For example, furfural is very important in agriculture as herbicides, fungicides, and insecticides. Furfural is also applied in food industry as flavouring agents. Its most beneficial

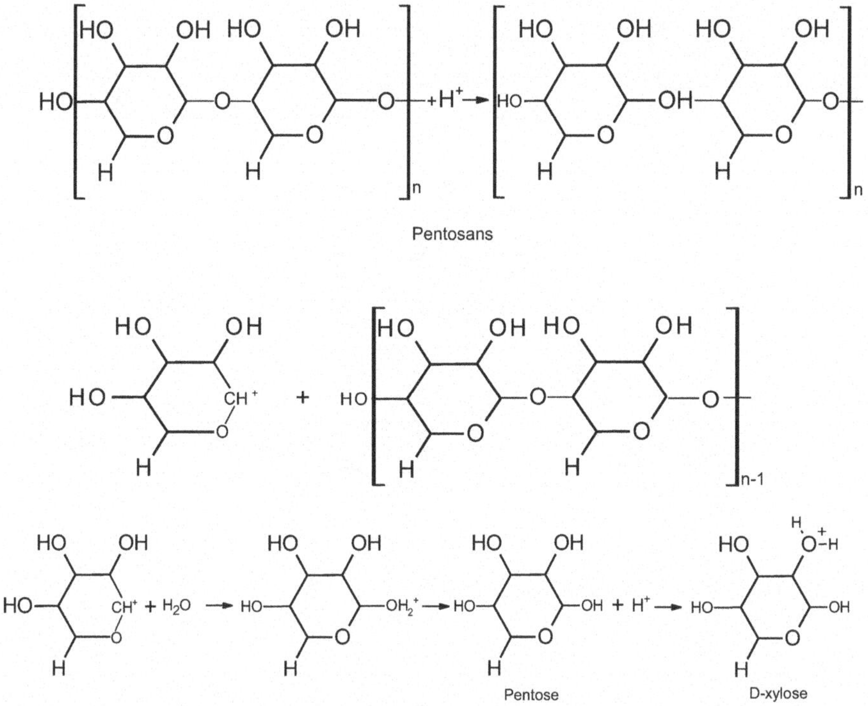

Fig. 4.3 Depolymerization and acid hydrolysis of pentosans in biomass to produce xylose

applications will be in medicine and pharmaceuticals as tuberculosis remedies and in plastics production as resins and synthesis fibres (López et al. 2014).

Figures 4.3 and 4.4 show the mechanism of biomass conversion to furfural through two main reactions (acid-catalysed and dehydration). The pentosan carbon bonds are broken down through depolymerization followed by the hyrolysis of pentose to release xylose via acid-hydrolysis reaction (Fig. 4.3). The elimination of three water molecules produces furfural (Fig. 4.4) through dehydration (Antal Jr et al. 1991; Nimlos et al. 2006). However, this mechanism may vary due to lignocellulosic material (particle size, shape, and structure) and the reaction medium (catalyst, concentration, temperature, and reaction time).

The reaction with real biomass is often complex and unpredictable due to diverse content of the biomass. The yield of furfural varies quite significantly due to the difference in carbohydrate content in the agricultural crops. The comparison of furfural yield obtained from various types of biomass is shown in Fig. 4.5. The highest yield obtained from oil palm frond was 37.0% comparable to other methods using the same feedstock. Riansa-ngawong and Prasertsan (2011) obtained 17.3% furfural yield with oil palm pressed fibre, comparable to de Carvalho et al. (2017) (wheat straw) and Marzialetti et al. (2008) (pine wood). Raman and Gnansounou (2015) reported furfural yield of 5.6% from empty fruit bunch using hydrothermal reactions. However, the yield was lower compared to studies by Zeitsch (2000) (bagasse),

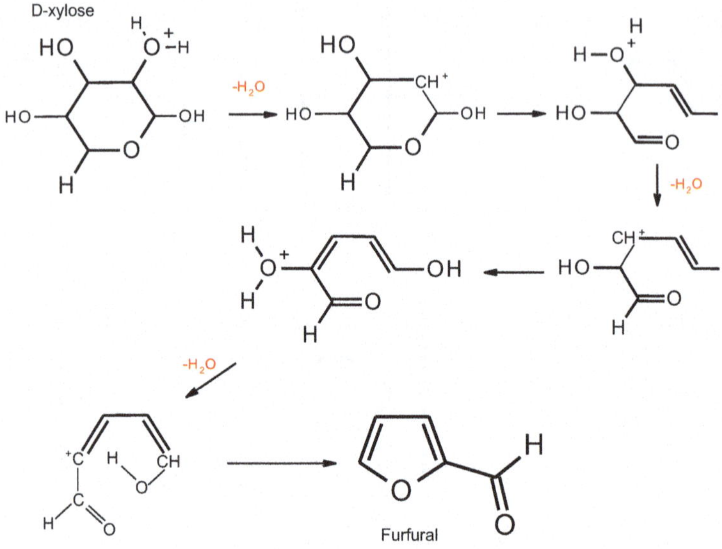

Fig. 4.4 Dehydration of xylose to form furfural

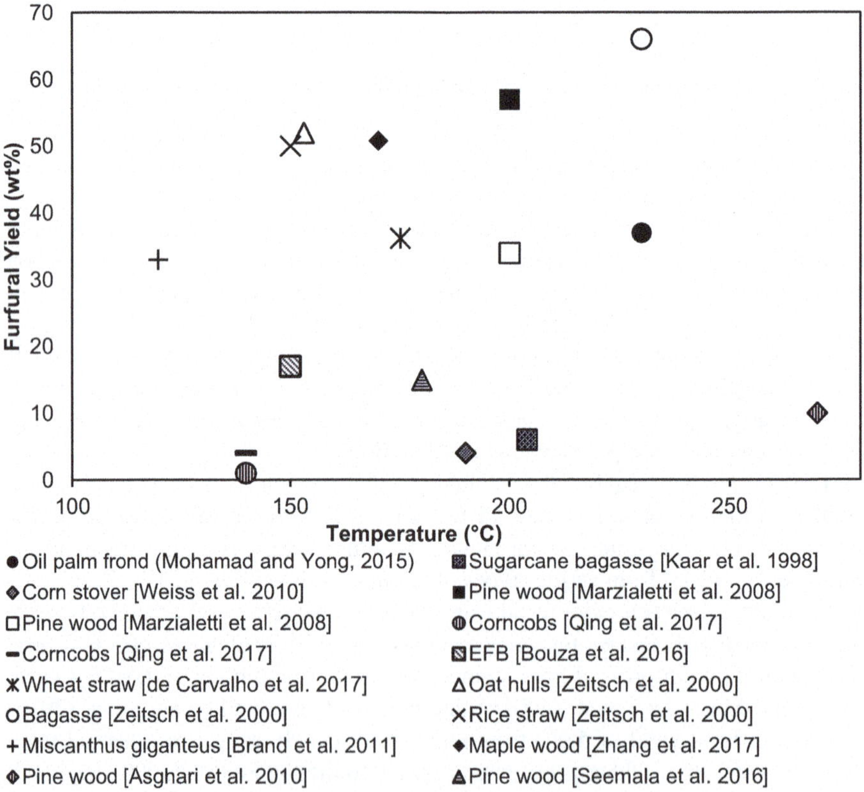

Fig. 4.5 Comparison of furfural yield using different biomass feedstocks

Marzialetti et al. (2008) (pine wood), and Zhang et al. (2013) (maple wood). The difference in the yield is due to different factors such as xylose content, types of catalyst, operating conditions, and reaction media.

Furfural production using supercritical CO_2 has shown significant outcome. It is particularly attractive as it can be used at high temperatures, has easy product separation and purification, as well as is environmentally friendly. Supercritical CO_2 offers acidifying properties that enhanced the biomass depolymerization and subsequently furfural formation (van Walsum and Shi 2004; Miyazawa and Funazukuri 2005; Sangarunlert et al. 2007; Gallina et al. 2017; Albarelli et al. 2018). Gairola and Smirnova (2012) studied the conversion of D-xylose and hemicellulose to furfural under hydrothermal conditions with simultaneous furfural extraction with supercritical CO_2 in catalyst-free conditions. The effect of the simultaneous supercritical CO_2 extraction is significant as it increases the furfural yield from xylose by 65% from 41 mol% to 68 mol%. The study was further extended using real biomass hydrolysates (straw and brewery wastes). Furfural yield from wheat straw and brewery waste increased from 20.3 mol% to 28.9 mol% and 4.8 mol% to 13.4 mol%, respectively. In another study, Sangarunlert et al. (2007) reported furfural yield of almost 90% from rice husks by acid hydrolysis with supercritical CO_2. Serna et al. (2016) in their study using supercritical CO_2 to produce furfural from rice husk further concluded that using the carbon dioxide as a green solvent for continuous processing of biomass depends on the composition characteristics of the lignocellulosic residues. Morais and Bogel-Lukasik (2016) investigated furfural formation from xylose using supercritical CO_2 and tetrahydrofuran (THF) as solvent. In their study, 43 mol% furfural yield was achieved in comparison to only 25 mol% in the absence of CO_2 (Morais and Lukasik 2017).

On the other hand, several studies concluded that furfural formation from biomass model compound (xylose and xylulose) and real biomass is enhanced under supercritical water conditions although the reported yield is relatively low (1–5%) compared to other supercritical fluids (Sasaki et al. 2003; Aida et al. 2010; Goodwin and Rorrer 2010; Paksung and Matsumura 2015; Paksung et al. 2016; Chalermsaktrakul and Matsumura 2018). On contrary, furfural is the most abundant liquid product when the temperature is just below the critical point of water (Paksung and Matsumura 2015; Paksung et al. 2016). Paksung and Matsumura (2015) further deduced the reactions occurring under supercritical water conditions are mainly retro-aldol condensation reactions as well as dehydration to produce furfural. The reaction pathway of D-xylose in supercritical water based on Aida et al. (2010) is shown in Fig. 4.6.

Mohamad and Yong (2015) and Yong et al. (2016) successfully demonstrated furfural production from oil palm frond using supercritical ethanol with catalyst (formic acid). The proposed method is self-sustainable as ethanol is produced from biomass and formic acid is a by-product of furfural production. The furfural yield reported from this study ranged from 4.9 to 35.8% (dry weight basis), comparable to conventional methods. The study shows that high furfural yield can be obtained using supercritical ethanol without the need of catalyst.

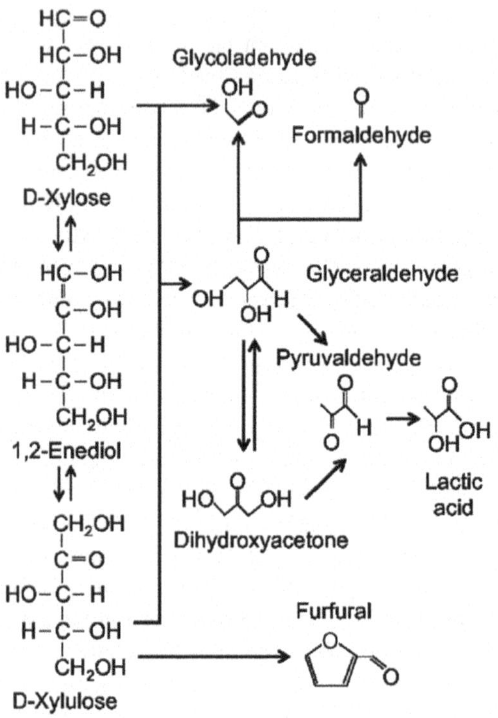

Fig. 4.6 Reaction pathway of D-xylose in supercritical water

4.3 Levulinic Acid

Levulinic acid is a top 12 value-added platform chemical determined by the US Department of Energy in 2004 (Werpy et al. 2004). Levulinic acid (4-oxopentanoic acid or γ-ketovaleric acid) is a 5-carbon chemical compound consisting of a short-chain fatty acid with an acidic carboxyl group (COOH) and a ketone carbonyl group (C=O). The two functional groups in levulinic acid provided its versatility as it can further react with other functional groups to produce numerous derivatives through condensation, esterification, oxydehydrogenation, halogenation, hydrogenation, and others (Zhang et al. 2012). Levulinic acid is an ideal platform chemical with diverse applications in areas such as biofuels, resins, polymers, herbicides, pharmaceuticals, flavours and fragrances, solvents, plasticizers, and antifreeze agents (Yan et al. 2015).

Levulinic acid is produced from C6 and C5 sugars in lignocellulosic biomass. In the reaction with C6 sugars, levulinic acid is produced through hydrolysis via intermediate formation of hydroxymethylfurfural with formic acid as by-product in the process (Chang et al. 2007). C5 sugars can be converted to levulinic acid as well through complex reactions. It involved the dehydration of xylose in the presence of

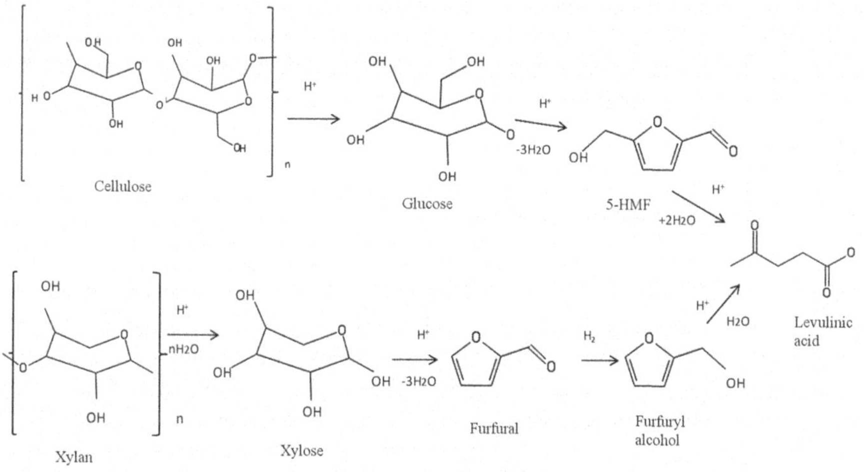

Fig. 4.7 Conversion routes of levulinic acid from lignocellulosic biomass

acid catalyst to produce furfural which will be further hydrogenated to furfuryl alcohol before being completely hydrolysed to levulinic acid using acid catalyst (Pileidis and Titirici 2016). Figure 4.7 shows the production routes of levulinic acid from lignocellulosic biomass.

Supercritical fluids with its both acid and basic properties are beneficial in the conversion of biomass to levulinic acid since it facilitates selective reactions and efficient separation of product from its medium. Jeong et al. (2017a) and Jang et al. (2018) studied the supercritical water hydrolysis of a lignocellulosic biomass (*Quercus mongolica*) at a pilot scale with varying acid catalyst loading. The reaction product was obtained within 1 s at 380 °C and 230 bar using 0.01–0.1% sulfuric acid with 0.1 g/L levulinic acid produced from the reaction. Levulinic acids were produced from fructose in supercritical water at 400 °C with 23–27 MPa (Cantero et al. 2015a). There has been reported study of levulinic acid production from municipal solid waste under supercritical water conditions in a plug flow reactor (210–230 °C, 25 bar, 12 s) and continuous stirred tank reactor (195–215 °C, 14 bar, 20 min) (Sadhukhan et al. 2016).

On the other hand, Kuznetsov et al. (2018) observed the formation of levulinic acid from aspen wood in supercritical ethanol with zeolite catalysts. Levulinic acid was produced at 270 °C with the pressure varying between 6.5 and 7.5 MPa. It is also interesting to note that the yield of levulinic acid increases two times more with the increase of zeolite catalyst loading. Zeng et al. (2015) obtained levulinic acid from the hydrolysation of 5-HMF in the decomposition products of both cellulose and hemicellulose from fir sawdust in supercritical ethanol with dissolved phosphotungstic acid.

Yu et al. (2018) carried out supercritical CO_2 extraction and subsequently thermochemical conversion to produce platform chemicals from food waste (corn, taro, lettuce, and bean sprout). The study observed lettuce and bean sprout produced a levulinic acid yield of ~7%, with glucose and fructose as the co-products. However, due to its high fibre content, it required a higher reaction temperature (170–190 °C) with sulfuric acid as catalyst.

4.4 5-Hydroxymethylfurfural

The conversion of lignocellulosic biomass to 5-HMF has gained interests due to its potential as substitute of fossil fuel-based monomers for various polymer productions. 5-HMF has been identified by US Department of Energy as a priority chemical deemed necessary for a successful biorefinery (Werpy et al. 2004). It is one of the very few petroleum-derived platform chemicals that can also be synthesized from biomass, a renewable resource (Lichtenthaler and Peters 2004). In addition, as a platform chemical, it can be used as a feedstock for various types of biofuels and chemicals in medicine and resin plastics and as diesel fuel additives. This can be achieved through numerous chemical reactions including hydrogenation, oxidative dehydrogenation, esterification, halogenation, polymerization, and hydrolysis with the aldehyde and hydroxymethyl group in 5-HMF (Román-Leshkov et al. 2006; Zheng et al. 2016).

Cellulose initially produced monosaccharides through hydrolysis. The subsequent dehydration (also termed as dehydrocyclization and cyclodehydration) and elimination of water molecules produced 5-HMF (Tong et al. 2017). The main reaction intermediates during these reactions are hexoses such as glucose and fructose. The reaction mechanism pathways to produce 5-HMF are shown in Fig. 4.8.

The synthesis of 5-HMF from biomass is often challenging as they are prone to side reactions, which included the rehydration of the 5-HMF to levulinic acid and formic acid. The reaction mechanism is shown in Fig. 4.9. In addition, they can undergo cross-polymerization to form soluble polymers and the undesired insoluble humin (Lewkowski 2001).

There are various studies on 5-HMF formation from cellulose (Sasaki et al. 1998; Cantero et al. 2015a; Martínez et al. 2015; Aspromonte et al. 2019) and glucose (Sasaki et al. 2002; Sınağ et al. 2003; Chuntanapum and Matsumura 2010a; Chuntanapum and Matsumura 2010b; Zhu et al. 2016) in supercritical water although it was concluded generally that the production of 5-HMF from cellulose and glucose in pressurized water is highly dependent on reaction temperature and the production of 5-HMF is suppressed once the reaction temperature is over the critical point of water. The proposed reaction mechanism of cellulose in sub- and supercritical water to produce 5-HMF according to Yu et al. (2007) is shown in Fig. 4.10.

Yang et al. (2016) obtained 5-HMF as the main product with yield of 14.6% using cellulose in supercritical ethanol-water solutions. Bicker et al. (2005) and Bicker et al. (2003) in their study of D-fructose dehydration to 5-HMF in supercritical acetone/water mixture (90:10) obtained 77% selectivity to 5-HMF and 99%

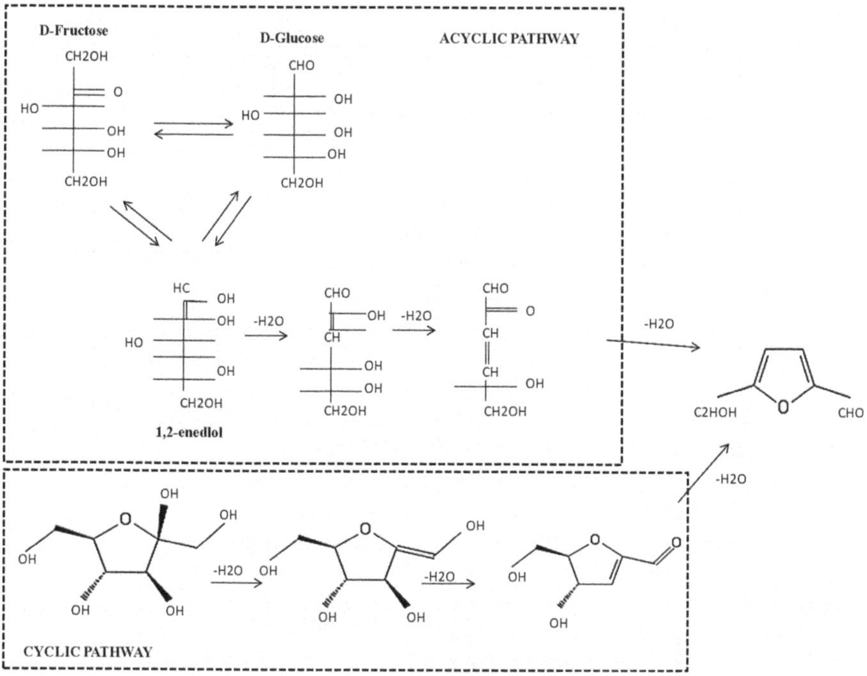

Fig. 4.8 Reaction mechanism of biomass sugars to 5-HMF

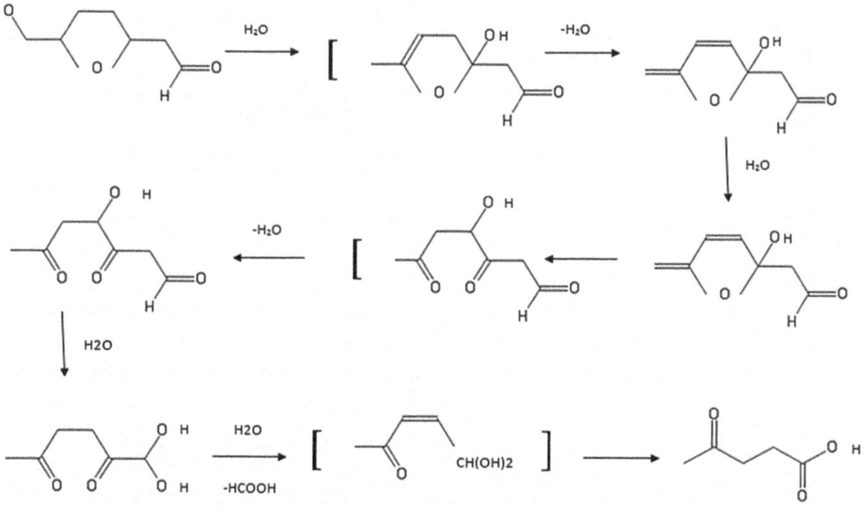

Fig. 4.9 Reaction mechanism of 5-*HMF* hydration to levulinic acid

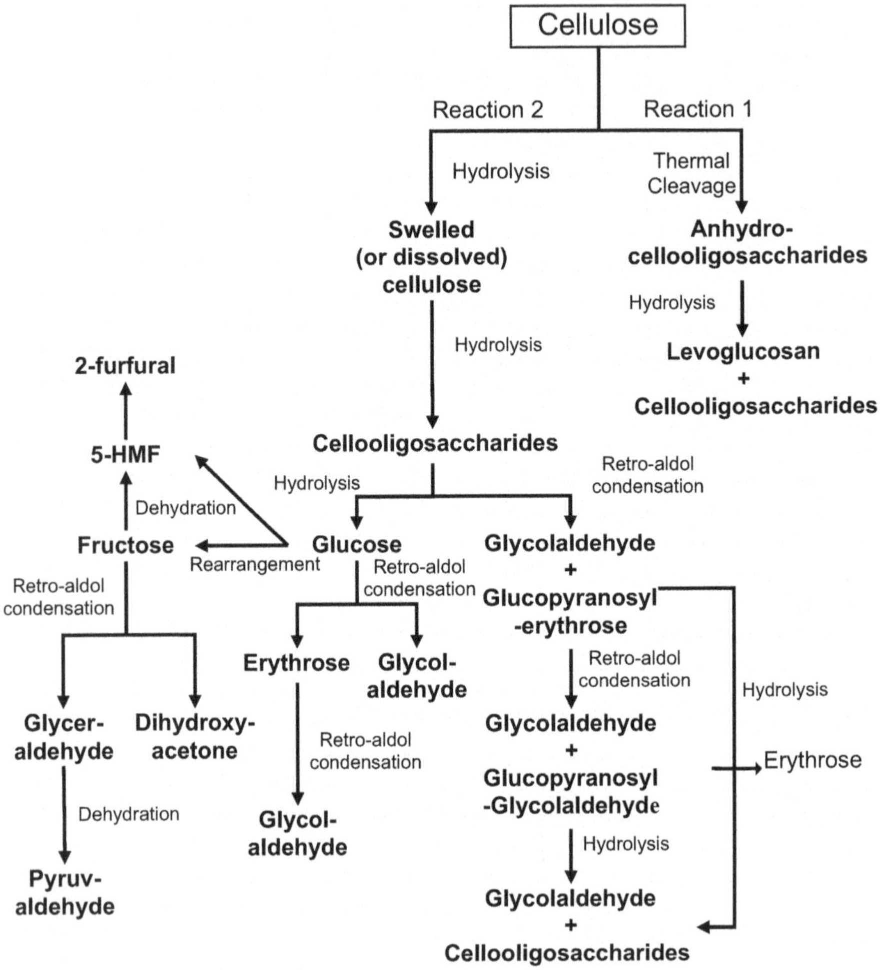

Fig. 4.10 The proposed reaction mechanism of cellulose in sub- and supercritical water

conversion at 180 °C and 120 s. The study on supercritical alcohol (methanol and ethanol) to yield 5-HMF also produced some promising results. For example, Galebach et al. (2018) obtained 5-HMF up to 1.5% carbon yield from cellulose without a catalyst. Brand and Kim (2015) observed the formation of 5-HMF in its liquid compounds from cellulose decomposition in supercritical ethanol.

There are several studies focusing on production of 5-HMF directly from biomass using supercritical fluids. Palma and Taylor (2001) obtained 5-HMF from raisins using supercritical CO_2 where they obtained average amount of 0.128 mg/g of raisin. The starch-rich biomass (taro and corn) resulted in 11–20% yield of 5-HMF in supercritical CO_2 after microwave heating at 140 °C for 5–10 min using $SnCl_4$

catalyst (Yu et al. 2018). Liquefaction of Japanese beech in supercritical methanol produced 5-HMF through the formation of methylated cellotriose and methylated cellobiose (Yamazaki et al. 2006). Martínez et al. (2018) obtained 2% yield of 5-HMF from sugar beet pulp under supercritical water conditions at 390 °C, 25 MPa, and reaction times between 0.11 and 1.15 s. 5-HMF were obtained from wheat bran by supercritical water hydrolysis although it was further deduced that 5-HMF production was highly inhibited in the experimented conditions to obtain yields lower than 0.5% (Cantero et al. 2015b). Serna et al. (2016) established the formation of 5-HMF from rich husk under supercritical CO_2 conditions. Studies on supercritical water hydrolysis of woody biomass (Jeong et al. 2017b) and holm oak (Piqueras et al. 2017) concluded that glucose are converted into their primary degradation products, 5-HMF, via dehydration above a certain level of hydrolysis severity.

4.5 Amino Acids

Protein and amino acids have a high commercial value in many industries which produce feedstock, foods, pharmaceuticals, and cosmetics. The high demand attributed to obtaining amino acids from foods or nutritional supplements makes them extremely valuable. There are at least 20 known standard amino acids, 8 of them are characterized and named as essential amino acids. Extracted protein from food wastes or biomass is an essential source of amino acids used for bulk platform chemical production. Globally, the production of amino acids is nearly five million tons within the few decades (Eggeling and Bott 2015). L-Lysine and L-glutamic acid can be hydrolysed from biomass, potentially serving as useful feedstock chemicals for a variety of commodity chemicals like N-methylpyrrolidone and N-vinylpyrrolidone which can be produced from glutamic acid. Similarly, lysine can be converted to 5-aminovaleric acid, 1,5-diaminopentane, and caprolactam (Clark and Deswarte 2015).

More recently supercritical fluid extraction using CO_2 as a solvent-free method has been used for the extraction of these value-added products. This is because it offers attractive features overcoming many of the limitations due to more conventional extraction methods and can preserve the quality of the bioactive compound. According to Chee et al. (2013), the optimum conditions for maximum recovery of protein from defatted rice bran fall within the intermediate range of 60 °C, 450 bar, and 90 min. There is minimal denaturation of protein that occurs within the temperature ranges of 35–80 °C and pressure between 10 and 30 MPa (Ackman 1991), although, in some conditions, for more viscous properties of waste, the temperature needs to be higher to extract amino acids. The waste refers to compounds that contain molasses, oil, tannin, fibres, starch, and minerals (Dividich et al. 1978; Steg and Van Der Meer 1985; Bhagavan and Ha 2011; Hoi and Martincigh 2013; Tanaka et al. 2017). Supercritical CO_2 was used to extract free amino acids from sugar beet and sugarcane molasses where it was observed that the extraction recoveries of amino acids were 42% and 31% for aspartic acid, 63% and 37% for glutamic acid, 46% and 48% for alanine, and 31% and 20% for lysine sequentially (Varaee et al. 2019).

Notwithstanding, according to Cheng et al. (2008), most amino acids provide maximum yield at the reaction temperature between 200 and 290 °C. Here, the quality of protein and amino acids was well preserved in various supercritical CO_2 extraction process conditions with certain types of amino acids like arginine. The protein damage observed was minor (Kasche et al. 1988; Lin et al. 1992; Kamat et al. 1995; Zheng and Tsao 1996; Tibbetts et al. 2015). Ali-Nehari et al. (2011) discovered that higher amino acid yield was produced from krill residue (537.78 mg/g) with higher temperature (280 °C). This studies proved that the process is feasible in low-pressure conditions due to the low solubility of the compounds in supercritical fluids.

4.6 L-Glutamic Acid and L-Lysine

L-Glutamic acid is an amino acid in which high yields can be obtained from different sources and is one of the first primary metabolites produced industrially through the process of fermentation. L-Glutamic acid is mainly used as a flavour enhancer. Notably, L-lysine is typically found among the important amino acids in animal feeds.

Given the continuous demand for amino acids, the use of natural sources from biomass as feedstock for L-glutamic acid and L-lysine is highly desirable. Contemporary lysine production is based on glucose and starch feedstocks found in Western parts of the world and on sucrose, raw sugar, and molasses found in Eastern regions, thus leading to unfavourable competition with human nutritional needs (Schneider et al. 2011). Accordingly, the need for more sustainable feedstocks seems inevitable given the growing world population. On the other hand, there is some biomass having high potential to produce these compounds using supercritical fluid technology. The use of inedible biomass as substrate for the microbial production of building block chemicals is considered ideal to prevent competition with food supply. Tibbetts et al. 2015 extracted L-lysine and L-glutamic acid at 70 °C and 35 MPa within 270 min from spirulina biomass. The study resulted in L-lysine and L-glutamic acid of 42.6 mg/g and 21.9 mg/g, respectively. Another study used deoiled palm kernel cake as the source of amino acids using a mild condition for SF extraction (60 °C, 44.6 MPa, 50 min) and found 21.11% and 5.23% of L-glutamic acid and L-lysine, respectively (Hossain et al. 2016). The decrease of amino acid yield may be attributed to the rapid decomposition rate compared to the extraction rate of amino acids by the hydrolysis of proteins.

4.7 Conclusion

The potential of lignocellulosic and protein biomass as an alternative platform to fossil fuel resources for the production of platform chemical has been demonstrated. Currently, the main technological barrier is to cost-effectively convert the biomass to the various types of platform chemicals due to its diverse competitions. In order to make the processes more competitive and economic, it will be a must to further produce value-added chemicals and to develop more efficient processes. The efficient utilization of biomass by conversion to platform chemicals is the backbone in the global economy shift towards a sustainable production. Plant and protein biomasses are valuable source for energy generation as well as chemical production in various industries. Efficient biomass separation is one of the major challenges in biorefineries due to the complex structure of the plant cell wall and the high crystallinity of cellulose causing it recalcitrant to separation. Although numerous studies as discussed previously have established the feasibility of producing various types of platform chemicals using supercritical fluid technology, the high initial cost often associated with the process as well as the large amount of energy needed is indeed the very last challenge and barrier for future expansion of the technology.

References

Ackman RG (1991) Application of gas-liquid chromatography to lipid separation and analysis. In: Perkins EG (ed) Analysis of fats, oils, and lipoproteins. American Oil Chemists' Society, Champaign, IL, p 270

Aida TM, Shiraishi N, Kubo M, Watanabe M, Smith RL Jr (2010) Reaction kinetics of d-xylose in sub-and supercritical water. J Supercrit Fluids 55:208–216. https://doi.org/10.1016/j.supflu.2010.08.013

Albarelli JQ, Santos DT, Ensinas AV, Marechal F, Cocero MJ, Meireles MAA (2018) Product diversification in the sugarcane biorefinery through algae growth and supercritical CO_2 extraction: thermal and economic analysis. Renew Energy 129:776–785. https://doi.org/10.1016/j.renene.2017.05.022

Ali-Nehari A, Kim SB, Lee YB, Chun BS (2011) Production of value added materials by subcritical water hydrolysis from krill residues extracted by supercritical carbon dioxide. Afr J Biotechnol 10(80):18450–18457. https://doi.org/10.5897/ajb10.2450

Antal MJ Jr, Leesomboon T, Mok WS, Richards GN (1991) Mechanism of formation of 2-furaldehyde from D-xylose. Carbohydr Res 217:71–85. https://doi.org/10.1016/0008-6215(91)84118-X

Asghari FS, Yoshida H (2010) Conversion of Japanese red pine wood (Pinus densiflora) into valuable chemicals under subcritical water conditions. Carbohydr Res 345:124–131. https://doi.org/10.1016/j.carres.2009.10.006

Aspromonte SG, Romero A, Boix AV, Alonso E (2019) Hydrolysis of cellulose to glucose by supercritical water and silver mesoporous zeolite catalysts. Cellulose 26:2471–2485. https://doi.org/10.1007/s10570-018-2221-5

Bhagavan NV, Ha CE (2011) Amino acids. In: Bhagavan NV, Ha CE (eds) Essentials of medical biochemistry: with clinical cases. Academic Press, pp 19–27

Bicker M, Hirth J, Vogel H (2003) Dehydration of fructose to 5-hydroxymethylfurfural in sub- and supercritical acetone. Green Chem 5:280–284. https://doi.org/10.1039/b211468b

Bicker M, Kaiser D, Ott L, Vogel H (2005) Dehydration of D-fructose to hydroxymethylfurfural in sub-and supercritical fluids. J Supercrit Fluids 36:118–126. https://doi.org/10.1016/j.supflu.2005.04.004

Boisen A, Christensen TB, Fu W, Gorbanev YY, Hansen TS, Jensen JS et al (2009) Process integration for the conversion of glucose to 2,5-furandicarboxylic acid. Chem Eng Res Des 87:1318–1327. https://doi.org/10.1016/j.cherd.2009.06.010

Bouza RJ, Gu Z, Evans JH (2016) Screening conditions for acid pretreatment and enzymatic hydrolysis of empty fruit bunches. Ind Crop Prod 84:67–71. https://doi.org/10.1016/j.indcrop.2016.01.041

Brand S, Kim J (2015) Liquefaction of major lignocellulosic biomass constituents in supercritical ethanol. Energy 80:64–74. https://doi.org/10.1016/j.energy.2014.11.043

Brandt A, Ray MJ, To TQ, Leak DJ, Murphy RJ, Welton T (2011) Ionic liquid pretreatment of lignocellulosic biomass with ionic liquid–water mixtures. Green Chem 13:2489–2499. https://doi.org/10.1039/c1gc15374a

Cai CM, Nagane N, Kumar R, Wyman CE (2014) Coupling metal halides with a co-solvent to produce furfural and 5-HMF at high yields directly from lignocellulosic biomass as an integrated biofuels strategy. Green Chem 16:3819–3829. https://doi.org/10.1039/c4gc00747f

Cantero DA, Vaquerizo L, Martinez C, Bermejo MD, Cocero MJ (2015a) Selective transformation of fructose and high fructose content biomass into lactic acid in supercritical water. Catal Today 255:80–86. https://doi.org/10.1016/j.cattod.2014.11.013

Cantero DA, Martínez C, Bermejo MD, Cocero MJ (2015b) Simultaneous and selectiverecovery of cellulose and hemicellulose fractions from wheat bran by supercritical water hydrolysis. Green Chem 17:610–618. https://doi.org/10.1039/c4gc01359j

Chalermsaktrakul T, Matsumura Y (2018) Effect of acetic acid addition on decomposition of xylose in supercritical water. Energy Fuel 32:1754–1760. https://doi.org/10.1021/acs.energyfuels.7b02720

Chang C, Cen P, Ma X (2007) Levulinic acid production from wheat straw. Bioresour Technol 98:1448–1453. https://doi.org/10.1016/j.biortech.2006.03.031

Chee FL, Iqbal S, Ismail M (2013) Effects of supercritical fluid extraction conditions on yield of protein from defatted rice bran. J Chem Soc Pak 35:192

Cheng H, Zhu X, Zhu C, Qian J, Zhu N, Zhao L, Chen J (2008) Hydrolysis technology of biomass waste to produce amino acids in sub-critical water. Bioresour Technol 99:3337–3341. https://doi.org/10.1016/j.biortech.2007.08.024

Chuntanapum A, Matsumura Y (2010a) Char formation mechanism in supercritical water gasification process: a study of model compounds. Ind Eng Chem Res 49:4055–4062. https://doi.org/10.1021/ie901346h

Chuntanapum A, Matsumura Y (2010b) Role of 5-HMF in supercritical water gasification of glucose. J Chem Eng Jpn 44:1011020091. https://doi.org/10.1252/jcej.10we146

Clark JH, Deswarte F (eds) (2015) Introduction to chemicals from biomass. Wiley, Chichester

de Carvalho, Martínez-Abad A, Evtuguin DV, Colodette JL, Lindström ME, Vilaplana F, Sevastyanova O (2017) Isolation and characterization of acetylated glucuronoarabinoxylan from sugarcane bagasse and straw. Carbohydr Polym 156:223–234

Dias AS, Lima S, Pillinger M, Valente AA (2010) Furfural and furfural-based industrial chemicals. In: Pignataro B (ed) Ideas in chemistry and molecular sciences: advances in synthetic chemistry. Wiley, pp 165–186. https://doi.org/10.1002/9783527630554.ch8

Dividich JL, Christon R, Peiniau J, Aumaitre A (1978) Proximate chemical analysis of final cane molasses and effect of feeding 30% molasses on intestinal sucrose and maltase activities in the rat. Anim Feed Sci Technol 3:15–22. https://doi.org/10.1016/0377-8401(78)90018-4

Eggeling L, Bott M (2015) A giant market and a powerful metabolism: L-lysine provided by Corynebacterium glutamicum. Appl Microbiol Biotechnol 99:3387–3394. https://doi.org/10.1007/s00253-015-6508-2

Gairola K, Smirnova I (2012) Hydrothermal pentose to furfural conversion and simultaneous extraction with SC-CO_2–kinetics and application to biomass hydrolysates. Bioresour Technol 123:592–598. https://doi.org/10.1016/j.biortech.2012.07.031

Galebach PH, McClelland DJ, Eagan NM, Wittrig AM, Buchanan JS, Dumesic JA, Huber GW (2018) Production of alcohols from cellulose by supercritical methanol depolymerization and hydrodeoxygenation. ACS Sustain Chem Eng 6:4330–4344. https://doi.org/10.1021/acssuschemeng.7b04820

Gallina G, Biasi P, Piqueras CM, García-Serna J (2017) Processing of lignocellulosic biomass derived monomers using high-pressure CO_2 and CO_2–H_2O mixtures. In: Łukasik RM (ed) High pressure technologies in biomass conversion, vol 48. Royal Society of Chemistry, p 115

Goodwin AK, Rorrer GL (2010) Reaction rates for supercritical water gasification of xylose in a micro-tubular reactor. Chem Eng J 163:10–21. https://doi.org/10.1016/j.cej.2010.07.013

Hoi LWS, Martincigh BS (2013) Sugar cane plant fibres: separation and characterisation. Ind Crop Prod 47:1–12. https://doi.org/10.1016/j.indcrop.2013.02.017

Hossain S, Nik Norulaini NA, Ahmad Naim AY, Mohd Zulkhairi AR, Massaud Bennama M, Mohd Omar AK (2016) Utilization of the supercritical carbon dioxide extraction technology for the production of deoiled palm kernel cake. J CO_2 Util 16:121–129

Huber GW, Iborra S, Corma A (2006) Synthesis of transportation fuels from biomass: chemistry, catalysts, and engineering. Chem Rev 106:4044–4098. https://doi.org/10.1021/cr068360d

Jana H, Ulla L (2011) Ionic liquids in the pretreatment of lignocellulosic biomass, ionic liquids: applications and perspectives. In: Kokorin A (ed) . InTech. http://www.intechopen.com/books/ionic-liquids-applications-and-perspectives/ionicliquids-in-thepretreatment-of-lignocellulosic-biomass; ISBN: 978-953-307-248-7

Jang SK, Kim JH, Jeong H, Choi JH, Lee SM, Choi IG (2018) Investigation of conditions for dilute acid pretreatment for improving xylose solubilization and glucose production by supercritical water hydrolysis from Quercus mongolica. Renew Energy 117:150–156. https://doi.org/10.1016/j.renene.2017.10.015

Jeong H, Jang SK, Hong CY, Kim SH, Lee SY, Lee SM et al (2017a) Levulinic acid production by two-step acid-catalyzed treatment of Quercus mongolica using dilute sulfuric acid. Bioresour Technol 225:183–190. https://doi.org/10.1016/j.biortech.2016.11.063

Jeong H, Park YC, Seong YJ, Lee SM (2017b) Sugar and ethanol production from woody biomass via supercritical water hydrolysis in a continuous pilot-scale system using acid catalyst. Bioresour Technol 245:351–357. https://doi.org/10.1016/j.biortech.2017.08.058

JSME steam tables (1983) In: Japanese Society of Mechanical Engineers (ed) Thermophysical properties of fluids. Japanese Society of Mechanical Engineers, Tokyo

Kaar WE, Gutierrez CV, Kinoshita CM (1998) Steam explosion of sugarcane bagasse as a pretreatment for conversion to ethanol. Biomass Bioenergy 14:277–287. https://doi.org/10.1016/S0961-9534(97)10038-1

Kamat EJ, Beckman A, Russell J (1995) Enzyme activity in supercritical fluids. Crit Rev Biotechnol 15:41–71. https://doi.org/10.3109/07388559509150531

Kang S, Li X, Fan J, Chang J (2013) Hydrothermal conversion of lignin: a review. Renew Sust Energ Rev 27:546–558. https://doi.org/10.1016/j.rser.2013.07.013

Kasche V, Schlothaucer R, Brunner G (1988) Enzyme denaturation in supercritical CO_2: stabilizing effects of SS bonds during depressurization step. Biotechnol Lett 10:569–574. https://doi.org/10.1007/BF01027131

Kruse A, Dinjus E (2007) Hot compressed water as reaction medium and reactant. J Supercrit Fluids 39(3):362–380

Kuznetsov BN, Sharypov VI, Beregovtsova NG, Baryshnikov SV, Pestunov AV, Vosmerikov AV, Djakovitch L (2018) Thermal conversion of mechanically activated mixtures of aspen wood-zeolite catalysts in a supercritical ethanol. J Anal Appl Pyrolysis 132:237–244. https://doi.org/10.1016/j.jaap.2018.01.022

Lewkowski J (2001) Synthesis, chemistry and applications of 5-hydroxymethyl-furfural and its derivatives. Arkivoc 1:17–54

Lichtenthaler FW, Peters S (2004) Carbohydrates as green raw materials for the chemical industry. C R Chim 7:65–90. https://doi.org/10.1016/j.crci.2004.02.002

Lin HM, Yang Z, Chen LF (1992) Inactivation of Saccharomyces cerevisiae by supercritical and subcritical carbon dioxide. Biotechnol Prog 8:458–461. https://doi.org/10.1021/bp00017a013

López F, García MT, Feria MJ, García JC, De Diego C, Zamudio MA, Díaz MJ (2014) Optimization of furfural production by acid hydrolysis of Eucalyptus globulus in two stages. Chem Eng J 240:195–201. https://doi.org/10.1016/j.cej.2013.11.073

Mamman AS, Lee JM, Kim YC, Hwang IT, Park NJ, Hwang YK et al (2008) Furfural: Hemicellulose/xylose derived biochemical. Biofuels Bioprod Bioref 2:438–454. https://doi.org/10.1002/bbb.95

Martínez CM, Cantero DA, Bermejo MD, Cocero MJ (2015) Hydrolysis of cellulose in supercritical water: reagent concentration as a selectivity factor. Cellulose 22:2231–2243. https://doi.org/10.1007/s10570-015-0674-3

Martínez CM, Cantero DA, Cocero MJ (2018) Production of saccharides from sugar beet pulp by ultrafast hydrolysis in supercritical water. J Cleaner Prod 204:888–895. https://doi.org/10.1016/j.jclepro.2018.09.066

Marzialetti T, Valenzuela Olarte MB, Sievers C, Hoskins TJ, Agrawal PK, Jones CW (2008) Dilute acid hydrolysis of loblolly pine: a comprehensive approach. Ind Eng Chem Res 47:7131–7140. https://doi.org/10.1021/ie800455f

Minami E, Saka S (2005) Decomposition behavior of woody biomass in water-added supercritical methanol. J Wood Sci 51:395–400. https://doi.org/10.1007/s10086-004-0670-y

Miyazawa T, Funazukuri T (2005) Polysaccharide hydrolysis accelerated by adding carbon dioxide under hydrothermal conditions. Biotechnol Prog 21:1782–1785. https://doi.org/10.1021/bp050214q

Mohamad N, Yong TLK (2015) Production of furfural from oil palm frond via green chemicals using supercritical ethanol and formic acid. J Jpn Inst Energy 94:835–840. https://doi.org/10.3775/jie.94.835

Mohan M, Banerjee T, Goud VV (2015) Hydrolysis of bamboo biomass by subcritical water treatment. Bioresour Technol 191:244–252. https://doi.org/10.1016/j.biortech.2015.05.010

Morais ARC, Bogel-Lukasik R (2016) Highly efficient and selective CO_2-adjunctive dehydration of xylose to furfural in aqueous media with THF. Green Chem 18:2331–2334. https://doi.org/10.1039/c5gc02863a

Morais ARC, Lukasik RM (2017) Hydrothermal pretreatment using supercritical CO_2 in the biorefinery context. In: Ruiz H, Hedegaard TM, Trajano H (eds) Hydrothermal processing in biorefineries. Springer, Cham. https://doi.org/10.1007/978-3-319-56457-9_14

Naik SN, Goud VV, Rout PK, Dalai AK (2010) Production of first and second generation biofuels: a comprehensive review. Renew Sust Energ Rev 14:578–597. https://doi.org/10.1016/j.rser.2009.10.003

Nimlos MR, Qian X, Davis M, Himmel ME, Johnson DK (2006) Energetics of xylose decomposition as determined using quantum mechanics modeling. J. Phys Chem A 110:11824–11838. https://doi.org/10.1021/jp0626770

Paksung N, Matsumura Y (2015) Decomposition of xylose in sub-and supercritical water. Ind Eng Chem Res 54:7604–7613. https://doi.org/10.1021/acs.iecr.5b01623

Paksung N, Nagano R, Matsumura Y (2016) Detailed mechanism of xylose decomposition in near-critical and supercritical water. Energy Fuel 30:7930–7936. https://doi.org/10.1021/acs.energyfuels.6b00918

Palma M, Taylor LT (2001) Supercritical fluid extraction of 5-hydroxymethyl-2-furaldehyde from raisins. J Agric Food Chem 49:628–632. https://doi.org/10.1021/jf001070s

Pileidis FD, Titirici MM (2016) Levulinic acid biorefineries: new challenges for efficient utilization of biomass. ChemSusChem 9:562–582. https://doi.org/10.1002/cssc.201501405

Piqueras CM, Cabeza A, Gallina G, Cantero DA, García-Serna J, Cocero MJ (2017) Online integrated fractionation-hydrolysis of lignocellulosic biomass using sub-and supercritical water. Chem Eng J 308:110–125. https://doi.org/10.1016/j.cej.2016.09.007

Qing Q, Guo Q, Zhou L, Wan Y, Xu Y, Ji H et al (2017) Catalytic conversion of corncob and corncob pretreatment hydrolysate to furfural in a biphasic system with addition of sodium chloride. Bioresour Technol 226:247–254. https://doi.org/10.1016/j.biortech.2016.11.118

Raman JK, Gnansounou E (2015) Furfural production from empty fruit bunch–a biorefinery approach. Ind Crops Prod 69:371–377. https://doi.org/10.1016/j.indcrop.2015.02.063

Riansa-ngawong W, Prasertsan P (2011) Optimization of furfural production from hemicellulose extracted from delignified palm pressed fiber using a two-stage process. Carbohydr Res 346(1):103–110

Román-Leshkov Y, Chheda JN, Dumesic JA (2006) Phase modifiers promote efficient production of hydroxymethylfurfural from fructose. Science 312:1933–1937. https://doi.org/10.1126/science.1126337

Sadhukhan J, Ng KS, Martinez-Hernandez E (2016) Novel integrated mechanical biological chemical treatment (MBCT) systems for the production of levulinic acid from fraction of municipal solid waste: a comprehensive techno-economic analysis. Bioresour Technol 215:131–143. https://doi.org/10.1016/j.biortech.2016.04.030

Sangarunlert W, Piumsomboon P, Ngamprasertsith S (2007) Furfural production by acid hydrolysis and supercritical carbon dioxide extraction from rice husk. Korean J Chem Eng 24:936–941. https://doi.org/10.1007/s11814-007-0101-z

Sasaki M, Kabyemela B, Malaluan R, Hirose S, Takeda N, Adschiri T, Arai K (1998) Cellulose hydrolysis in subcritical and supercritical water. J Supercrit Fluids 13:261–268. https://doi.org/10.1016/S0896-8446(98)00060-6

Sasaki M, Goto K, Tajima K, Adschiri T, Arai T (2002) Rapid and selective retro-aldol condensation of glucose to glycolaldehyde in supercritical water. Green Chem 4(3):285–287

Sasaki M, Hayakawa T, Arai K, Adschiri T (2003) Measurement of the rate of retro-aldol condensation of D-xylose in subcritical and supercritical water. In: Hydrothermal reactions and techniques, pp 169–176. https://doi.org/10.1142/9789812705228_0022

Savage PE (1999) Organic Chemical Reactions in Supercritical Water. Chem Rev 99(2):603–622

Schneider J, Niermann K, Wendisch VF (2011) Production of the amino acids l-glutamate, l-lysine, l-ornithine and l-arginine from arabinose by recombinant Corynebacterium glutamicum. J Biotechnol 154(2–3):191–198

Seemala B, Haritos V, Tanksale A (2016) Levulinic acid as a catalyst for the production of 5-hydroxymethylfurfural and furfural from lignocellulose biomass. ChemCatChem 8:640–647. https://doi.org/10.1002/cctc.201501105

Serna LD, Alzat CO, Alzate CC (2016) Supercritical fluids as a green technology for the pretreatment of lignocellulosic biomass. Bioresour Technol 199:113–120. https://doi.org/10.1016/j.biortech.2015.09.078

Sınag A, Kruse A, Schwarzkopf V (2003) Key compounds of the hydropyrolysis of glucose in supercritical water in the presence of K_2CO_3. Ind Eng Chem Res 42:3516–3521. https://doi.org/10.1021/ie030079r

Steg A, Van Der Meer JM (1985) Differences in chemical composition and digestibility of beet and cane molasses. Anim Feed Sci Technol 13:83–91. https://doi.org/10.1016/0377-8401(85)90044-6

Tanaka F, Tanaka A, Uchino T (2017) Effect of high temperature drying on amino acids decomposition in feed rice. Eng Agric Environ Food 10:1–3. https://doi.org/10.1016/j.eaef.2016.06.003

Tibbetts SM, Bjornsson WJ, McGinn PJ (2015) Biochemical composition and amino acid profiles of *Nannochloropsis granulata* algal biomass before and after supercritical fluid CO_2 extraction at two processing temperatures. Anim Feed Sci Technol 204:62–71. https://doi.org/10.1016/j.anifeedsci.2015.04.006

Tong X, Yu L, Chen H, Zhuang X, Liao S, Cui H (2017) Highly efficient and selective oxidation of 5-hydroxymethylfurfural by molecular oxygen in the presence of cu-MnO_2 catalyst. Catal Commun 90:91–94. https://doi.org/10.1016/j.catcom.2016.11.024

Van Walsum GP, Shi H (2004) Carbonic acid enhancement of hydrolysis in aqueous pretreatment of corn stover. Bioresour Technol 93:217–226. https://doi.org/10.1016/j.biortech.2003.11.009

Varaee M, Honarvar M, Eikani MH, Omidkhah MR, Moraki N (2019) Supercritical fluid extraction of free amino acids from sugar beet and sugar cane molasses. J Supercrit Fluids 144:48–55. https://doi.org/10.1016/j.supflu.2018.10.007

Weiss ND, Farmer JD, Schell DJ (2010) Impact of corn stover composition on hemicellulose conversion during dilute acid pretreatment and enzymatic cellulose digestibility of the pretreated solids. Bioresour Technol 101:674–678. https://doi.org/10.1016/j.biortech.2009.08.082

Werpy T, Petersen G, Aden A, Bozell J, Holladay J, White J, Manheim A, Elliot D, Lasure L, Jones S, Gerber M, Ibsen K, Lumberg L, Kelley S (2004) Top value added chemicals from biomass – volume 1: results of screening for potential candidates from sugars and synthesis gas. Pacific Northwest National Laboratory, National Renewable Energy Laboratory and Department of Energy, Washington, DC. 76 pp

Yamazaki J, Minami E, Saka S (2006) Liquefaction of beech wood in various supercritical alcohols. J Wood Sci 52:527–532. https://doi.org/10.1007/s10086-005-0798-4

Yan K, Wu G, Lafleur T, Jarvis C (2014) Production, properties and catalytic hydrogenation of furfural to fuel additives and value-added chemicals. Renew Sust Energ Rev 38:663–676. https://doi.org/10.1016/j.rser.2014.07.003

Yan K, Jarvis C, Gu J, Yan Y (2015) Production and catalytic transformation of levulinic acid: a platform for speciality chemicals and fuels. Renew Sust Energ Rev 51:986–997. https://doi.org/10.1016/j.rser.2015.07.021

Yang L, Yang X, Tian E, Lin H (2016) Direct conversion of cellulose into ethyl lactate in supercritical ethanol–water solutions. ChemSusChem 9:36–41. https://doi.org/10.1002/cssc.201500855

Yokoyama S, Matsumura Y (2008) Thermochemical conversion of biomass. In: The Asian biomass handbook: a guide for biomass production and utilization, 1st edn, part 4, p 71

Yong TLK, Matsumura Y (2012) Reaction kinetics of the lignin conversion in supercritical water. Ind Eng Chem Res 51:11975–11988. https://doi.org/10.1021/ie300921d

Yong TLK, Matsumura Y (2013) Kinetic analysis of guaiacol conversion in sub- and supercritical water. Ind Eng Chem Res 52:9048–9059. https://doi.org/10.1021/ie4009748

Yong TLK, Mohamad N, Yusof NNM (2016) Furfural production from oil palm biomass using a biomass-derived supercritical ethanol solvent and formic acid catalyst. Procedia Eng 148:392–400. https://doi.org/10.1016/j.proeng.2016.06.495

Yu Y, Lou X, Wu H (2007) Some recent advances in hydrolysis of biomass in hot-compressed water and its comparisons with other hydrolysis methods. Energy Fuel 22:46–60. https://doi.org/10.1021/ef700292p

Yu IK, Attard TM, Chen SS, Tsang DC, Hunt AJ, Jérôme F et al (2018) Supercritical carbon dioxide extraction of value-added products and thermochemical synthesis of platform chemicals from food waste. ACS Sustain Chem Eng 7:2821–2829. https://doi.org/10.1021/acssuschemeng.8b06184

Zeitsch KJ (2000) The chemistry and technology of furfural and its many by-products. Elsevier, Amsterdam/Oxford, p 13

Zeng C, Zheng H, Lv J, Chen X, Huang B (2015) Liquefaction of fir sawdust in supercritical ethanol with dissolved phosphotungstic acid. Bioresources 10:7738–7751

Zhang J, Wu S, Li B, Zhang H (2012) Advances in the catalytic production of valuable levulinic acid derivatives. ChemCatChem 4:1230–1237. https://doi.org/10.1002/cctc.201200113

Zhang L, Yu H, Wang P, Dong H, Peng X (2013) Conversion of xylan, d-xylose and lignocellulosic biomass into furfural using AlCl3 as catalyst in ionic liquid. Bioresour Technol 130:110–116

Zhang L, Xi G, Zhang J, Yu H, Wang X (2017) Efficient catalytic system for the direct transformation of lignocellulosic biomass to furfural and 5-hydroxymethylfurfural. Bioresour Technol 224:656–661. https://doi.org/10.1016/j.biortech.2016.11.097

Zheng Y, Tsao GT (1996) Avicel hydrolysis by cellulase enzyme in supercritical CO_2. Biotechnol Lett 18:451–454. https://doi.org/10.1007/BF00143469

Zheng X, Gu X, Ren Y, Zhi Z, Lu X (2016) Production of 5-hydroxymethyl furfural and levulinic acid from lignocellulose in aqueous solution and different solvents. Biofuels Bioprod Biorefin 10:917–931. https://doi.org/10.1002/bbb.1720

Zhu C, Guo L, Jin H, Huang J, Li S, Lian X (2016) Effects of reaction time and catalyst on gasification of glucose in supercritical water: detailed reaction pathway and mechanisms. Int J Hydrog Energy 41:6630–6639. https://doi.org/10.1016/j.ijhydene.2016.03.035

Chapter 5
Supercritical Carbon Dioxide: A Glimpse from the Modern Era of Green Chemistry

Atul Kumar Varma and Santanu Ghosh

Contents

A. K. Varma (✉) · S. Ghosh
Coal Geology and Organic Petrology Laboratory, Department of Applied Geology, Indian Institute of Technology (Indian School of Mines), Dhanbad 826004, Jharkhand, India
e-mail: atul@iitism.ac.in

Inamuddin, A. M. Asiri (eds.), *Advanced Nanotechnology and Application of Supercritical Fluids*, Nanotechnology in the Life Sciences,
https://doi.org/10.1007/978-3-030-44984-1_5

Abbreviations

AOT	Aerosol OT
CCS	Carbon capture and sequestration
CNT	Carbon nanotubes
CO_2	Carbon dioxide
EO	Ethylene oxide
EOR	Enhanced oil recovery
F7H7	A hybrid surfactant containing seven fluorine and seven hydrogen atoms
FFV	Fraction free volume
FTIR	Fourier transform infrared spectroscopy
ILs	Ionic liquids
MPa	Megapascal
n–C7	Straight-chain normal alkane with seven carbon atoms
n–CF_2H	Normal alkane in which a carbon atom is covalently bonded with two fluorine atoms and one hydrogen atom
n–CF_3	Normal alkane in which a carbon atom is covalently bonded with three fluorine atoms
NMR	Nuclear magnetic resonance spectroscopy
PET	Positron emission tomography
P_{trans}	Cloud point pressure
RESOLV	Rapid expansion of a supercritical solution into a liquid solvent
RESS	Rapid expansion of supercritical solutions
SAA	Supercritical assisted atomization
SAS	Supercritical anti-solvent precipitation
sCO_2	Supercritical carbon dioxide
SIS1	Sodium 2-(4,4-dimethylpentan-2yl)-5,7,7-trimethyl sulfate
STP	Standard temperature and pressure condition
TBP	Tri-*n*-butyl phosphate
TOA	Tri-*n*-octylamine
VOCs	Volatile organic compounds
W	Surfactant ratio
Φ_{surf}	Surfactant coverage at the interface

5.1 Introduction

In modern timeframe, science is looking toward renewability and sustainability, which encompasses the reduction in the use of toxic chemical substances and a sharp turn toward the applications of eco-friendly chemically efficient products with an increasing trend in grasping the philosophy of green chemistry (Anastas and Warner 1998; Peach and Eastoe 2014). Several types of operations, reactions, extraction and separation procedures in chemical processing industries require toxic and flammable organic solvents. Besides having adverse handling and disposal issues, those organic solvents, reasonably, toxicate the environment and its dwellers. Often, many organic solvents are used as volatile organic compounds (VOCs), and many others are restricted because of their significant contributions in ozone layer depletion. Progressively, use of supercritical fluids like supercritical carbon dioxide (supercritical CO_2) as solvents (Eckert et al. 1996) in chemical processing industrial appliances can result in the minimization of the applications of the traditional hazardous organic reagents and VOCs (Noyori 2005). Supercritical carbon dioxide comes out to be a potential alternative solvent that is not that much hazardous to the environment, and thus, it is considered as a green medium in modern chemical industrial processes. This is non-toxic, non-flammable unlike many other organic solvents, and inexpensive with low critical constants and has good availability in pure form and suitable transport properties, and these characteristics have made the supercritical carbon dioxide as a green solvent (Henczka and Djas 2016). Although the supercritical carbon dioxide is not a volatile organic compound, it is a potential greenhouse gas. However, if this is captured, processed, and then returned to the environment, the greenhouse effect can be pacified. The reduction in applications of toxic volatile organic compounds will enhance the sustainability of many chemical processes and also of the environment (Peach and Eastoe 2014).

This chapter takes a look on the chemical properties of the supercritical carbon dioxide, its applications in the chemical reactions, uses in material synthesis, roles in the synthesis of nanoparticles, and its influences in hydrocarbon exploration from the shale and tight sandstone reservoirs. As carbon dioxide is a non-polar compound, it faces difficulties in dissolving polar compounds, ionic complexes, and high molecular weight moieties. Therefore, the solubility of the compounds in carbon dioxide can be enhanced through the addition of fluorinated surfactants (Hoefling et al. 1991) and partially fluorinated moieties (Harrison et al. 1994; Sagisaka et al. 2003). However, for avoiding the adverse effects of the fluorinated compounds, non-fluorinated surfactants are receiving potential attention (da Rocha et al. 2003; Hollamby et al. 2009; Mohamed et al. 2010; Sarbu et al. 2000; Sagisaka et al. 2013). These compounds include the trisiloxanes (da Rocha et al. 2003), hydrocarbon surfactants (Mohamed et al. 2010; Pitt et al. 1996; Sagisaka et al. 2013; Trickett et al. 2010), and oxygenated moieties (Kazarian et al. 1996; Raveendran and Wallen 2002; Reilly et al. 1995). Addition of fluorinated polymers (Beyer et al. 2000; Loeker et al. 2003; Rindfleisch et al. 1996) and non-fluorinated polymers (Hong et al. 2008; Potluri et al. 2003; Sarbu et al. 2000; Tapriyal et al.

2008) had been reported to enhance the solubility of the carbon dioxide. Additionally, the carbon dioxide solubility can be quantified employing models like fraction free volume (FFV) and surfactant coverage at the interface (Φ_{surf}) (Peach and Eastoe 2014). Depending on these parameters, the affinity of any surfactant and polymer to the carbon dioxide can be estimated. The solvent property of the supercritical carbon dioxide can also be enhanced by tuning the viscosity. Addition of the carbon dioxide thickeners like polymeric viscosifiers (Heller et al. 1985; Huang et al. 2000) and salts (Hatzopoulos et al. 2013; James et al. 2014; Xu et al. 1993) may thicken the carbon dioxide. Moreover, the supercritical carbon dioxide can be applied as a green reaction medium in many chemical reactions to reduce the toxicity of the organic solvents and the by-products. Carbon dioxide acts as Lewis acid and interacts with strong bases like phosphenes, amines, alkyl anions, etc. (Inoue and Yamazaki 1982; West et al. 2001). Many chemical reactions like hydrogenation (Clark et al. 2007; Chouchi et al. 2001; Fujita et al. 2011; Han and Poliakoff 2012; Hiyoshi et al. 2006, 2007, 2009, 2012; Kainz et al. 1999; Lange et al. 2000; Seki et al. 2007, 2008; Stevens et al. 2009), hydroformylation (Bektesevic et al. 2006; Haumann and Riisagar 2008; Niessen and Woelk 2007; Sellin et al. 2001; Webb et al. 2003), oxidation (Beckman 2004; Bourne et al. 2008; Caravati et al. 2006; Han and Poliakoff 2012; Mayadevi 2012), C–C bonding (Ikariya and Kayaki 2000; Amandi et al. 2007; Early et al. 2001; Galia et al. 2011; Goddard et al. 2000; Marathe et al. 2002; Muljana et al. 2011), esterification reactions (Jackson et al. 2006), and transesterifications (Ciftci and Temelli 2011; Galia et al. 2011; Rodrigues et al. 2011; Shin et al. 2012; Varma and Madras 2007; Varma et al. 2010) employ the supercritical carbon dioxide as a reactant medium. Further, the supercritical carbon dioxide is used as a green medium for the synthesis of nanomaterials (Hu et al. 1999a, b; Kameo et al. 2003; McLeod et al. 2004; Ohde et al. 2000; Reverchon 2002; Reverchon and Adami 2006; Reverchon et al. 2003; Shah et al. 2000; Sun et al. 2001), reverse micelles (Chattopadhyay and Gupta 2003; Hong et al. 2003), nanowires (Coleman et al. 2001; Hanrath and Korgel 2002; Holmes et al. 2000, 2003), nanofibers (Meziani et al. 2005; Sun et al. 2005), carbon nanotubes (Motiei et al. 2001), nanofilms (Ye et al. 2003a, b, 2005), composite nanoparticles (Liu et al. 2004; Wang et al. 2004; Ye et al. 2003a, b; Zhang et al. 2004a, b, c), and drug nanoparticles (Padrela et al. 2018), among others. Supercritical carbon dioxide is also used as a green foaming agent to produce microcellular cross-linked polyethylene foams (Xing et al. 2008), polypropylene foams (Antunes et al. 2012; Bao et al. 2016; Rizvi and Park 2014; Yang et al. 2017), semicrystalline microcellular polymers (Tang and Wang 2017), etc. In complementary, the supercritical carbon dioxide has been used as a green solvent in reactive extraction procedures of carboxylic acid (Djas and Henczk 2016, 2018; Hartl and Marr 1993; Kapucu et al. 1999; Rahmanian and Ghaziaskar 2008; Wang et al. 2005), citric acid (Djas and Henczk 2016, 2018), and acetic and propionic acids (Henczka and Djas 2016). Supercritical carbon dioxide possesses the ability to extract organic matter from shale (Jiang et al. 2016). They had observed the increment in the density and solvent strength with an increase in temperature, pressure, and treatment time and, hence, greater extraction of organic matter from the pores and the fractures of shales, which would have enhanced the shale gas exploration.

Additionally, supercritical carbon dioxide is considered as an alternate fracturing fluid in shale gas exploration (Ishida et al. 2012; Jia et al. 2018; Middleton et al. 2015; Pei et al. 2015) to minimize the adverse effects of water, such as excessive use of water in dry areas (Scanlon et al. 2014; Vengosh et al. 2014), mechanical alterations of the shale gas resource plays (Jia et al. 2017, 2018; Lu et al. 2016), and groundwater contamination (Myers 2012; Osborn et al. 2011; Vengosh et al. 2014). Supercritical carbon dioxide can stimulate better propagation of the fractures (Ishida et al. 2012; Li et al. 2016) and induces small resistance in flow and desorption (Middleton et al. 2015). However, the supercritical carbon dioxide has also been observed to induce coal matrix swelling and decrease the coal seam permeability (Perera et al. 2011a) and, therefore, imparts adverse effects on shale gas exploration. Hence, this chapter covers a myriad of attributes associated with the supercritical carbon dioxide in terms of its advantages, disadvantages, efficiencies, and deficiencies and the plausible solutions to overcome those hindrances to place it as a suitable green medium replacing the traditional hazardous organic solvents in the industries.

5.2 Supercritical Fluids

The critical state of a substance is marked by a critical pressure and a critical temperature, where the thermodynamic equilibrium of the phase boundaries comes to an end. Above the critical point, the boundary between the liquid and vapor phase of a fluid vanishes, and the fluid properties can be regulated by adjusting the temperature and pressure conditions. Supercritical fluids possess density like the liquid but reveal diffusivity, viscosity, and surface tension like the gas. The gas-like low viscosity provides large mass transfer property, and due to this low viscosity compounded by low surface tension, supercritical fluids can easily penetrate into pores within the solid materials. On the other hand, the solvent strength of these fluids can be comparable to that of liquids. The strongly polar compounds exhibit higher critical temperatures compared to that of weakly polar substances due to the requirement of a huge amount of energy to overcome the polar attractive strength. That's why the water, toluene, isopropanol, etc. show a higher critical temperature than carbon dioxide, ethylene, etc. The molecular attraction of a supercritical fluid is scaled by the kinetic energy. The density of the fluids at the supercritical stage is crucially influenced by the temperature and pressure, and change in density due to changes in temperature and pressure conditions controls the solvent strength of the fluid. Hence, the supercritical fluids are considered as adjustable or tunable solvents as their properties are strongly regulated by pressure and temperature modifications, unlike the conventional solvents.

5.3 Characteristics of Supercritical Carbon Dioxide

Carbon dioxide behaves like gas in standard temperature and pressure condition (STP), and it also freezes as dry ice. However, it acts as a supercritical fluid (Fig. 5.1), when it is kept above the critical temperature of 304.2 K or 31.05 °C and the critical pressure of 7.38 megapascal (MPa) or 73.8 bar (Henczka and Djas 2016; Song et al. 2016; Span and Wagner 1996; Wang and Ni 2013; Yang et al. 2019). Like other supercritical fluids, the supercritical carbon dioxide shows the gas-like diffusivity and liquid-like density. The solvent power of the supercritical carbon dioxide can be adjusted through changing the pressure-temperature conditions.

However, low surface tension, viscosity, and dielectric constant unlike other common organic solvents, often, restrict the supercritical carbon dioxide from becoming an effective solvent (Harrison et al. 1994; Hoefling et al. 1991). Due to linear molecular geometry with sp hybridization, carbon dioxide lacks any net dipole moment and, hence, struggles to dissolve polar and ionic compounds (Hoefling et al. 1991). The polarity of the compounds also depends on the determination of the point groups (Peach and Eastoe 2014). When the dipole moment, individual, within a molecule pacifies one another, the molecule becomes non-polar. Additionally, a molecule having multiple C axes or a horizontal mirror plane or σ does not show any net dipole moment. Carbon dioxide is characterized by the presence of the $D_{\infty h}$, a point group, and, therefore, is non-polar (Peach and Eastoe 2014).

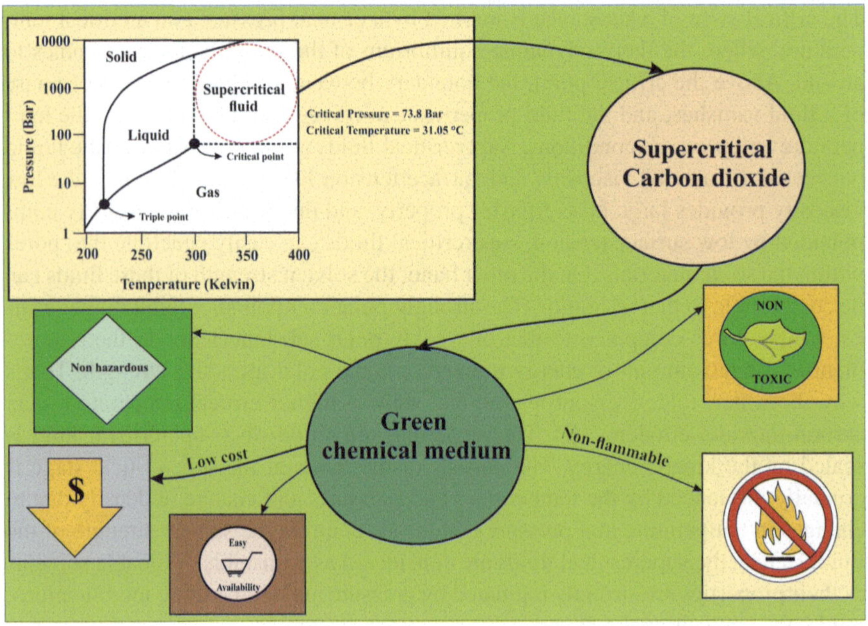

Fig. 5.1 Phase diagram of carbon dioxide with triple point and critical points along with the characteristics of supercritical carbon dioxide for considering it as a green medium

The supercritical carbon dioxide may be the good solvent for low molecular weight compounds but not for high molecular weight compounds. The physicochemical characteristics of the supercritical carbon dioxide require modifications to enable it for atmospheric carbon dioxide sequestration, storage, and enhanced oil recovery (EOR) procedures (Orr and Taber 1984; Peach and Eastoe 2014; Plasynski et al. 2009). Due to the enhanced concentration of greenhouse gases and consequent global warming, carbon capture and sequestration (CCS) offers one of the most viable practices to placate the situation. The effectiveness of the CCS and EOR requires control in fluid properties, especially an increase in fluid viscosity of the supercritical carbon dioxide (Peach and Eastoe 2014). Addition of surfactants and polymers may modify the fluid properties like solubility and viscosity of the supercritical carbon dioxide. Consani and Smith (1990) had observed that most of the commercially available surfactants are insoluble in supercritical carbon dioxide, while very few non-ionic compounds were marginally soluble. Therefore, there was an urge for developing carbon dioxide-soluble additives, especially surfactants and polymers in water-in-CO_2 systems (Peach and Eastoe 2014). However, nowadays, the focus is broadened toward other systems like ionic liquid-in-carbon dioxide systems (Chandran et al. 2010; Eastoe et al. 2005; Liu et al. 2007) and the development of lamellar and bicontinuous phases (Klostermann et al. 2012; Sagisaka et al. 2013)

5.3.1 Addition of Fluorinated Surfactants

Due to high electronegativity and large electron affinity of fluorines, fluorocarbons with carbon number equal to or greater than (≥) 4 have boiling points lower than the corresponding hydrocarbons (Kissa 1994). Moreover, in comparison to the hydrocarbons, fluorocarbons possess larger molecular volume and, hence, show lower polarizability per volume and Hildebrand[1] solubility parameter. Therefore, fluorocarbons appear more compatible with carbon dioxide compared to the hydrocarbons. Owing to large solubility of the fluorocarbon chains in the carbon dioxide, these compounds were synthesized and used extensively. Hoefling et al. (1991) had introduced those fluorocarbon compounds through synthesizing fluorinated Aerosol-OT such as sodium dioctylsulfosuccinate (Fig. 5.2) equivalents and an eye on Windsor II microemulsions.

However, because of high cost and adverse environmental issues, the use of fluorocarbons has been reduced in recent time (Butenhoff et al. 2002; Falandysz et al. 2006; Houde et al. 2006; Olsen et al. 2003). To minimize the use of fluorocarbons, investigators had focused to develop partially fluorinated hybrid compounds through

[1] Hildebrand solubility parameter: $\partial = \sqrt{\frac{(H_v - RT)}{V_m}}$ ΔH_v = Change in enthalpy due to vaporization; R = Universal gas constant; T = Temperature; V_m = Molar volume.

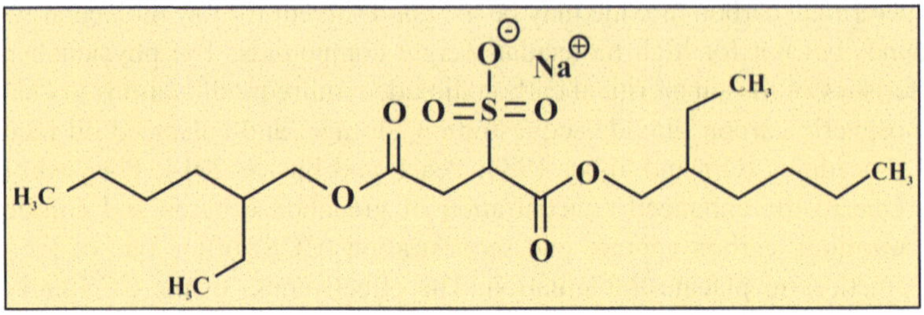

Fig. 5.2 Chemical structure of sodium dioctylsulfosuccinate (an Aerosol-OT surfactant)

Fig. 5.3 F7H7 (a combination of n–C7 fluorocarbon chain and an n–C7 hydrocarbon chain) (after Guo et al. 1992a, b; Harrison et al. 1994)

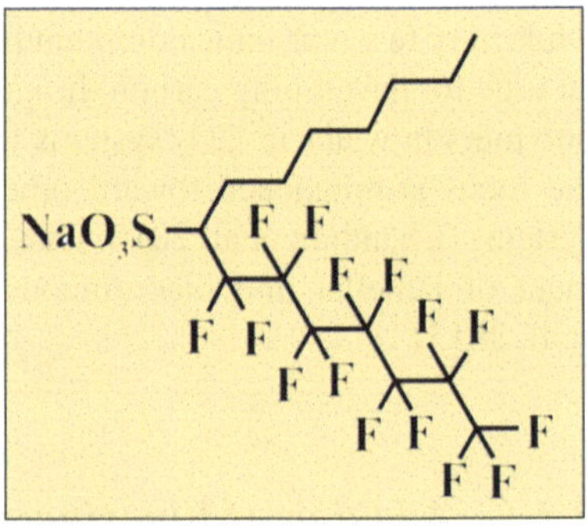

the production of F7H7, which is a combination of n–C7 fluorocarbon and n–C7 hydrocarbon chains (Fig. 5.3) (Guo et al. 1992a, b; Harrison et al. 1994). The solubility of the surfactants had been expressed through temperature-pressure phase behavior investigations. The cloud point pressure or P_{trans} denotes the minimum pressure when a dispersion becomes a single transparent and stable phase (Peach and Eastoe 2014). Considering this parameter in the case of the surfactant solubility within carbon dioxide, it had been demonstrated that a low cloud point pressure corresponds to a more efficient surfactant. The water-to-surfactant ratio (w) quantifies the water molecule solubilized by a specimen of surfactant in carbon dioxide (Eq. 5.1; Peach and Eastoe 2014).

P_{trans} has been observed to show a positive trend with the w value. Further, the stability of microemulsions is also influenced by the chain length of the surfactants (Peach and Eastoe 2014). It was observed that the surfactants having longer chain length became more stable in the microemulsions (Peach and Eastoe 2014), plausibly because of the enhanced rate of interfacial activities and transformations from $-CF_3$ to $-CF_2H$ at terminal position, which would have declined the simplicity to form microemulsion (Eastoe et al. 1996, 2000; Sagisaka et al. 2003). Mohamed et al. (2012) had, however, observed that replacement of terminal fluorine by a hydrogen atom (n–CF_3 to n–CF_2H) had introduced dipole moment, which might have a good chance in declining the solubility in carbon dioxide.

$$w = [\text{Water}] / [\text{Surfactant}] \quad (5.1)$$

5.3.2 *Non-fluorinated Surfactants*

Although the harsh effects of fluorocarbons can be minimized by the production of partially fluorinated surfactants, the modern trends are inclining toward the development of non-fluorinated compounds, which are based on siloxane, hydrocarbons, and oxygenated surfactants (da Rocha et al. 2003; Hollamby et al. 2009; Mohamed et al. 2010; Sarbu et al. 2000; Sagisaka et al. 2013).

5.3.2.1 Trisiloxanes

Trisiloxanes can stabilize emulsions within carbon dioxide-in-water and water-in-carbon dioxide systems, but those have not been observed to stabilize the microemulsions (Peach and Eastoe 2014). A range of ethylene oxide (EO) repeat units comprising a reversed emulsion morphology from water-in-carbon dioxide to carbon dioxide-in-water system has also been studied, which exhibited an increase of EO repeat units from 2 to 7, and the EO7 was found to be very stable due to strong degree of solvation of trisiloxanes by the carbon dioxide (da Rocha et al. 2003).

5.3.2.2 Hydrocarbon Surfactants

As the majority of the hydrocarbon surfactants are not soluble in carbon dioxide (Consani and Smith 1990), the structural aspects of the hydrocarbons have been modified through increasing tail branching and methylation in order to enhance the solubility of the hydrocarbon surfactants in carbon dioxide (Trickett et al. 2010). This was observed to produce more stable water-in-carbon dioxide microemulsions due to an improvement in partition coefficient through weak communications between tails of surfactants and comparatively weaker affinity of the surfactants to water (Peach and Eastoe 2014). Hydrocarbon surfactants having a tert-butyl chain at the terminal end exhibited the lowest surface energy (Pitt et al. 1996). Mohamed et al. (2010) had also observed that the hydrocarbons with a larger degree of branching at the chain tip exhibited smaller P_{trans} than the structures containing a lesser degree of branching. Sagisaka et al. (2013) had synthesized and solubilized SIS1 (sodium 2-(4,4-dimethylpentan-2yl)-5,7,7-trimethyl sulfate), a hydrocarbon-based surfactant, in carbon dioxide. The design of this surfactant was constructed based on the fact that a tail with strong hydrophobicity was required to solubilize the surfactant within the carbon dioxide. This was compounded by the discovery of a highly methylated isostearyl unit that exhibited affinity to dissolve in the carbon dioxide

(Peach and Eastoe 2014). SIS1 achieved high solubilizing efficiency through lowering of interfacial tension (Peach and Eastoe 2014).

5.3.2.3 Oxygenated Surfactants

Oxygenated surfactants are, also, the potential replacements of the fluorinated compounds. Kazarian et al. (1996) had employed the Fourier transform infrared (FTIR) spectroscopy of the polymer compounds incorporating carbonyl groups to investigate the interactions between carbonyl groups (electron-donating group) and carbon dioxide. The polymers revealed splitting of the band attributed to CO_2 *v2* mode, the phenomenon that was absent in the Fourier transform infrared (FTIR) spectra of the polymers lacking electron-donating groups (Kazarian et al. 1996). The donation of an electron from the lone electron pair of oxygen in the carbonyl group to the carbon atom in carbon dioxide would have brought about Lewis acid-base interactions resulting in the splitting of the band (Peach and Eastoe 2014). Reilly et al. (1995), after studying the interactions between the carbon dioxide and the *d*-methanol, had suggested that carbon dioxide acted as an acceptor of electron rather as a donor. They had concluded that a CO_2-*d*-methanol compatibility was formed through Lewis acid-base interactions dominating over the hydrogen bonding mechanisms. Raveendran and Wallen (2002) had designed oxygenated surfactants, where some selected carbohydrates with alternative functional groups having an affinity toward carbon dioxide were solubilized. Eastoe et al. (2005) had successfully stabilized spherical reverse micellar microemulsions with the aid of designed Aerosol-OT analogs, one consisting of a carbonyl group in every chain with a *tert*-butyl group at the end of the chains and another comprising twin chains of vinyl acetate.

5.3.3 Addition of Fluorinated Polymers

Amphiphilic block copolymers exhibit large capability to adsorb in the interfaces and synthesize aggregation structures (Peach and Eastoe 2014). The molecular structures of the polymers are required to be modified to enhance their affinity toward carbon dioxide. Krytox 16,350 (Fig. 5.4), a perfluoropolyether oil (Enick et al. 1998) with 11,350 g/mol molecular weight, was one of the initial polymers that were reported to get solubilized in the carbon dioxide. Enick et al. (1998) had also revealed that the solubility of the polymers declined with an increase in the molecular weight and the polymers with low molecular weight (lower than 15,000 g/mol) exhibited a larger degree of solubility in carbon dioxide. The formation of microemulsions from the polymers was observed to be influenced by the headgroup composition (Loeker et al. 2003). When the perfluoropolyether compounds were in ammonium carboxylate form, water-in-carbon dioxide emulsions were formed, but when these polymers were in carboxylic acid form, microemulsion formation was culminated (Peach and Eastoe 2014). In addition, within the polymer compounds

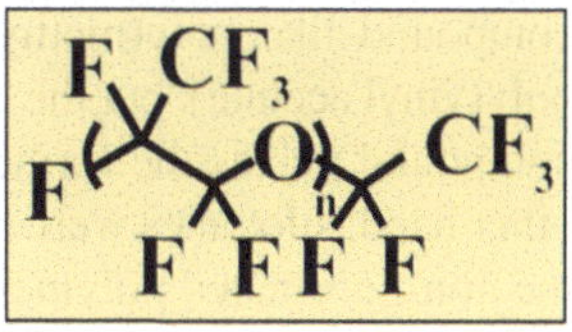

Fig. 5.4 Krytox 16,350 (after Enick et al. 1998)

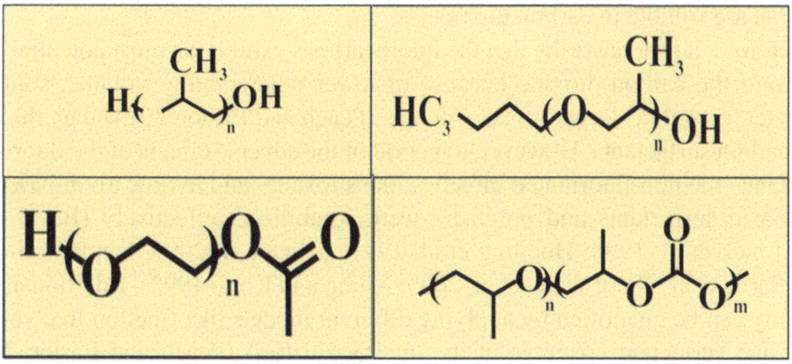

Fig. 5.5 (**a**) Poly(propylene glycol)-diol; (**b**) poly(propylene glycol)-monobutyl ether; (**c**) poly(propylene glycol) acetate; (**d**) poly(ether carbonate)

having an optimum molecular weight (2500 g/mol), the longer-chain homologs were reported to show affinity toward carbon dioxide, but the shorter-chain counterparts exhibited hydrophilic nature (Loeker et al. 2003). On the other hand, partially fluorinated polymers like ethylene-based copolymer and fluorinated Teflon analogs had been found to be soluble in carbon dioxide (Beyer et al. 2000; Rindfleisch et al. 1996).

5.3.4 *Non-fluorinated Polymers*

Among the non-fluoropolymers, poly(propylene glycol)-diol (Fig. 5.5a), poly(propylene glycol)-monobutyl ether (Fig. 5.5b), poly(propylene glycol) acetate (Fig. 5.5c), poly(ether carbonate) (Fig. 5.5d), etc. were designed and solubilized (Sarbu et al. 2000). There are some factors like chain flexibility, free volume, functional groups, low crystallinity, etc. that can influence the solubility of these polymers (Kilic et al. 2009). Ether linkage addition would raise the chain flexibility at the backbone of the polymer structures. Also, the company of functional groups having favorable thermodynamic interactions with the carbon dioxide would increase the solubility of the polymers within the carbon dioxide. In contrast, an association of amine groups as well as allyl polymers with a $-CH_2$ spacer between the functional group and the polymer backbone was reported to decline the solubility of the polymers in carbon dioxide (Kilic et al. 2003, 2009). In addition, a compound like poly(methyl acrylate) shows a larger degree of acetylation than poly(vinyl acetate), but the higher melting point of the former compared to the latter compound makes the former compound to be insoluble in carbon dioxide. On the other hand, siloxanes were reported to exhibit the largest degree of solubility within the non-fluorinated polymers in carbon dioxide (Kilic et al. 2003; Xiong and Kiran 1995). Cellulose triacetate (Hong et al. 2008), cyclodextrins (Potluri et al. 2003), and glucopyranoside (Tapriyal et al. 2008) have been observed as promising polymers that are soluble in carbon dioxide.

Therefore, it is noteworthy that the fluorocarbons exhibit the most potential affinity toward the carbon dioxide because of lower polarizability volume, solubility parameter, as well as larger molar volumes (Peach and Eastoe 2014) than the other hydrocarbon surfactants. However, to get rid of the adverse effects of the fluorinated surfactants, the non-fluorinated moieties like siloxane- and hydrocarbon-based and oxygenated surfactants and polymers were solubilized, effectively (Beyer et al. 2000; Enick et al. 1998; Hoefling et al. 1991; Kilic et al. 2009; Rindfleisch et al. 1996; Sarbu et al. 2000; Wang et al. 2009; Xiong and Kiran 1995). Additionally, the solubility can be quantified by applying different models like fraction free volume[2] (FFV) and surfactant coverage at the interface[3] (Φ_{surf}) (Peach and Eastoe 2014). FFV considers the geometry of the surface tails, the surface coverage measured using tail length and volume, and the surfactant headgroup. Chemical moieties with smaller fraction free volume values were believed to show greater affinity toward carbon dioxide with respect to that of larger fraction free volume due to enhanced interface stability resulted from the compact interfacial film (Peach and Eastoe 2014). The fluorinated surfactant moieties possess greater tail volume than the hydrocarbon surfactants and, therefore, exhibit more affinity toward the carbon dioxide. On the other hand, surfactant coverage at the interface counts on the fractional fragmental volume of the surfactant (computed from the area of surfactant headgroups at the critical concentration of the micelle and interfacial thickness) at the interface in relation to the total volume of the surfactant in the system (Peach and Eastoe 2014). Larger surfactant coverage at the interface has been attributed to the larger parting between the two phases at the interface.

[2] Fraction free volume: $FFV = 1 - \frac{V}{tA_{cmc}}$ (after, Stone et al. 2004)

V = Surfactant tail volume; t = Surfactant tail length; A = Area of the surfactant headgroup at the critical concentration of micelle.

[3] Surfactant coverage at the interface: $\Phi_{surf} = \frac{V_{cal}}{V_{meas}}$

V_{means} = Fragment volume of the surfactant at an interface; V_{cal} = Total volume of the surfactant
$V_{meas} = A_{cmc} - \tau$ (after Mohamed et al. 2011, 2012)
A_{cmc} = Area of the surfactant headgroups; τ = Interfacial thickness.

5.3.5 *Viscosifiers*

The solvent property of the supercritical carbon dioxide can also be modified by adjusting the viscosity. Enhancement in viscosity results from the increase in the internal structure of the solutes (Peach and Eastoe 2014). Most of the structures associated with the polymers and surfactants were the reversed spherical micelles (Eastoe et al. 1996). Ellipsoidal and rod-like, lamellar structures, bicontinuous phases, and gel structures can undergo self-assembly and aggregation. These structures can also be used to enhance the viscosity of the polymers and surfactants (Acharya and Kunieda 2006; Cummings et al. 2012; James et al. 2014; Nagarajan 1993; Tata et al. 1994; Trickett et al. 2010; Trickett and Eastoe 2008).

5.3.5.1 Polymeric Viscosifiers

Among the polymeric carbon dioxide viscosifiers, Heller et al. (1985) observed that poly(1-decane) could enhance the viscosity of the carbon dioxide due to its large degree of solubility (Bae and Irani 1993; Heller et al. 1985). Huang et al. (2000) had boosted the viscosity of the carbon dioxide by a factor ~5–400 through adding styrene and fluoroacrylate copolymers at a range of polymer concentrations (1–5 w/w%) and styrene-to-fluoroacrylate molar ratios. The π-π stacking between the phenyl groups was expected to enhance the viscosity in the system. However, polymer solubility was noticed to get declined with an inclination in styrene concentrations within the polymer chain because of its poor solubility in carbon dioxide (Huang et al. 2000).

5.3.5.2 Addition of Salts

Addition of salt to the systems was observed to form the elongated micelles that led to increase in the surfactant headgroup screening and subsequent declination in effective size. That, in turn, favored a minimized surface curvature (Peach and Eastoe 2014). Moreover, Hatzopoulos et al. (2013), James et al. (2014), and Xu et al. (1993) had noticed the increase in the viscosity of the water-in-carbon dioxide system through adding of phenols and hydrotropic salts to microemulsions.

5.4 Application of Supercritical CO_2 in Chemical Reactions

The supercritical carbon dioxide was used mainly for extraction and separation in the early days (Mayadevi 2012). Kurt Zosel had decaffeinated the roasted coffee beans using the supercritical carbon dioxide, which was its first industrial application (Zosel 1978). Supercritical carbon dioxide is used as a solvent in the chemical reactions as a C1 building block or as both solvent and reactant (Jessop and Leitner 1999; Oakes et al. 2001). Moreover, the supercritical carbon dioxide is used in hydrogenations, hydroformylations, controlled polymerization reactions (Cooper 2000; de Vries et al. 2000), enzyme-catalyzed reactions (Randolph et al. 1988), radical reactions (Hadida et al. 1997), cycloaddition reactions (Lsaacs and Keating 1992), etc.

As mentioned above, supercritical fluids possess almost similar dissolving power to that of liquids and can be compressed more than the dilute gases. Supercritical fluids benefit with a faster rate of chemical reactions and better selectivity (Mayadevi 2012). These two properties are considered to be the consequences of better solubility of gases in supercritical fluids; faster diffusion of solutes into, out, of, and within the supercritical stage; weak degree of solvation around the reacting species; localized clustering of reactants or the solvents; and larger negative activation volumes near the critical point and declined cage effect in radical reactions (Mayadevi 2012). As the density of the supercritical fluids like the supercritical carbon dioxide can be adjusted through changes in temperature and pressure at or above the critical state, these provide additional advantages over the traditional organic solvents (Mayadevi 2012). Additionally, the supercritical carbon dioxide renders inertness to most of the reactive compounds and, hence, minimizes the side reactions and by-products (Mayadevi 2012).

There are two principal schemes of the organic reactions in the supercritical carbon dioxide: (a) as a reactant and (b) as solvent or co-solvent.

5.4.1 Supercritical CO_2 as a Reactant

Although being inert, carbon dioxide acts as Lewis acid and reacts with strong bases like phosphenes, amines, alkyl anions, etc. (Inoue and Yamazaki 1982; West et al. 2001). In hydrogenation reactions, carbon dioxide can be reduced to carbon monoxide at temperatures in the presence of catalysts greater than 69.85 °C, and with longer residence time (Jessop et al. 1999), it can poison the palladium catalyst (Mayadevi 2012). Depending on the catalyst and conditions of reactions, high-temperature catalytic hydrogenation of carbon dioxide synthesizes carbon monoxide, methane, methanol, and derivatives of the formic acid (Jessop et al. 1995). Production of poly(ether carbonate) through copolymerization of carbon dioxide with oxiranes has been considered as a green alternative route to the conventional phosgene route (Cheng et al. 2001; Kuran 1998). Hydrogenation of methanol, ammonia, and amines through carbon dioxide yield methyl formate and derivatives of formamide, respectively (Jessop et al. 1995).

5.4.2 Reactions in Supercritical CO_2 Medium

As discussed above, the supercritical carbon dioxide acts as a green solvent, unlike many other common organic solvents that are toxic to the environment. Supercritical carbon dioxide has been reported to boost up the miscibility of the gaseous solvents and, hence, accelerates the catalyst activity. The reactions can be adjusted by controlling the pressure, which increases the catalytic activities, recovery, recycle, and production yields (Mayadevi 2012). These phenomena have been observed in homogeneous and heterogeneous catalytic reactions (Inoue and Yamazaki 1982) like oxidation, alkylation, esterification, hydroformylation, hydrogenation, disproportionation, etc. (Mayadevi 2012).

5.4.2.1 Hydrogenation Reactions

Supercritical carbon dioxide has been, widely, used as a reaction medium in hydrogenation reactions in continuous/batch modes and in homogeneous/heterogeneous phases (Mayadevi 2012). Both fluid-/gas- and liquid-phase hydrogenation reactions are carried out in industrial scale (Mayadevi 2012). The fluid/gas phase employs solid catalyst in a two-phase mode, while the liquid-phase reactions take place in two or even in three modes (Mayadevi 2012). A representative hydrogenation reaction using the supercritical carbon dioxide as a medium of reaction is presented in Fig. 5.6 following Fujita et al. (2011). In liquid-phase hydrogenation reactions, palladium, platinum, ruthenium, iridium, and rhodium are used as catalysts (Kainz et al. 1999; Lange et al. 2000). Neutral transition metals dissolve easily in the supercritical carbon dioxide, while the cationic complexes do not dissolve easily due to the non-polar nature of the carbon dioxide. Perfluorination can be employed to enhance the stability of the cationic complexes. Often, the substrate can also act as co-solvent that can enhance the catalyst solubility and in turn step up the reaction rate in the supercritical carbon dioxide (Niessen and Woelk 2007). Liquid-phase hydrogenation eradicates the diffusion rate of hydrogen gas at the gas-liquid interface enhancing solubility of the hydrogen gas and minimizes total transport resistance accelerating the rate of reactions and improving the production yields (Mayadevi 2012). Selectivity was reported to increase in asymmetric hydrogenation reactions by using carbon dioxide replacing the traditional organic solvents

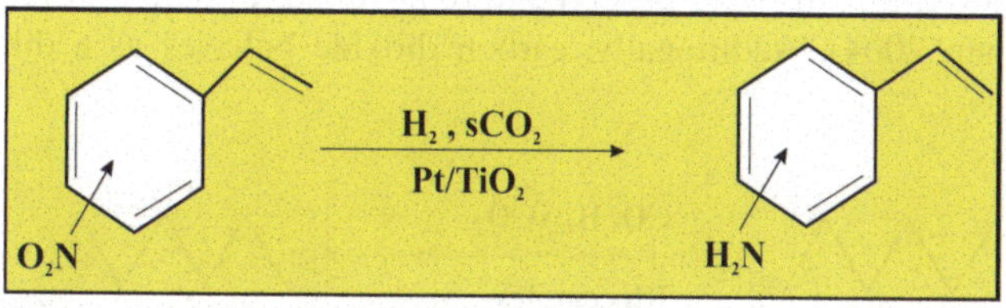

Fig. 5.6 Representative hydrogenation reaction of nitrostyrene using supercritical carbon dioxide as a reactant following Fujita et al. (2011)

(Mayadevi 2012). In complementary, stereoselective hydrogenation in the supercritical carbon dioxide prefers the cis-product that has large selectivity and yield (Clark et al. 2007; Hiyoshi et al. 2006, 2007, 2009, 2012; Kainz et al. 1999). During the exothermic hydrogenation reactions of alkanes and alkynes, the control in the concentration of H_2 is a crucial factor for hydrogenation reactions in the supercritical carbon dioxide (Chouchi et al. 2001; Jessop et al. 1999). The inventory and compression cost can be reduced through the in situ generation of carbon dioxide (Mayadevi 2012). Two reactions within the same reactor under the supercritical carbon dioxide in a sequence are the other achievements in hydrogenation reactions (Mayadevi 2012). Aldol condensation reaction followed by hydrogenation in presence of bi-functional palladium catalyst was reported to enhance the activity and lengthen the life of the catalyst (Han and Poliakoff 2012; Seki et al. 2007, 2008; Stevens et al. 2009).

5.4.2.2 Hydroformylation

Hydroformylation or oxosynthesis corresponds to the production of oxygenated compounds (Fig. 5.7) through the addition of equivalent amounts of hydrogen and carbon monoxide to alkene (Mayadevi 2012). Besides using rhodium and cobalt, ruthenium and manganese are used as catalysts in hydroformylation in supercritical carbon dioxide (Bektesevic et al. 2006). Oxosynthesis reactions in supercritical fluids and the use of nuclear magnetic resonance (NMR) spectroscopy to get detailed insights into the reactions were reviewed by Niessen and Woelk (2007). Sellin et al. (2001) and Webb et al. (2003) had carried out hydroformylation reactions in association with ionic liquids (ILs). Haumann and Riisagar (2008) had reviewed oxosynthesis in room-temperature ionic liquids. Those systems were reported to be valid for gas-phase reactions and reactions in which carbon monoxide was produced from carbon dioxide through in situ processes.

5.4.2.3 Oxidation

At supercritical state, carbon dioxide and oxygen are miscible at all ranges of concentration (Musie et al. 2001). In complementary, catalytic oxidation cannot oxidize carbon dioxide (Mayadevi 2012). Therefore, oxidation reactions in the supercritical carbon dioxide (Fig. 5.8) were observed to remove the formation of the side products (Beckman 2004). Additionally, carbon dioxide behaves as a diluent where

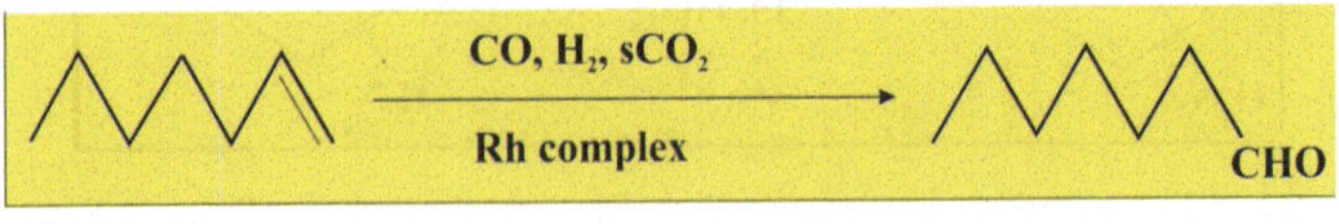

Fig. 5.7 Hydroformylation of 1-hexene (after Tadd et al. 2003)

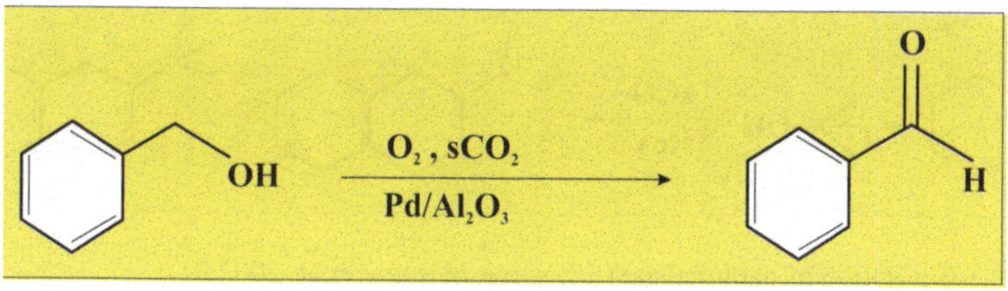

Fig. 5.8 The oxidation reaction of benzyl alcohol to benzaldehyde using the supercritical carbon dioxide as a reactant (following Jenzer et al. 2001)

molecular oxygen is used as the oxidizing agent and, thus, offers a safer reaction environment (Mayadevi 2012). A number of oxides and metal oxide catalysts are required for semi-continuous/continuous oxidation reactions in heterogeneous phase in supercritical carbon dioxide (Mayadevi 2012). These reactions are performed at high temperature (≥199.85 °C) and involve oxides of molybdenum, cobalt, nickel, silica, tungsten, manganese, and copper in variable combinations (Mayadevi 2012). Influence of pressure had been observed to enhance the rate of the reaction that used carbon dioxide as a solvent (Caravati et al. 2006; Han and Poliakoff 2012). Photocatalytic oxidation in the supercritical carbon dioxide, additionally, serves as a potential topic of interest (Bourne et al. 2008).

5.4.2.4 C–C Bonding, Alkylation, Acetylation, and Esterification Reactions

Ikariya and Kayaki (2000) had reviewed the earlier studies on metal-catalyzed C–C production in the supercritical carbon dioxide, which encompassed cyclotrimerization of alkynes, Heck-Mizoroki olefination of aryl halides, and Suzuki, Sonogashira, and Stille coupling reactions that used cobalt or palladium catalysts.

- The Stille reaction is a chemical reaction widely used in organic synthesis that involves the coupling of an organotin compound with a variety of organic electrophiles via palladium-catalyzed coupling reaction.
- Suzuki coupling is the reaction of vinyl or aryl boronic acids with aryl and vinyl halides or triflates using a palladium catalyst.
- Sonogashira coupling is the coupling of terminal alkynes with aryl and vinyl halides in presence of a palladium catalyst, a copper co-catalyst, and an amine.

The reactions exhibited better conversion and selectivity when the supercritical carbon dioxide was used as the medium for reaction (Amandi et al. 2007; Caballero et al. 2011; Early et al. 2001; Galia et al. 2011; Goddard et al. 2000; Marathe et al. 2002; Muljana et al. 2011). The *tert*-butylation of naphthalene is presented in Fig. 5.9 following Marathe et al. (2002). Production of biodiesel through esterification/transesterification of fatty acids or triglycerides in the supercritical carbon dioxide is continuously attaining interests of the researchers (Mayadevi 2012). Further, Jackson et al. (2006) had compared the esterification reaction of oleic acid

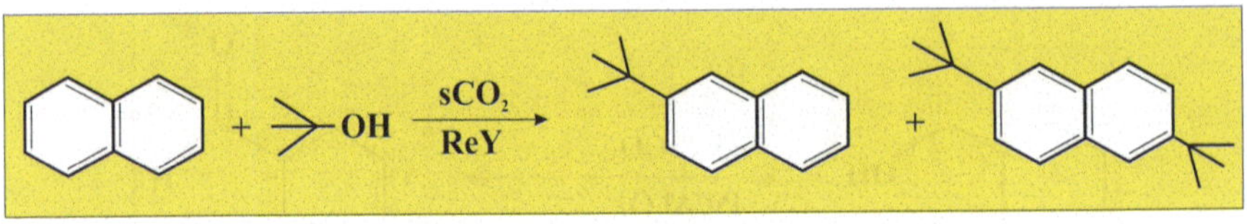

Fig. 5.9 *tert*-Butylation of naphthalene (following Marathe et al. 2002)

with methanol in the supercritical carbon dioxide over several catalysts and found that immobilized lipase exhibited high activity under the supercritical carbon dioxide in comparison to sulfonic acid-functionalized silica, standard acidic resin, and Amberlyst 15. Enzyme-catalyzed and non-catalytic transesterification reactions were performed on vegetable oils (Ciftci and Temelli 2011; Galia et al. 2011; Rodrigues et al. 2011; Shin et al. 2012; Varma and Madras 2007; Varma et al. 2010) with methanol or ethanol, and it was observed that catalytic activity and reaction rate were enhanced, when the reactions were carried out in the supercritical carbon dioxide. Better yields and faster kinetics have been observed in biodiesel production through the continuous synthesis of fatty acid methyl esters with the use of immobilized Novozym 435 catalyst in the supercritical carbon dioxide (Ciftci and Temelli 2011; Rodrigues et al. 2011). Moreover, an enhanced degree of substitution was noticed during the acetylation reaction of starch with acetic anhydride and sodium acetate carried out in the supercritical carbon dioxide (Muljana et al. 2011). Jackson et al. (2006) had reported catalytic functionalization of methane in the supercritical carbon dioxide at 39.85 °C. They had employed perfluorinated ligands containing a silver catalyst to activate the methane and formations of C–C bonding between methane and ethyl diazoacetate for the production of ethyl propionate, and these mechanisms got success as a result of the use of the supercritical carbon dioxide.

5.4.3 *Supercritical CO_2 in Material Synthesis*

For the production of nanoparticles, polymers, foams, and porous organic materials, the supercritical carbon dioxide has been used as a green solvent. Cooper (2001) had reviewed the use of the supercritical carbon dioxide as a foaming agent or as a solvent in synthesis of porous organic materials, as developing green solvent for lithography, in the synthesis of metal nanoparticles, and for controlled delivery of drugs. Du et al. (2009) had reviewed the synthesis of fluorinated polymers in supercritical carbon dioxide. The synthesized materials were reported to possess superior characteristics in comparison to that produced using common organic solvents.

Although supercritical carbon dioxide has been widely applied as a green medium in several organic-chemical reactions, this requires high energy costs to become compressed, which is a great hindrance in the way of getting commercially used. Multiple reaction sequencing in the same reactor as well as in situ generation of carbon dioxide through proper decomposition procedures may overcome that hindrance (Mayadevi 2012).

5.5 Evolving Areas Using Supercritical Carbon Dioxide

5.5.1 Bicontinuous Phase and Foams

Klostermann et al. (2012) had designed a balanced supercritical carbon dioxide microemulsion system characterized by equal volumes of carbon dioxide and water. These systems were reported to use a polyfluoroether surfactant compounded with ethoxylated surfactants. Here, the systems were interpreted to follow a bicontinuous phase, where the microemulsion domain size increased with increment in pressure (Klostermann et al. 2012). Further, carbon dioxide-in-water foams that have potential applications in carbon dioxide sequestration and enhanced oil recovery can be produced with branched non-ionic hydrocarbon surfactants (Adkins et al. 2010; Johnston and da Rocha 2009). Compared to the water-in-carbon dioxide microemulsions, carbon dioxide-in-water foams were documented to provide flexibility in surfactant designing because of lower surface tension values required to stabilize water-in-carbon dioxide microemulsions (Stone et al. 2003). In addition, surfactant tail branching was reported to increase the stability of the foams compared to the linear structures by minimizing the contact between water and carbon dioxide (Peach and Eastoe 2014).

5.5.2 Ionic Liquids-in-Supercritical Carbon Dioxide

Ionic liquids, the organic salts comprising both inorganic and organic ions and having melting point lower than 100 °C, are also known as green solvents because of their low toxicity, non-flammability, chemical stability, and small volatility (Rogers and Seddon 2003). Investigations are going on combining the ionic liquids with the supercritical carbon dioxide to synthesize microemulsions for getting advantages of both of these green media (Peach and Eastoe 2014). Liu et al. (2007) had formed the ionic liquids-in-carbon dioxide microemulsions employing *N*-ethyl perfluorooctylsulfonamide, a fluorinated surfactant. Additionally, ionic liquids like 1,1,3,3-tetramethylguanidinium acetate, 1,1,3,3-tetramethylguanidinium lactate, and 1,1,3,3-tetramethylguanidinium trifluoroacetate had been solubilized (Peach and Eastoe 2014). Through computer stimulation technique, Chandran et al. (2010) had observed the formation of reverse ionic liquid micelles, which supports the production of ionic liquid-stable droplets in a continuous carbon dioxide phase by the addition of surfactant with amphiphilic property. This work also indicated that the cations of the ionic liquid have the least influence and the stability of the emulsions is controlled by the interactions between ionic liquid anions and surfactant headgroup (Peach and Eastoe 2014).

Hence, several investigations have been carried out for the last two decades regarding the application of the supercritical carbon dioxide as a green solvent replacing traditional toxic solvents and volatile organic compounds, which have been also associated with the works related to modifications in carbon dioxide handling in order to enhance the carbon sequestration and enhanced oil recovery processes. The latter aspects include designing of surfactants and polymers with high affinity toward carbon dioxide as well as the manufacture of the carbon dioxide thickeners. A lot of structures had been observed in water-in-carbon dioxide emulsions encompassing the reverse micelles, lamellar structures, and bicontinuous phases. The solvent properties of the supercritical carbon dioxide have been, further, enhanced by introduction and combination of ionic liquids for getting benefited from the combined good effects of both of these green solvents.

5.6 Supercritical Carbon Dioxide and Nanomaterials

Nanometer-sized particles exhibit a new myriad of properties of industrial interest (Reverchon and Adami 2006). These properties correspond to the ratio of the surface to the volume at the nanoscale, where surficial properties gain relevance with respect to the volume properties (Reverchon and Adami 2006). The surface molecules were witnessed to introduce large hardness to the metals and higher energy to the explosives and propellants, as well as these would help in the production of pharmaceuticals and electronic devices with improved performances (Reverchon and Adami 2006). The production methodologies of the nanomaterials, usually, follow two principal approaches: (a) top-down and (b) bottom-up (Reverchon and Adami 2006). The top-down process is related to the manufacturing of nanoparticles through reducing the size of the initial material, while the bottom-up approach is attributed to the synthesis of nanoparticles from scale of molecular level (Reverchon and Adami 2006). Supercritical fluids owing to their gas-like diffusivity, tunable solvent properties, and lower toxicity, have been used as a green medium for the production of nanoparticles. Especially, the liquid- and gas-like properties were reviewed to be useful in nanoparticle synthesis (Reverchon and Adami 2006). Among other media like alcohol, ammonia, low molecular weight hydrocarbons, and water at supercritical state, the supercritical carbon dioxide is found to be, usually, cheap and non-toxic for the synthesis of nanomaterials (Ohde et al. 2000; Reverchon and Adami 2006). Nanoparticles and nanostructured materials are the two main explored areas of all the nanomaterials (Reverchon and Adami 2006). The supercritical fluid-based classification of the nanoparticle generation techniques can be carried out depending on the role of supercritical fluid in the generation procedures.

5.6.1 Rapid Expansion of Supercritical Solution Method

The rapid expansion of supercritical solutions (RESS) comprises the saturation of a supercritical fluid in presence of a solid substrate followed by depressurization of the solution with the help of a hot nozzle to a chamber having low pressure, which yields a swift substrate nucleation in the form of tiny particles, which are accumulated from the stream of gas (Reverchon and Adami 2006). This process is very interesting because no organic solvents are used in this case. Turk et al. (2002) had applied this process for the production of sitosterol nanomaterials (average diameter ~200 nm). They had tested this process at variable pre-expansion pressure and temperature ranges. A typical version of the RESS procedure is the rapid expansion of a supercritical solution into a liquid solvent (RESOLV) (Reverchon and Adami 2006). This method has been considered to improve the RESS performance through quenching the particle growth within the precipitator (Reverchon and Adami 2006). Sun et al. (2001) had modified the RESOLV process by using water in supercritical carbon dioxide microemulsion that was employed as a modified solvent for dissolving silver nitrate ($AgNO_3$). The water-in-carbon dioxide microemulsion consisting of silver nitrate was found to be expanded into sodium borohydride solution stable at the room temperature. However, these two processes were found to be limited within the compounds showing reasonable solubility in the supercritical carbon dioxide, while the majority of high molecular weight compounds and ionic groups are less soluble in the supercritical carbon dioxide due to its non-polar property.

5.6.2 Supercritical Anti-solvent Precipitation Technique

Supercritical anti-solvent precipitation (SAS) is another process to synthesize the nanoparticles (Reverchon and Adami 2006). A liquid solution consists of a solute to be micronized, and during this process, the liquid solvent should entirely be miscible with a supercritical fluid, while the solute ought to be insoluble in that fluid. Hence, the liquid solution upon contact with the supercritical fluid forms a solution resulting in supersaturation and precipitation of the solute (Reverchon and Adami 2006). Owing to the presence of supercritical fluid, the mass transfer of the reaction is enhanced, and therefore, the liquid solution forms at a rapid rate, which in turn synthesizes the nanoparticles. Vapor-liquid equilibria at high-pressure and the transfer of mass between the supercritical fluid and liquid play a crucial role in the SAS technique (Reverchon and Adami 2006). The synthesis of a single supercritical phase is the most important step for the entire production of the nanomaterial, successfully (Reverchon et al. 2003).

5.6.3 *Supercritical Assisted Atomization Procedure*

Supercritical assisted atomization (SAA) utilizes supercritical fluids as an atomizing medium (Reverchon 2002). The basis of this method lies in the solubilization of the supercritical carbon dioxide in a liquid solution and subsequent atomization with the use of a thin-wall nozzle. During this conduction, a couple of atomization procedures occur, where the first one is related to the primary droplet synthesis at the exit point of the nozzle by pneumatic atomization followed by the destruction of those droplets through rapid expulsion of carbon dioxide from the inner part of the droplets (Reverchon and Adami 2006).

5.6.4 *Drying of Sol/Gel*

Supercritical fluids, moreover, allow the drying of gels without any surface tension, which prevents the gels from collapsing (Reverchon and Adami 2006). Hu et al. (1999b) suggested that the gels should be prepared in aqueous solution followed by replacement of water with a mixture of benzene and *n*-propanol in a precipitate and thereafter the removal of the organic solution through the use of the supercritical carbon dioxide. This procedure got finalized through the synthesis of copper borate nanomaterials.

5.6.5 *Synthesis in Supercritical CO_2*

5.6.5.1 Synthesis of Silver Nanoparticles

Silver nanocrystals can be stabilized in the supercritical carbon dioxide in presence of opportune surfactants like alkanethiols (Shah et al. 2000). They coined the name of this process as arrested growth, which is related to the reduction of organometallic precursors soluble in the supercritical carbon dioxide with hydrogen in the presence of perfluoro-octanediol ligand, which binds at the surface of nanomaterials and reduces the growth of the particles. In this procedure, the precursors should be soluble in the supercritical carbon dioxide, while the polar compounds are not. In addition, through this process, Shah et al. (2002) had investigated the role of process parameters on the particle diameter and the polydispersity, where they suggested that the density of carbon dioxide had a major influence on particle size and distribution. They obtained larger silver crystals in the lower-density carbon dioxide as the growth of the particles took place before the surfactant coverage barred their coagulation, while smaller silver crystals were found in higher-density supercritical carbon dioxide because of strong hindrance imparted by the surfactants. Silver nanoparticles were synthesized by Korgel et al. (2002) by arrested precipitation in the supercritical carbon dioxide through the reduction of silver acetylacetonate in the presence of organic ligands. In complementary, Kameo et al. (2003) had synthesized palladium and silver nanoparticles through the reduction of palladium acetate and silver acetylacetonate in the supercritical carbon dioxide medium. Further, metallic nanomaterials of silver and palladium were produced by McLeod et al. (2004) through spraying the supercritical carbon dioxide solution consisting of metal precursors soluble in carbon dioxide into another carbon dioxide solution comprising a reducing agent and a stabilizer compound required for limiting the nanoparticle growth. Furthermore, Reverchon et al. (2003) had produced nanoparticles of titanium hydroxide in the supercritical carbon dioxide through the hydrolysis of titanium *tetra*-isopropoxide with the use of a continuous large-scale plant that was operated at 80–140 bar pressure and 40–60 °C temperature range. Stallings and Lamb (2003) had, also, carried out the hydrolysis of titanium *tetra*-isopropoxide, where they had injected the pure titanium *tetra*-isopropoxide into water-in-supercritical carbon dioxide dispersion medium employing a batch reactor.

5.6.5.2 Reverse Micelles

The thermodynamically stable water-in-oil microemulsions are produced by surfactants of amphiphilic property with a headgroup of hydrophilic characteristics that surrounds the nanometric-sized water core as well as a hydrophobic tail extended into the non-polar continuous phase (Johnston et al. 1999). Micelles are the cylindrical or spherical objects synthesized by surfactants that separate the oil and water (Reverchon and Adami 2006). Oil drops in water are called micelles, whereas water drops in oil are known as reverse micelles (Reverchon and Adami 2006). Reverse micelles are employed as nanoreactors for producing inorganic and organic nanoparticles. Supercritical carbon dioxide possesses high affinity toward many organic solvents, and hence, this can be employed in recovery steps of reactions as antisolvent for extraction of liquid solvents to synthesize dry non-coalescing nanomaterials in one step. Chattopadhyay and Gupta (2003) had injected the water-in-oil microemulsion into a batch reactor comprising the supercritical carbon dioxide, which acted either as a solvent or as a solvent catcher removing the organic solvents. They had carried out this technique during the synthesis of silica nanoparticles. Ohde et al. (2001), during the production of the nanomaterials, had suggested that the step that determines the rate for the synthesis of nanoparticles depends on the diffusion between the micellar core and the supercritical carbon dioxide. Hong et al. (2003) had synthesized the nanoparticles of titanium oxide through controlled hydrolysis reaction of titanium *tetra*-isopropoxide within water-in-carbon dioxide reverse micelles that were stabilized by the polymeric surfactants. They witnessed an increase in the average particle size with an inclination of water-to-surfactant ratio.

5.6.5.3 Nanowire and Nanofiber Synthesis

Nanowires are useful materials for building blocks like in the microelectronics industry, where these can play roles both as an electrical contact and as devices (Reverchon and Adami 2006). Different techniques have been suggested for the synthesis of nanowires, such as laser ablation, liquid crystal templating process, and vapor-liquid-solid growth methods (Wu and Yang 2001). On the other hand, the synthesis of nanowires depends on two methodologies, which are (a) a supercritical fluid-based vapor-liquid-solid formation procedure (Hanrath and Korgel 2002; Holmes et al. 2000) and (b) a mesoporous template filling supercritical fluid-aided process (Coleman et al. 2001; Holmes et al. 2003). The first procedure utilizes alkanethiol-capped gold nanocrystals as the seed particles for continuing the single-dimensional crystallization of the silicon and germanium in supercritical hexane. The second method corresponds to the supercritical carbon dioxide-dependent variation of mesoporous template filling of pores because of limitations in the mass transfer procedure. The lowered surface tension coupled with enhanced coefficients of diffusion helps to weaken this hindrance (Reverchon and Adami 2006). The rapid expansion of a supercritical solution into a liquid solvent (RESOLV) method has been, also, considered for synthesizing polymeric nanofibers of poly(heptadecafluorodecyl acrylate), poly(methyl methacrylate), etc. (Meziani et al. 2005). The poly(heptadecafluorodecyl acrylate) is soluble in the supercritical carbon dioxide, and it expands in aqueous sodium chloride solution that synthesizes the nanomaterials (Sun et al. 2005) or nanofibers with <100 nm diameter on the basis of the concentration of the polymer within the expanding solution (Reverchon and Adami 2006). Louvier-Hernandez et al. (2005) had synthesized chitin microfibers through dissolving in hexafluoroisopropanol followed by processing of the solution by supercritical anti-solvent precipitation technique using the supercritical carbon dioxide as the anti-solvent.

Fukushima et al. (1999), Wakayama et al. (2001, 2002, 2003, 2004), and Wakayama and Fukushima (1999, 2000a, b) had synthesized nanoparticles employing nanoscale casting using templates. This technique includes the soluble precursor dissolutions within the supercritical carbon dioxide followed by their deposition on the templates of activated carbon. After this process, the activated carbon is eliminated, and the deposited matter develops nanometer-sized porous structure.

5.6.5.4 Carbon Nanotubes

Owing to structural, electrical, mechanical, and technological characteristics, carbon nanotubes (CNT) are in limelight (Reverchon and Adami 2006). They are, widely, used in scanning probe microscopes, semiconductor machines, quantum wires, gas storage devices, chemical filters, fuel cells, batteries, etc. (Reverchon and Adami 2006). Motiei et al. (2001) suggested a reaction of magnesium with carbon dioxide at 1000 °C for the production of magnesium oxide and carbon nanotubes, and during this process, the CO_2 remains at supercritical state.

5.6.5.5 Nanofilms

Nanofilms can be employed as gas sensors, catalysts, and permselective membranes in several technological appliances (Reverchon and Adami 2006). Ye et al. (2003a, 2005) had proposed a nanofilm deposition procedure aided by the supercritical carbon dioxide, and they coined the name of that procedure as supercritical fluid immersion deposition. This procedure was a type of the classical galvanic displacement procedures carried out, commonly, in aqueous hydrofluoric acid solutions. This reaction yielded nanofilms of lead, copper, silver, etc. on silica-based substrates in the supercritical carbon dioxide solutions. The supercritical carbon dioxide, there, was reported to enhance the performance of the procedure.

5.6.5.6 Composite Nanomaterials

Composite nanoparticles are classified as nanocapsules and nanospheres (Reverchon and Adami 2006). Nanocapsules are produced from a shell belonging to one component with a core that belongs to an active component, while the nanospheres are built up from the dispersion of two or more compounds (Reverchon and Adami 2006). These composite nanoparticles serve as semi- and superconductors, medical instruments, controlled release of pharmaceuticals, etc. (Reverchon and Adami 2006). Wang et al. (2004) had suggested the synthesis of nanocapsules composed of silica and Eudragit (a type of polymer) through supercritical anti-solvent precipitation procedure. Through dissolving Eudragit in acetone, a polymer solution was prepared, and the silica nanomaterials were suspended within that solution at the anticipated polymer-silica ratio. After that, the entire solution was treated with the supercritical carbon dioxide resulting in the nucleation of heterogeneous polymer with the silica nanomaterials of spherical shape serving as the nuclei for the polymer growth. Silica nanoparticles had the original diameter of 16–30 nm, and the resulting nanocapsules had the diameter ranging from 50 to 100 nm. The composite nanomaterials had also been observed to be synthesized from varying water-in-oil reverse micelle procedures (Reverchon and Adami 2006). Zhang et al. (2004a) had introduced a process, where, initially, a couple of micellar solutions with precursor compounds were formed and then, those were mixed for producing zinc sulfide or cadmium sulfide composite nanomaterials within the reverse micelles because of the content exchange. Here, the supercritical carbon dioxide was employed as a solvent catcher for removing the surfactant and the liquid solvent. They (Zhang et al. 2004b) had also synthesized silver nanomaterials in a water-in-oil reverse micellar system and employed the supercritical carbon dioxide for the elimination of the cyclohexane and the surfactant and also to stimulate the polymer precipitation on the silver nanoparticles wrapped within the polymer.

Zhang et al. (2004c) had proposed the production of polymer-encapsulated silver nanofibers with the use of water-in-oil reverse micelles and supercritical carbon dioxide combined with an ultrasonic treatment. They had also solubilized polystyrene within the solution, and at the final stage, the supercritical carbon dioxide was added to stimulate the precipitation of polystyrene on silver nanoparticles, and the entire procedure ended up with the production of silver nanowires coated with a polymer. Liu et al. (2004) had tried to encapsulate the polystyrene within carbon nanotubes with the help of the supercritical carbon dioxide. Initially, the carbon nanotubes were dispersed in the styrene monomer, ultrasonically, followed by addition of the supercritical carbon dioxide to induce the penetration of monomer inside the carbon nanotubes. Then, following the release of carbon dioxide, the vessel had been transferred to an oven at the temperature of 100 °C to begin the polymerization, and the carbon nanotubes were found to be filled up by the polystyrene. Ye et al. (2003a) had suggested the deposition of metal within the carbon nanotubes employing a supercritical carbon dioxide-based procedure. Ye et al. (2003b) introduced the precipitation of palladium nanoparticles on carbon nanotubes through reduction of a palladium precursor in the supercritical carbon dioxide.

5.6.5.7 Drug Nanoparticles and Supercritical CO_2

Padrela et al. (2018) had reviewed the supercritical carbon dioxide-based procedures on formation and control of physicochemical and morphological characteristics of crystalline or amorphous drug nanoparticles. They had found versatility in these procedures as the supercritical carbon dioxide can play triple roles as a solvent, an anti-solvent, and an additive with high mobility. Introduction of some simple changes to the standard procedures, it is possible to distinguish the hydrophilic and hydrophobic compounds of low to high molecular weights and to obtain a myriad of crystalline or amorphous drug nanoparticles (Padrela et al. 2018). Simple optimization toward the final products and product quality and integrity, additionally, reflect the versatility of the supercritical carbon dioxide-induced processes (Padrela et al. 2018).

5.7 Application of Supercritical Carbon Dioxide as a Foaming Agent

5.7.1 Production of Microcellular Cross-Linked Polyethylene Foams

Supercritical carbon dioxide can be used as a foaming agent for the production of microcellular cross-linked polyethylene foams (Xing et al. 2008). Microcellular polymeric foams offer more advantages over solid polymers like reduction in material cost, chemical resistance, low electrical and thermal conductivity, and better sound insulation (Collias et al. 1994; Goel and Beckman 1994; Matuana et al. 1998). Polyethylene foams provide high sound and heat insulation benefits along with the outstanding capability to absorb shocks, and these foams are used in packaging and automotive industries, in the buildings, and in sports as well as in leisure products (Xing et al. 2008). However, the foaming agents, traditionally, employed for the production of these polyethylene foams include chlorofluorocarbons and hydrochlorofluorocarbons (Xing et al. 2008), which are responsible for ozone layer depletion. Alternatively, carbon dioxide is considered as a green solvent for its non-toxic, non-flammable, and moderate critical constraints. Moreover, carbon dioxide is able to plasticize most of the polymers and reduces the glass transition temperature (Cooper 2003; Nalawade et al. 2006). Hence, the supercritical carbon dioxide can be used as an alternative environment-friendly foaming agent and has been applied as a clean blowing agent to produce polyethylene foams (Xing et al. 2008).

5.7.2 Polypropylene Foams

Polypropylene materials with micro-pores exhibit outstanding performances (Acebo et al. 2014; Hochleitner et al. 2014; Mori et al. 2009; Moutsatsou et al. 2015; Nofar et al. 2015; Pientka et al. 2009; Sai et al. 2013) like they can resist high temperature as well as can provide superior mechanical strength than other polyolefins. The supercritical carbon dioxide has been widely used as a green blowing agent in foaming of many polymers (Ameli et al. 2015; Keshtkar et al. 2014; Nofar et al. 2015; Nofar and Park 2014). However, owing to the narrow foaming temperature window, the continuous synthesis of polypropylene foams is difficult (Yang et al. 2017). Additionally, polypropylene has low melt strength hindering it to produce prominent cellular structures (Yang et al. 2017). On the other hand, cross-linked polypropylenes have higher melt strength (Bouhelal et al. 2007). Radiation cross-linking also strengthens the melt of the polypropylenes, and the addition of the cross-linking compounds is necessary to minimize the polypropylene decomposition upon their exposure to irradiation (Bouhelal et al. 2007; Kubo et al. 1997). Antunes et al. (2012) and Rizvi and Park (2014) had reported preparation of polypropylene foams through foaming by the supercritical carbon dioxide. Yang et al. (2017), in their study, set the range of the foaming temperature window between 8 and 12 °C that was extremely useful for foaming of polypropylene through the supercritical carbon dioxide in industrial procedures.

Li et al. (2014), Merlet et al. (2007), Krause et al. (2001a, b, 2002), Martina and Hutmacher (2007), Rezwan et al. (2006), Wang et al. (2014), Zhong and Dehghani (2012), Rizvi and Park (2014), Rodeheaver and Colton (2001), Keshtkar et al. (2014), Tomasko et al. (2003), Lee et al. (2005, 2006), Liao and Nawaby (2012), and Pintado-Sierra et al. (2014) had designed a myriad of methodologies to employ carbon dioxide as a foaming agent for the manufacture of open-cell foams. Isotactic polypropylene foam provides many advantages in comparison to the polyolefin foams (Colton 1989; Vasile 2000; Zhai et al. 2010). Bao et al. (2016) had, also, prepared polypropylene foams of different cell sizes and relative densities through solid-state foaming of carbon dioxide in order to develop relationships between mechanical properties of the foams with cell morphological characteristics.

5.7.3 *Semicrystalline Polymers*

Semicrystalline polymers have been effectively used because of their high stiffness and large mechanical strength (Tang and Wang 2017). Moreover, semicrystalline microcellular polymers are cheap and possess better mechanical strength (Tang and Wang 2017). However, the crystalline parts of the semicrystalline polymers may influence the behavior of the foams, and the effects on the crystalline region depend on the foaming processes (Tang and Wang 2017). Tang and Wang (2017) had investigated the influence of foaming on the poly(vinylidene fluoride-*co*-hexafluoropropylene) employing a supercritical carbon dioxide batch process for a saturated timeframe, where the foaming temperature was kept below the poly(vinylidene fluoride-*co*-hexafluoropropylene) melting temperature. The degree of dispersion of the supercritical carbon dioxide was considered to affect the foaming tendency and structure of the semicrystalline polymers, significantly, in that study. The pressure of carbon dioxide in the polymer was interpreted to affect its melting point (Tang and Wang 2017). The melting point was found to decrease with increment in carbon dioxide pressure in the case of isotactic polypropylene foaming (Jiang et al. 2009). Additionally, the foaming temperature should not be too much lowered because if that is the case, then the solubility of carbon dioxide would diminish the cellular structures (Tang and Wang 2017). In their study, they kept the foaming temperature at 130 °C and foaming pressure at 35 MPa for documenting the foaming influence on the poly(vinylidene fluoride-*co*-hexafluoropropylene).

5.8 Supercritical CO_2 and Polyhedral Oligomeric Silsesquioxanes

Phase behavior investigations of the supercritical fluids with the modern-age materials help to develop environment-friendly materials employing the supercritical carbon dioxide (Demirtas and Dilek 2019). Polyhedral oligomeric silsesquioxanes with trifluoropropyl isobutyl, methacryl, and isooctyl functional groups have been observed to be soluble within the supercritical carbon dioxide (Dilek 2013; Kanya and Dilek 2015). Polyhedral oligomeric silsesquioxanes are the organic-inorganic hybrid cage structures containing 8, 10, or even 12 Si atoms with chemical formula $R_nSi_nO_{1.5n}$, where R represents hydrogen or alkyl, alkylene, acrylate, aryl, hydroxyl, amine, and other groups (Demirtas and Dilek 2019). These serve as nanofillers in improving the thermal resistance of the polymers, oxygen and fire resistance, and electric insulation capabilities (Ayandele et al. 2012; Joshi and Butola 2004; Wu and

Mather 2009). The functional groups of these compounds enable their interaction with the polymer chains for increasing the compatibility with the matrix of the polymers that in turn enhance the distribution of polyhedral oligomeric silsesquioxanes within the polymers (Demirtas and Dilek 2019). Polyhedral oligomeric silsesquioxanes are non-volatile, non-toxic, biocompatible, and environment-friendly (Demirtas and Dilek 2019). Like other cases, the supercritical carbon dioxide serves as the preferred solvent and more accurately as a green solvent among the other supercritical fluids. Although lacking polarity, carbon dioxide possesses quadrupole moment, and through the electron-deficient carbon, it can take part in Lewis acid-base reactions (Demirtas and Dilek 2019). As all the polyhedral oligomeric silsesquioxanes are not soluble in CO_2 (Kanya and Dilek 2015), those are required to be functionalized for enabling the Lewis acid-base interactions coupled with weak hydrogen bonding with carbon dioxide (Tezsevin et al. 2017). Trifluoropropyl polyhedral oligomeric silsesquioxanes exhibited the strongest solubility (Dilek 2013), whereas the methacrylate variety had shown the lowest solubility in the supercritical carbon dioxide (Kanya and Dilek 2015). Eris et al. (2013) had formed microchannels within the silica aerogel for optofluidic uses through extracting the fine rods of trifluoropropyl polyhedral oligomeric silsesquioxanes from the aerogel.

5.9 Application of Supercritical CO_2 as a Green solvent in Extraction Procedures

5.9.1 Extraction of Carboxylic Acid

Carboxylic acids are used in large scale at chemical, pharmaceutical, and food industries (Djas and Henczk 2018). Liu et al. (2012) and Soccol et al. (2006) had reviewed that the carboxylic acids can be applied in the production of textile, dyes, perfumes, detergents, pharmaceuticals, etc. Dhillon et al. (2011) and Song and Lee (2006) had reported some advanced uses of this acid in tissue engineering, drug delivery, biopolymer production, etc. Usually, the extraction of a carboxylic acid is carried out with the help of the organic solvents, which are toxic to the environment and can potentially pollute the environment, as well as these solvents are difficult to be regenerated. To avoid these disadvantages, green solvents like supercritical fluids (especially, the supercritical carbon dioxide) are gaining attention for the extraction process of the carboxylic acid (Djas and Henczk 2018). After carrying out an extensive review, they had suggested that the supercritical fluid-driven extraction procedure is a very effective way of separating carboxylic acid from aqueous solution. This process is a very cost-effective and simple reaction procedure and has a potentially large yield. However, more extensive studies and investigations at length are required in this field to overcome the challenges imparted by the transformation of this method from laboratory scale to industrial scale.

5.9.2 Citric Acid Extraction

Citric acid is produced on an industrial scale through fermentation (Djas and Henczk 2016). Citric acid is applied as a preservative and also as an additive to introduce a pleasant essence to foods. It is also used as a pH adjuster, a buffer, and a chelating medium (Djas and Henczk 2016). The application of non-polar supercritical carbon dioxide in the extraction of polar compounds like carboxylic acids requires the use of reactants that build up the complexes with the acid and boost up the acid solubility in the supercritical fluid (Kapucu et al. 1999; Rahmanian and Ghaziaskar 2008). Carboxylic acid extraction requires longer chain-length tertiary aliphatic amines and phosphorus-bonded oxygen-containing reactants (Kertes and King 1986). The reaction between the tertiary aliphatic amines and carboxylic acid is a reversible procedure (Djas and Henczk 2016).

Hartl and Marr (1993), Kapucu et al. (1999), Rahmanian and Ghaziaskar (2008), Wang et al. (2005), and Henczka and Djas (2016) had reviewed the use of supercritical carbon dioxide in reactive extractions of carboxylic acids. Kapucu et al. (1999) and Wang et al. (2005) had observed that the citric acid had poor solubility in supercritical carbon dioxide. However, the application of Alamine 336 or Hostarex A327 and tri-*n*-butyl phosphate (TBP), respectively, were found to increase the solubility of the citric acid. Djas and Henczk (2016) had carried out the citric acid extraction from aqueous solutions applying the supercritical carbon dioxide and tri-*n*-octylamine (TOA) as reactants that formed a complex with acid. The efficiency of reactive extraction was found to be increased with increment in pressure, while it declined with increase in temperature. Additionally, reactive extraction process operating in semi-continuous mode was observed to separate citric acid more efficiently in comparison to the batch mode. They had concluded that citric acid can be recovered in an efficient way with the use of the supercritical carbon dioxide flow saturated with tri-*n*-octylamine at low temperature (35 °C) but high pressure (16 MPa).

5.9.3 Acetic and Propionic Acid Extraction

Further, Henczka and Djas (2016) investigated the application of the supercritical carbon dioxide and tri-*n*-octylamine in reactive extraction of acetic and propionic acids from aqueous solution. They had suggested that the efficiency of the acetic acid and propionic acid recovery could be enhanced by reactive extraction of these two acids using the tri-*n*-octylamine compared to the physical extraction process. They had also recommended to carry out the extraction process by applying the supercritical carbon dioxide in semi-continuous mode. Additionally, this reactive extraction process was observed to become more fruitful with high pressure and a bit longer process time.

5.10 Brief Applications of Supercritical CO_2 in Hydrocarbon Exploration: Pros and Cons

Apart from the applications in chemical industries, the supercritical carbon dioxide has also been employed in hydrocarbon exploration from many potential geological formations. Here, the applications of the supercritical carbon dioxide in hydrocarbon exploration are briefly penned down as they are a bit diverse from its use as a green medium in chemical industries.

5.10.1 Extraction of Organic Matter from Source Rock

Supercritical carbon dioxide can extract organic matter from shale and has the ability to dissolve primary pores as well as fractures (Jiang et al. 2016). Owing to low interconnected porosity and permeability, hydrofracturing and horizontal well fracturing of the reservoir are required for commercial exploration of the shale gas. Research is also going on the supercritical carbon dioxide-induced fracturing of the shale gas reservoirs to enhance the gas pumping rate (Jiang et al. 2016). Ebrahim and Akkutlu (2014), Faye et al. (2013), Hongru et al. (2015), Jing et al. (2015), Jinxing et al. (2014), Roozbeh et al. (2014), Svetlana et al. (2015), Xiangxin et al. (2014), and Yuan et al. (2015), among others, had witnessed that the injection of the supercritical carbon dioxide in the oil or gas field had replaced the methane and accelerated the mobility ratio to boost up the hydrocarbon harvest rate as well as it sealed some amounts of underground carbon dioxide. Usually, the supercritical carbon dioxide is capable to extract lipophilic low molecular weight compounds with a low boiling point at pressure less than 10 MPa like ether, ester, hydrocarbons, etc. (Jiang et al. 2016). Jiang et al. (2016) had treated shale samples from Longmaxi Formation in the Sichuan Basin, China, at a different temperature, pressure, and time to document the influence of the supercritical carbon dioxide on the shale microstructure. They had observed that clay minerals within the shale samples released crystal water after treatment. The specific surface area of the shales and porosity were observed to increase with pressure and time. At low pressure, the treatment of the shales with the supercritical carbon dioxide did not bring about any change in the microstructure. The porosity of the shales was found to exhibit a linear relationship with the specific surface area. Additionally, they had also observed that the density and solvent power of the supercritical carbon dioxide increased with increase in pressure, temperature, and treatment time and, consequently, the supercritical carbon dioxide started to extract a greater amount of organic matter from the fractures and pores of the shales. These, in turn, augmented the shale gas seepage channel numbers and broadened them, which would have enhanced the shale gas development.

5.10.2 Application of Supercritical Carbon Dioxide as a Fracturing Fluid in Shale Gas Reservoir

Further, in hydraulic fracturing for shale gas exploration, the role of fracturing fluid is very crucial as this fluid influences the permeability through inducing fractures, which are the way-outs of the hydrocarbons (Odumabo et al. 2014). The most popular fracturing fluid comprises the water-based fluid system (Ciriello et al. 2016; Federico et al. 2017; Jia et al. 2018; Longo and Di Federico 2015). However, the water in the fracturing fluid may alter the mechanical properties of the shale gas reservoirs (Jia et al. 2017, 2018; Lu et al. 2016). Additionally, in dry areas, excessive use of water may result in its deficiency (Scanlon et al. 2014; Vengosh et al. 2014). Hydrofracturing may contaminate the groundwater with the additives within the fluid (Myers 2012; Osborn et al. 2011; Vengosh et al. 2014). To circumvent these cons of hydrofracturing, supercritical carbon dioxide has been suggested as a substitute for the traditional fracturing fluid (Ishida et al. 2012; Jia et al. 2018; Middleton et al. 2015; Pei et al. 2015). Supercritical carbon dioxide can induce better propagation of fractures (Ishida et al. 2012; Li et al. 2016) with low hindrance in flow and desorption (Middleton et al. 2015) and provide better potential for carbon dioxide sequestration (Dahaghi 2010). Li et al. (2016) and Zhang et al. (2016) had observed that using the supercritical carbon dioxide as a fracturing fluid in Green River and Longmaxi shale deposits had induced better-interconnected fracture networks than the water-based system. Wang et al. (2018) had studied the simulation behavior and found that fracturing with the supercritical carbon dioxide had comparatively smaller breakdown pressure and generated fractures with higher complexity. Fernø et al. (2015) had applied positron emission tomography (PET) to observe the flow of carbon dioxide in shale at the time of injection experiments and found that the enhanced oil recovery factor had a significant possibility to reach up to 55% without any fracturing process. Jia et al. (2018) had investigated the fracturing tests with both water and supercritical carbon dioxide in shale, and they used profilometry for estimating complexity and surface roughness of the induced fractures and to calculate the permeability of those fractures. They found that the supercritical carbon dioxide-induced fracturing introduced fractures with greater tortuosity in comparison to the hydrofracturing. They had also noticed that the supercritical carbon dioxide-induced fracture surfaces appeared more complex and rougher than the hydrofractured surfaces. Additionally, the large minerals were found to be repositioned from the fracture surfaces during the supercritical carbon dioxide fracturing, which acted as micro-proppants, resulting in the larger aperture of the fractures. In complementary, during supercritical carbon dioxide fracturing, the permeability of the shale fractures raised in magnitude nearly about 5 orders, and the magnitude of this augmentation was about 3 orders larger than that induced by the hydrofracturing. Therefore, they suggested that the supercritical carbon dioxide was feasible enough as fracturing fluid for simulations of shale gas reservoirs.

5.10.3 Use of Supercritical Carbon Dioxide in the Tight Gas/ Oil Reservoir

Supercritical carbon dioxide-induced fracturing is also gaining attention in the exploration of tight gas or oil reservoirs as well (Zou et al. 2018). Zou et al. (2018) had investigated the growth behavior of supercritical carbon dioxide-induced fractures in tight sandstones. They had found that the supercritical carbon dioxide stimulated shearing of the sandstone bedding planes and promoted natural fracture systems leading to a complex network of fractures under large horizontal differential stress. A large number of branching fractures were found to be tremendously low, and therefore, they had impeded the proppant transport during hydrofracturing. The growth of the fractures was increased with an increase in injection pressure. Hence, a large rate of pumping was essential in order to maintain the high injection pressure that would open and widen the branching fractures in the sandstone for exploring tight hydrocarbons (Zou et al. 2018).

5.10.4 Disadvantages of Supercritical Carbon Dioxide: Coal Matrix Swelling and Plausible Solutions

However, supercritical carbon dioxide does also have some cons as it has been reported to reduce the permeability of the coal seam through coal matrix swelling (Perera et al. 2011a). Perera et al. (2011a) had carried out an investigation on the influence of the swelling of coal mass on the permeability of naturally fractured black coal. They had conducted the permeability tests for carbon dioxide and nitrogen gas injections at 2–20 MPa injection pressures at 33 °C and under 10–24 MPa confining pressure. Within an hour of carbon dioxide injection, they had observed a significant reduction in permeability of the black coal samples owing to matrix swelling. Coal matrix swelled to its peak within 3–4 h of the carbon dioxide injection. They had also reported that the coal matrix swelling induced by supercritical carbon dioxide adsorption was about twice greater than swelling induced by subcritical carbon dioxide adsorption.

Besides, carbon dioxide sequestration-enhanced coal bed methane recovery by carbon dioxide is another way to reduce the carbon dioxide concentration in the atmosphere for reducing global warming. The carbon dioxide-enhanced coal bed methane recovery represents an injection of carbon dioxide into deep coal seams where carbon dioxide displaces the adsorbed methane through desorption, which is collected through recovery well (Ranathunga et al. 2017). However, the coal matrix swelling due to adsorption of carbon dioxide alters the petrophysical characteristics, and as a consequence, there is a reasonable possibility of carbon dioxide leakage into the adjacent aquifers (Masoudian 2016; Pan et al. 2010; Perera and Ranjith 2012; Vishal et al. 2013). The desorption of methane during the carbon dioxide-enhanced recovery takes place due to stronger attraction and van der Walls bonds between the coal mass and carbon dioxide compared to that in between coal mass and methane (Harpalani and Schraufnagel 1990; Levine 1993). However, the polymer-like structure of coal is deeply affected by the adsorption of carbon dioxide, which results in matrix swelling. This coal matrix swelling develops strains between the macro-pores (Perera et al. 2011b) and the adsorbing carbon dioxide gas layer, which results in declination of pore spaces for transportation of carbon dioxide, and in turn it reduces the permeability of the coal (Gathitu et al. 2009; Perera et al. 2011a). The permeability of the coal is transformed by high pressure and temperature existing deep in the coal seams. At that pressure and temperature, carbon dioxide exists at the supercritical state, and it possesses a higher adsorption efficiency in comparison to the subcritical carbon dioxide and, hence, has greater ability to make the coal matrix swelled (Day et al. 2010; Perera et al. 2011a).

Perera et al. (2011b), Vishal and Singh (2015), and Kiyama et al. (2011) had reported that in high-rank coals, nitrogen gas can enhance the permeability of carbon dioxide in coal mass through reversing the coal matrix swelling caused upon carbon dioxide adsorption. Owing to lower adsorption potential of nitrogen gas within the fracture space of coal mass compared to that of carbon dioxide (Day et al. 2010), it remains in a free gas state, which leads to an imbalance between the free gas and sorbed gas phases. Consequently, this phenomenon lowers the partial pressure of carbon dioxide, and hence, carbon dioxide is released from the coal mass (Reeves 2003), which may, in turn, reduce the coal matrix swelling and enhance the coal seam permeability.

5.11 Conclusions

The authors, therefore, have, briefly, reviewed the chemical signatures of the supercritical carbon dioxide and its utilization as a green medium in different industrial applications. They have collected the stems of the published literature and agglomerated them in this chapter. The traditional organic solvents and volatile organic compounds, used for material synthesis and extraction procedures, are the potential harms to the environment that cannot be overlooked. Therefore, there was a strong urge to adopt the green philosophy for enhancing the sustainability of the material synthesis in the industrial scale. Supercritical fluid, especially the supercritical carbon dioxide, comes out to be a green medium, which can be available at ease, transported without any hazard, and economic and can pacify the toxicity and wreck harms to the environment. With its low critical pressure and temperature compared to the other fluids, the supercritical carbon dioxide turns out to be the suitable alternate green solvent in the chemical industries. The fluorinated, partially fluorinated, and non-fluorinated surfactants as well as fluorinated and non-fluorinated polymers were solubilized to enhance the solvent strength of the carbon dioxide. Moreover, the addition of polymeric viscosifiers and hydrotropic salts may thicken the carbon dioxide and, therefore, enhance its solvent strength. In complementary, the authors have reviewed different organic chemical reactions that employ supercritical carbon dioxide as a green reactant and synthesize the products at low cost and with high yield. Scientists had, successfully, synthesized the nanoparticles, nanowires, nanofilms, nanocapsules, carbon nanotubes, drug nanoparticles, etc. through adopting green technologies. Additionally, supercritical carbon dioxide has been, widely, used as a foaming agent in the preparation of different types of foams in industrial scale as well as in reactive extractions of carboxylic acid, citric acid, and acetic and propionic acids from the aqueous solutions. The supercritical carbon dioxide is, also, capable of extracting organic matter from the source rocks to enhance the shale gas production. Further, the supercritical carbon dioxide has also been considered to be a potential fracturing fluid in shale gas exploration as it minimizes the adverse effects of hydrofracturing. Hence, in a nutshell, the supercritical carbon dioxide has a myriad of advantages as a green medium in several chemical reactions as well as in different types of industries, which can potentially outrun the disadvantages of the traditional toxic organic solvents. But, there is a long way to go before it turns out to be the most ideal green solvent for fulfilling most of the desired objectives in all the fields. By far, supercritical carbon dioxide is a large advancement toward the green philosophy that the world requires, now, the most, in the true sense.

References

Acebo C, Fernandez-Francos X, Messori M, Ramis X, Serra A (2014) Novel epoxysilica hybrid coatings by using ethoxysilyl-modified hyperbranched poly(ethyleneimine) with improved scratch resistance. Polymer 55:5028–5035. https://doi.org/10.1016/j.polymer.2014.08.021

Acharya DP, Kunieda H (2006) Wormlike micelles in mixed surfactant solutions. Adv Colloid Interface Sci 123–126:401–413. https://doi.org/10.1016/j.cis.2006.05.024

Adkins SS, Chen X, Chan I, Torino E, Nguyen QP, Sanders AW, Johnston KP (2010) Morphology and stability of CO_2-in-water foams with nonionic hydrocarbon surfactants. Langmuir 26:5335–5348. https://doi.org/10.1021/la903663v

Amandi R, Scovell K, Licence P, Lotz TJ, Poliakoff M (2007) The synthesis of o-cyclohexylphenol in supercritical carbon dioxide: towards a continuous two-step reaction. Green Chem 9:797–801. https://doi.org/10.1039/B618727G

Ameli A, Nofar M, Jahani D, Rizvi G, Park CB (2015) Development of high void fraction polylactide composite foams using injection molding: crystallization and foaming behaviors. Chem Eng J 262:78–87. https://doi.org/10.1016/j.cej.2014.09.087

Anastas PT, Warner JC (1998) Green chemistry: theory and practice. Oxford University Press, New York. ISBN: 9780198506980

Antunes M, Realinho V, Ignacio Velasco J, Solorzano E, Rodriguez-Perez M-A, Antonio de Saja J (2012) Thermal conductivity anisotropy in polypropylene foams prepared by supercritical CO_2 dissolution. Mater Chem Phys 136:268–276. https://doi.org/10.1016/j.matchemphys.2012.07.001

Ayandele E, Sarkar B, Alexandridis P (2012) Polyhedral oligomeric silsesquioxane (POSS)-containing polymer nanocomposites. Nano 2:445–475. https://doi.org/10.3390/nano2040445

Bae J, Irani CA (1993) Laboratory investigation of viscosified CO_2 process. Soc Pet Eng Adv Technol Ser 1:166–171. https://doi.org/10.2118/20467-PA

Bao JB, Junior AN, Weng GS, Wang J, Fang YW, Hu GH (2016) Tensile and impact properties of microcellular isotactic polypropylene (PP) foams obtained by supercritical carbon dioxide. J Supercrit Fluids 111:63–73. https://doi.org/10.1016/j.supflu.2016.01.016

Beckman EJ (2004) Supercritical and near-critical CO_2 in green chemical synthesis and processing. J Supercrit Fluids 28:121–191. https://doi.org/10.1016/S0896-8446(03)00029-9
Bektesevic S, Kleman AM, Marteel-Parrish AE, Abraham MA (2006) Hydroformylation in supercritical carbon dioxide: catalysis and benign solvents. J Supercrit Fluids 38:232–241. https://doi.org/10.1016/j.supflu.2006.01.019
Beyer C, Oellrich LR, McHugh MA (2000) Effect of copolymer composition and solvent polarity on the phase behavior of mixtures of poly(ethylene-co-vinyl acetate) with cyclopentane and cyclopentene. Chem Eng Technol 23:592–595. https://doi.org/10.1002/cite.330711116
Bouhelal S, Cagiao ME, Benachour D, Calleja FJB (2007) Structure modification of isotactic polypropylene through chemical crosslinking: toughening mechanism. J Appl Polym Sci 103:2968–2976. https://doi.org/10.1002/app.25406
Bourne RA, Han X, Chapman AO, Arrowsmith NJ, Kawanami H, Poliakoff M, George MW (2008) Homogeneous photochemical oxidation via singlet O_2 in supercritical CO_2. Chem Commun 37:4457–4459. https://doi.org/10.1039/b806063k
Butenhoff JL, Costa G, Elcombe C, Farrar D, Hansen K, Iwai H, Jung R, Kennedy G Jr, Lieder P, Olsen G, Thomford P, Butenhoff J (2002) Toxicity of ammonium perfluorooctanoate in male Cynomolgus monkeys after oral dosing for 6 months. Toxicol Sci 69:244–257. https://doi.org/10.1016/j.plantsci.2016.04.008
Caballero A, Despagnet-Ayoub E, Mar Diaz-Requejo M, Diaz-Rodriguez A, Gonzalez-Nunez ME, Mello R, Munoz BK, Ojo W-S, Asensio G, Etienne M, Perez PJ (2011) Silver-catalyzed C-C bond formation between methane and ethyl diazoacetate in supercritical CO_2. Science 332(6031):835–838. https://doi.org/10.1126/science.1204131
Caravati M, Grunwaldt JD, Baiker A (2006) Solvent-modified supercritical CO_2 a beneficial medium for heterogeneously catalyzed oxidation reactions. Appl Catal A Gen 50:298–303. https://doi.org/10.1016/j.apcata.2005.09.037
Chandran A, Prakash K, Senapati S (2010) Self-assembled inverted micelles stabilize ionic liquid domains in supercritical CO_2. J Am Chem Soc 132:12511–12516. https://doi.org/10.1021/ja1055005
Chattopadhyay P, Gupta RB (2003) Supercritical CO_2-based formation of silica nanoparticles using water-in-oil microemulsions. Ind Eng Chem Res 42:465–472. https://doi.org/10.1021/ie020261c
Cheng M, Moore DR, Reczek JJ, Chamberlain BM, Lobkovsky EB, Coates GW (2001) Single-site β-diiminate zinc catalysts for the alternating copolymerization of CO_2 and epoxides: catalyst synthesis and unprecedented polymerization activity. J Am Chem Soc 123:8738–8749. https://doi.org/10.1021/ja003850n
Chouchi D, Gourgouillon D, Couret M, Vital J, Ponte MN (2001) The influence of phase behavior on reactions at supercritical conditions: the hydrogenation of α-pinene. Ind Eng Chem Res 40:2551–2554. https://doi.org/10.1021/ie000859p
Ciftci ON, Temelli F (2011) Continued production of fatty acid methyl esters from corn oil in a supercritical carbon dioxide bioreactor. J Supercrit Fluids 58:79–87. https://doi.org/10.1016/j.supflu.2011.05.011
Ciriello V, Longo S, Chiapponi L, Di Federico V (2016) Porous gravity currents: a survey to determine the joint influence of fluid rheology and variations of medium properties. Adv Water Resour 92:105–115. https://doi.org/10.1016/j.advwatres.2016.03.021
Clark P, Poliakoff M, Wells A (2007) Continuous flow hydrogenation of a pharmaceutical intermediate 4-(3,4-dichlorophenyl)-3,4-dihydro-2H-naphthalenydene methylamine in supercritical carbon dioxide. Adv Synth Catal 349:2655–2659. https://doi.org/10.1002/adsc.200700395
Coleman NRB, Morris MA, Spalding TR, Holmes JD (2001) The formation of dimensionally ordered silicon nanowires within mesoporous silica. J Am Chem Soc 123:187–188. https://doi.org/10.1021/ja005598p
Collias DI, Baird DG, Borggreve RJM (1994) Impact toughening of polycarbonate by microcellular foaming. Polymer 35:3978–3983. https://doi.org/10.1016/0032-3861(94)90283-6

Colton JS (1989) The nucleation of microcellular foams in semicrystalline thermoplastics. Mater Manuf Process 4:253–262. https://doi.org/10.1080/10426918908956288

Consani KA, Smith RD (1990) Observations on the solubility of surfactants and related molecules in carbon dioxide at 50 °C. J Supercrit Fluids 3:51–65. https://doi.org/10.1016/0896-8446(90)90008-A

Cooper AI (2000) Polymer synthesis and processing using supercritical carbon dioxide. J Mater Chem 10:207–234. https://doi.org/10.1039/A906486I

Cooper AI (2001) Recent developments in materials synthesis and processing using supercritical CO_2. Adv Mater 13:1111–1114. https://doi.org/10.1002/1521-4095(200107)13:143.0.CO;2-L

Cooper AI (2003) Porous materials and supercritical fluids. Adv Mater 15:1049–1059. https://doi.org/10.1002/adma.200300380

Cummings S, Enick R, Rogers S, Heenan R, Eastoe J (2012) Amphiphiles for supercritical CO_2. Biochimie 94:94–100. https://doi.org/10.1016/j.biochi.2011.06.021

da Rocha SRP, Dickson J, Cho D, Rossky PJ, Johnston KP (2003) Stubby surfactants for stabilization of water and CO_2 emulsions: trisiloxanes. Langmuir 19:3114–3120. https://doi.org/10.1021/la026608y

Dahaghi AK (2010) Numerical simulation and modeling of enhanced gas recovery and CO_2 sequestration in shale gas reservoirs: a feasibility study. In: SPE international conference on CO_2 capture, storage, and utilization, 10-12 November, New Orleans, Louisiana, USA

Day S, Fry R, Sakurovs R, Weir S (2010) Swelling of coals by supercritical gases and its relationship to sorption. Energy Fuel 24:2777–2783. https://doi.org/10.1021/ef901588h

de Vries TJ, Duchateau R, Vortsman MAG, Keurentjes JTF (2000) Polymerisation of olefins catalysed by a palladium complex in supercritical carbon dioxide. Chem Commun 2000:263–264. https://doi.org/10.1039/A908966G

Demirtas C, Dilek C (2019) Enhanced solubility of siloxy-modified polyhedral oligomeric silsesquioxanes in supercritical carbon dioxide. J Supercrit Fluids 143:358–364. https://doi.org/10.1016/j.supflu.2018.09.015

Dhillon GS, Brar SK, Verma M, Tyagi RD (2011) Recent advances in citric acid bioproduction and recovery. Food Bioprocess Technol 4:505–529. https://doi.org/10.1007/s11947-010-0399-0

Dilek C (2013) Supercritical carbon dioxide-soluble polyhedral oligomeric silsesquioxane (POSS) nanocages and polymer surface modification. J Supercrit Fluids 73:171–177. https://doi.org/10.1016/j.supflu.2012.10.012

Djas M, Henczk M (2016) Reactive extraction of citric acid using supercritical carbon dioxide. J Supercrit Fluids 117:59–63. https://doi.org/10.1016/j.supflu.2016.05.005

Djas M, Henczk M (2018) Reactive extraction of carboxylic acids using organic solvents and supercritical fluids: a review. Sep Purif Technol 201:106–119. https://doi.org/10.1016/j.seppur.2018.02.010

Du L, Kelly JY, Roberts GW, DeSimone JM (2009) Fluoropolymer synthesis in supercritical carbon dioxide. J Supercrit Fluids 47:447–457. https://doi.org/10.1016/j.supflu.2008.11.011

Early TR, Gordon RS, Carrol MA, Holmes AB, Shute RE, McConvey IF (2001) Palladium-catalysed cross-coupling reactions in supercritical carbon dioxide. Chem Commun (19):1966–1967. https://doi.org/10.1039/B106808C

Eastoe J, Bayazit Z, Martel S, Steytler DC, Heenan RK (1996) Droplet structure in a water-in-CO_2 microemulsion. Langmuir 12:1423–1424. https://doi.org/10.1021/la950546c

Eastoe J, Downer A, Paul A, Steytler DC, Rumsey E, Penfold J, Heenan RK (2000) Fluorosurfactants at air/water and water/CO_2 interfaces. Phys Chem Chem Phys 2:5235–5242. https://doi.org/10.1039/b005858k

Eastoe J, Gold S, Rogers SE, Paul A, Welton T, Heenan RK, Grillo IJ (2005) Ionic liquid-in-oil microemulsions. J Am Chem Soc 127:7302–7303. https://doi.org/10.1021/ja051155f

Ebrahim F, Akkutlu IY (2014) Multi-component gas transport and adsorption effects during CO_2 injection and enhanced shale gas recovery. Int J Coal Geol 123:52–61. https://doi.org/10.1016/j.coal.2013.07.021

Eckert CA, Knutson BL, Debenedetti PG (1996) Supercritical fluids as solvents for chemical and materials processing. Nature 383:313–318. https://doi.org/10.1038/383313a0

Enick R, Beckman E, Yazdi A, Krukonis V, Schonemann H, Howell J (1998) Phase behavior of CO_2–perfluoropolyether oil mixtures and CO_2–perfluoropolyether chelating agent mixtures. J Supercrit Fluids 13:121–126. https://doi.org/10.1016/S0896-8446(98)00043-6

Eris G, Sanli D, Ulker Z, Bozbag SE, Jonás A, Kiraz A, Erkey C (2013) Three-dimensional optofluidic waveguides in hydrophobic silica aerogels via supercritical fluid processing. J Supercrit Fluids 73:28–33. https://doi.org/10.1016/j.supflu.2012.11.001

Falandysz J, Taniyasu S, Gulkowska A, Yamashita N, Schulte-Oehlmann U (2006) Is fish a major source of fluorinated surfactants and repellents in humans Living on the Baltic coast? Environ Sci Technol 40:748–751. https://doi.org/10.1021/es051799n

Faye L, Kevin E, Yitian X, John AR (2013) Assessing the feasibility of CO_2 storage in the New Albany Shale (Devonian-Mississippian) with potential enhanced gas recovery using reservoir simulation. Int J Greenh Gas Control 17:111–126. https://doi.org/10.1016/j.ijggc.2013.04.018

Federico V, Di Longo S, King SE, Chiapponi L, Petrolo D, Ciriello V (2017) Gravity-driven flow of Herschel – bulkley fluid in a fracture and in a 2D porous medium. J Fluid Mech 821:59–84. https://doi.org/10.13140/RG.2.2.32496.46084

Fernø MA, Hauge LP, Uno Rognmo A, Gauteplass J, Graue A (2015) Flow visualization of CO_2 in tight shale formations at reservoir conditions. Geophys Res Lett 42:7414–7419. https://doi.org/10.1002/2015GL065100

Fujita S-I, Yoshida H, Asai K, Meng X, Arai M (2011) Selective hydrogenation of nitrostyrene to aminostyrene over Pt/TiO_2 catalysts: effects of pressurized carbon dioxide and catalyst preparation conditions. J Supercrit Fluids 60:106–112. https://doi.org/10.1016/j.supflu.2011.02.016

Fukushima Y, Wakayama H (1999) Nanoscale casting using supercritical fluid. J Phys Chem B 103:3062–3064. https://doi.org/10.1021/jp984649q

Galia A, Scialdone O, Tortorici E (2011) Transesterification of rapeseed oil over acid resins promoted by supercritical carbon dioxide. J Supercrit Fluids 56:186–193. https://doi.org/10.1016/j.supflu.2010.12.005

Gathitu BB, Chen W-Y, McClure M (2009) Effects of coal interaction with supercritical CO_2: physical structure. Ind Eng Chem Res 48:5024–5034. https://doi.org/10.1021/ie9000162

Goddard R, Bosley J, Al-Duri B (2000) Lipase-catalysed esterification of oleic acid and ethanol in a continuous packed bed reactor, using supercritical CO_2 as solvent: approximation of system kinetics. J Chem Technol Biotechnol 75:715–721. https://doi.org/10.1002/1097-4660(200008)75:8<715::AID-JCTB253>3.0.CO;2-J

Goel SK, Beckman EJ (1994) Generation of microcellular polymeric foams using supercritical carbon dioxide. I: effect of pressure and temperature on nucleation. Polym Eng Sci 34:1137–1147. https://doi.org/10.1002/pen.760341408

Guo W, Li Z, Fung BM, O'Rear EA, Harwell JH (1992a) Hybrid surfactants containing separate hydrocarbon and fluorocarbon chains. J Phys Chem 96:6738–6742. https://doi.org/10.1021/j100195a038

Guo W, Fung BM, O'Rear EA (1992b) Exchange of hybrid surfactant molecules between monomers and micelles. J Phys Chem 96:10068–10074. https://doi.org/10.1021/j100203a088

Hadida S, Super MS, Beckman EJ, Curran DP (1997) Radical reactions with alkyl and fluoroalkyl (fluorous) tin hydride reagents in supercritical CO_2. J Am Chem Soc 119:7406–7407. https://doi.org/10.1021/ja971120i

Han X, Poliakoff M (2012) Continuous reactions in supercritical carbon dioxide: problems, solutions and possible ways forward. Chem Soc Rev 41:1428–1436. https://doi.org/10.1039/c2cs15314a

Hanrath T, Korgel BA (2002) Nucleation and growth of germanium nanowires seeded by organic monolayer-coated gold nanocrystals. J Am Chem Soc 124:1424–1429. https://doi.org/10.1021/ja016788i

Harpalani S, Schraufnagel RA (1990) Shrinkage of coal matrix with release of gas and its impact on permeability of coal. Fuel 69:551–556. https://doi.org/10.1016/0016-2361(90)90137-F

Harrison K, Goveas J, Johnston KP, O'Rear EA (1994) Water-in-carbon dioxide microemulsions with a fluorocarbon-hydrocarbon hybrid surfactant. Langmuir 10:3536–3541. https://doi.org/10.1021/la00022a028

Hartl J, Marr R (1993) Extraction processes for bioproduct separation. Sep Sci Technol 28:805–819. https://doi.org/10.1080/01496399308019522

Hatzopoulos MH, Eastoe J, Dowding PJ, Grillo I (2013) Cylinder to sphere transition in reverse microemulsions: the effect of hydrotropes. J Colloid Interface Sci 392:304–310. https://doi.org/10.1016/j.jcis.2012.09.078

Haumann M, Riisagar A (2008) Hydroformylation in room temperature ionic liquids (RTILs): catalysts and process developments. Chem Rev 108:1474–1497. https://doi.org/10.1021/cr078374z

Heller JP, Dandge DK, Card RJ, Donaruma LG (1985) Direct thickeners for mobility control of CO_2 floods. Soc Pet Eng J 25:679–686. https://doi.org/10.2118/11789-PA

Henczka M, Djas M (2016) Reactive extraction of acetic acid and propionic acid using supercritical carbon dioxide. J Supercrit Fluids 110:154–160. https://doi.org/10.1016/j.supflu.2015.11.018

Hiyoshi N, Mine E, Rode CV, Sato O, Shirai M (2006) Low temperature hydrogenation of tetralin over supported rhodium catalysts in supercritical carbon di oxide solvent. Appl Catal A Gen 310:194–198. https://doi.org/10.1016/j.apcata.2006.05.040

Hiyoshi N, Rode CV, Sato O, Tetsuka H, Shirai M (2007) Stereoselective hydrogenation of tert-butylphenols over charcoal-supported rhodium catalyst in supercritical carbon dioxide solvent. J Catal 252:57–68. https://doi.org/10.1016/j.jcat.2007.08.011

Hiyoshi N, Bando KK, Sato O, Yamaguchi A, Rode CV, Shirai M (2009) Stereoselective hydrogenation of 4-alkylphenols over carbon-supported rhodium catalyst in supercritical carbon dioxide solvent. Catal Commun 10:1702–1705. https://doi.org/10.1016/j.catcom.2009.05.011

Hiyoshi N, Sato O, Yamaguchi A, Rode CV, Shirai M (2012) Kinetic analysis of 4-isopropylphenol hydrogenation over activated carbon-supported rhodium catalysts in supercritical carbon dioxide solvent. Green Chem 14:633–638. https://doi.org/10.1039/C1GC16038A

Hochleitner G, Huemmer JF, Luxenhofer R, Groll J (2014) High definition fibrous poly(2-ethyl-2-oxazoline) scaffolds through melt electrospinning writing. Polymer 55:5017–5023. https://doi.org/10.1016/j.polymer.2014.08.024

Hoefling TA, Enick RM, Beckman EJ (1991) Microemulsions in near-critical and supercritical carbon dioxide. J Phys Chem 95:7127–7129. https://doi.org/10.1021/j100172a006

Hollamby MJ, Trickett K, Mohamed A, Cummings S, Tabor RF, Myakonkaya O, Gold S, Rogers S, Heenan RK, Eastoe J (2009) Tri-chain hydrocarbon surfactants as designed micellar modifiers for supercritical CO_2. Angew Chem Int Ed 48:4993–4995. https://doi.org/10.1002/anie.200901543

Holmes JD, Johnston KP, Doty RC, Korgel BA (2000) Control of thickness and orientation of solution-grown silicon nanowires. Science 287:1471–1473. https://doi.org/10.1126/science.287.5457.1471

Holmes JD, Lyons DM, Ziegler KJ (2003) Supercritical fluid synthesis of metal and semiconductor nanomaterials. Chem Eur J 9:2144–2150. https://doi.org/10.1002/chem.200204521

Hong S-S, Lee MS, Hwang H-S, Lim K-T, Park SS, Ju C-S, Lee G-D (2003) Preparation of titanium dioxides in the w/c microemulsions and their photocatalytic activity. Sol Energy Mater 80:273–282. https://doi.org/10.1016/j.solmat.2003.08.003

Hong L, Fisher M, Enick R, Beckman E (2008) Cellulose triacetate oligomers exhibit high solubility in dense CO_2. Green Chem 10:756–761. https://doi.org/10.1039/B800812D

Hongru L, Dongdong J, Ruihua L, Bingqian S (2015) Correlating and estimating the solubilities of solid organic compounds in supercritical CO_2 based on the modified expanded liquid model and the group contribution method. Fluid Phase Equilib 385:10–24. https://doi.org/10.1016/j.fluid.2014.10.036

Houde M, Bujas TAD, Small J, Wells RS, Fair PA, Bossart GD, Solomon KR, Muir DCG (2006) Biomagnification of perfluoroalkyl compounds in the bottlenose dolphin (Tursiops truncatus) food web. Environ Sci Technol 40:4138–4414. https://doi.org/10.1021/es060233b

Hu ZS, Dong JX, Chen GX (1999a) Preparation of nanometer titanium oxide with n-butanol supercritical drying. Powder Technol 101:205–210. https://doi.org/10.1016/S0032-5910(98)00153-3
Hu ZS, Dong JX, Chen GX, Lou F (1999b) Preparation of nanometer copper borate with supercritical carbon dioxide drying. Powder Technol 102:171–176. https://doi.org/10.1016/S0032-5910(98)00190-9
Huang Z, Shi C, Xu J, Kilic S, Enick RM, Beckman EJ (2000) Enhancement of the viscosity of carbon dioxide using styrene/fluoroacrylate copolymers. Macromolecules 33:5437–5442. https://doi.org/10.1021/ma992043+
Ikariya T, Kayaki Y (2000) Supercritical fluids as reaction media for molecular catalysis. Catal Surv 4(1):39–50. https://doi.org/10.1023/A:1019032004130
Inoue S, Yamazaki N (1982) Organic and bioorganic chemistry of carbon dioxide, vol 13. Wiley, New York. ISBN 9780470273098
Ishida T, Aoyagi K, Niwa T, Chen Y, Murata S, Chen Q, Nakayama Y (2012) Acoustic emission monitoring of hydraulic fracturing laboratory experiment with supercritical and liquid CO_2. Geophys Res Lett 39:16309. https://doi.org/10.1029/2012GL052788
Jackson MA, Mbaraka IK, Shank BH (2006) Esterification of oleic acid in supercritical carbon dioxide catalyzed by functionalized mesoporous silica and an immobilized lipase. Appl Catal A Gen 310:48–53. https://doi.org/10.1016/j.apcata.2006.05.019
James C, Hatzopoulos MH, Yan C, Smith GN, Alexander S, Rogers SE, Eastoe J (2014) Shape transitions in supercritical CO_2 microemulsions induced by hydrotropes. Langmuir 30:96–102. https://doi.org/10.1021/la404144a
Jenzer G, Mallat T, Baiker A (2001) Continuous oxidation of benzyl alcohol in "supercritical" carbon dioxide. Catal Lett 73:5–8. https://doi.org/10.1023/A:1009014000630
Jessop PG, Leitner W (1999) Chemical synthesis using supercritical fluids. Wiley-VCH, Weinheim. ISBN:9783527613687
Jessop PG, Ikaria T, Noyori R (1995) Homogeneous hydrogenation of carbon dioxide. Chem Rev 95:259–272. https://doi.org/10.1021/cr00034a001
Jessop PG, Ikariya T, Noyori R (1999) Homogeneous catalysis in supercritical fluids. Chem Rev 99:475–494. https://doi.org/10.1021/cr970037a
Jia Y, Lu Y, Elsworth D, Fang Y, Wang C, Tang J (2017) Hydro-mechanical-chemical effects on permeability evolution of fractures in Longmaxi shale. In: Proceedings of the 51st US rock mechanics/geomechanics symposium (San Francisco, CA, USA). Doc ID: ARMA-2017-0296
Jia Y, Lu Y, Elsworth D, Fang Y, Tang J (2018) Surface characteristics and permeability enhancement of shale fractures due to water and supercritical carbon dioxide fracturing. J Pet Sci Eng 165:284–297. https://doi.org/10.1016/j.petrol.2018.02.018
Jiang XL, Liu TZ, Xu M, Zhao L, Hu GH, Yuan WK (2009) Effects of crystal structure on the foaming of isotactic polypropylene using supercritical carbon dioxide as a foaming agent. J Supercrit Fluids 48:167–175. https://doi.org/10.1016/j.supflu.2008.10.006
Jiang Y, Luo Y, Lu Y, Qin C (2016) Effects of supercritical CO_2 treatment time, pressure, and temperature on microstructure of shale. Energy 97:173–181. https://doi.org/10.1016/j.energy.2015.12.124
Jing P, Haijian Y, Hongwei Y (2015) Solubilities and partial molar volumes of three bis(2-ethoxyethyl) ethanedioate derivatives in supercritical carbon dioxide. J Chem Thermodyn 83:33–42. https://doi.org/10.1016/j.jct.2014.11.009
Jinxing D, Caineng Z, Shimeng L (2014) Geochemistry of the extremely high thermal maturity Longmaxi shale gas southern Sichuan Basin. Org Geochem 74:3–12. https://doi.org/10.1016/j.orggeochem.2014.01.018
Johnston KP, da Rocha SRP (2009) Colloids in supercritical fluids over the last 20 years and future directions. J Supercrit Fluids 47:523–530. https://doi.org/10.1016/j.supflu.2008.10.024
Johnston KP, Jacobson GB, Lee CT, Meredith C, da Rocha SRP, Matt ZY, DeGrazia J (1999) Microemulsions, emulsions and latexes. In: Jessop PG, Leitner W (eds) Chemical synthesis using supercritical fluids. Wiley-VCH, Weinheim, p 127. ISBN: 9783527613694

Joshi M, Butola BS (2004) Polymeric nanocomposites-polyhedral oligomeric silsesquioxanes (POSS) as hybrid nanofiller. J Macromol Sci Polym Rev 44:389–410. https://doi.org/10.1081/MC-200033687

Kainz S, Brinkmann A, Leitner W, Pfaltz AJ (1999) Iridium-catalyzed enantioselective hydrogenation of imines in supercritical carbon dioxide. Am Chem Soc 121:6421–6429. https://doi.org/10.1021/ja984309i

Kameo A, Yoshimura T, Esumi K (2003) Preparation of noble metal nanoparticles in supercritical carbon dioxide. Colloids Surf A Physicochem Eng Asp 215:181–189. https://doi.org/10.1016/S0927-7757(02)00439-9

Kanya B, Dilek C (2015) Effects of functional groups on the solubilities of polyhedral oligomeric silsesquioxanes (POSS) in supercritical carbon dioxide. J Supercrit Fluids 102:17–23. https://doi.org/10.1016/j.supflu.2015.03.024

Kapucu N, Güvenç A, Mehmetoğlu Ü, Çalimli A (1999) Reactive extraction of organic acids by supercritical CO_2. Rev Chem Eng 15:233–243. https://doi.org/10.1515/REVCE.1999.15.4.233

Kazarian SG, Vincent MF, Bright FV, Liotta CL, Eckert CAJ (1996) Specific intermolecular interaction of carbon dioxide with polymers. Am Chem Soc 118:1729–1736. https://doi.org/10.1021/ja950416q

Kertes AS, King CJ (1986) Extraction chemistry of fermentation product carboxylic acids. Biotechnol Bioeng 28:269–282. https://doi.org/10.1002/bit.260280217

Keshtkar M, Nofar M, Park CB, Carreau PJ (2014) Extruded PLA/clay nanocomposite foams blown with supercritical CO_2. Polymer 55:4077–4090. https://doi.org/10.1016/j.polymer.2014.06.059

Kilic S, Michalik S, Wang Y, Johnson JK, Enick RM, Beckman E (2003) Effect of grafted Lewis base groups on the phase behavior of model poly(dimethyl siloxanes) in CO_2. J Ind Eng Chem Res 42:6415–6424. https://doi.org/10.1021/ie030288b

Kilic S, Wang Y, Johnson JK, Beckman EJ, Enick RM (2009) Influence of tert-amine groups on the solubility of polymers in CO_2. Polymer 50:2436–2444. https://doi.org/10.1016/j.polymer.2009.03.012

Kissa E (1994) Fluorinated surfactants: synthesis, properties and applications. Marcell Dekker Inc., New York, p 50. ISBN: 978-0824790110

Kiyama T, Nishimoto S, Fujioka M, Xue Z, Ishijima Y, Pan Z, Connell LD (2011) Coal swelling strain and permeability change with injecting liquid/supercritical CO_2 and N_2 at stress-constrained conditions. Int J Coal Geol 85:56–64. https://doi.org/10.1016/j.coal.2010.09.010

Klostermann M, Strey R, Sottmann T, Schweins R, Lindner P, Holderer O, Monkenbusch M, Richter D (2012) Structure and dynamics of balanced supercritical CO_2-microemulsions. Soft Matter 8:797–807. https://doi.org/10.1039/C1SM06533E

Korgel BA, Shah PS, Pell LE, Johnston KP (2002) Nanocrystal stabilization and synthesis in supercritical solvents. In: Bertucco A (ed) Proceedings fourth international symposium on high pressure technology and chemical engineering, vol 2, Venice, Italy, p 633

Krause B, Boerrigter ME, van der Vegt NFA, Strathmann H, Wessling M (2001a) Novel open-cellular polysulfone morphologies produced with trace concentrations of solvents as pore opener. J Membr Sci 187:181–192. https://doi.org/10.1016/S0376-7388(01)00329-5

Krause B, Sijbesma HJP, Munuklu P, van der Vegt NFA, Wessling M (2001b) Bicontinuous nanoporous polymers by carbon dioxide foaming. Macromolecules 34:8792–8801. https://doi.org/10.1021/ma010854j

Krause B, Diekmann K, van der Vegt NFA, Wessling M (2002) Open nanoporous morphologies from polymeric blends by carbon dioxide foaming. Macromolecules 35:1738–1745. https://doi.org/10.1021/ma011672s

Kubo J, Otsuhata K, Ikeda S, Seguchi T (1997) Electron-induced crosslinking of polypropylene with the addition of hydrogen-donating hydrocarbons. J Appl Polym Sci 64:311–319. https://doi.org/10.1002/(SICI)1097-4628(19970411)64:2<311::AID-APP11>3.0.CO;2-3

Kuran W (1998) Coordination polymerization of heterocyclic and heterounsaturated monomers. Prog Polym Sci 23:919–992. https://doi.org/10.1016/S0079-6700(97)00031-2

Lange S, Brinkmann A, Trautner P, Woelk K, Bargon J, Leitner W (2000) Mechanistic aspects of dihydrogen activation and transfer during asymmetric hydrogenation in supercritical carbon dioxide. Chirality 12:450–457. https://doi.org/10.1002/(SICI)1520-636X(2000)12:5/6<450::AID-CHIR26>3.0.CO;2-H

Lee PC, Naguib HE, Park CB, Wang J (2005) Increase of open-cell content by introducing a secondary blowing agent. Polym Eng Sci 45:1445–1451. https://doi.org/10.1002/pen.20422

Lee PC, Wang J, Park CB (2006) Extruded open-cell foams using two semicrystalline polymers with different crystallization temperatures. Ind Eng Chem Res 45:175–181. https://doi.org/10.1021/ie050498j

Levine JR (1993) Coalification: the evolution of coal as source rock and reservoir. In: Law BE, Rice DD (eds.), Hydrocarbons from Coal. American Association of Petroleum Geologists, AAPG Studies in Geology, 38:39–77. ISBN: 9781629811048.

Li Z, Zhou Z, Armstrong SR, Baer E, Paul DR, Ellison CJ (2014) Multilayer coextrusion of rheologically modified main chain liquid crystalline polymers and resulting orientational order. Polymer 55:4966–4975. https://doi.org/10.1016/j.polymer.2014.08.009

Li X, Feng Z, Han G, Elsworth D, Marone C, Saffer D, Cheon D-S (2016) Breakdown pressure and fracture surface morphology of hydraulic fracturing in shale with H_2O, CO_2 and N_2. Geomech Geophys Geo Energy Geo Resour 2:63–76. https://doi.org/10.1007/s40948-016-0022-6

Liao X, Nawaby AV (2012) Solvent free generation of open and skinless foam in poly(L-lactic acid)/poly(D,L-lactic acid) blends using carbon dioxide. Ind Eng Chem Res 51:6722–6730. https://doi.org/10.1021/ie3000997

Liu Z, Dai X, Xu J, Han B, Zhang J, Wang Y, Huang Y, Yang G (2004) Encapsulation of polystyrene within carbon nanotubes with the aid of supercritical CO_2. Carbon 42:458–460. https://doi.org/10.1016/j.carbon.2003.12.031

Liu J, Cheng S, Zhang J, Feng X, Fu X, Han B (2007) Reverse micelles in carbon dioxide with ionic-liquid domains. Angew Chem Int Ed 46:3313–3315. https://doi.org/10.1002/anie.200605019

Liu L, Zhu Y, Li JH, Wang M, Lee P, Du GC, Chen J (2012) Microbial production of propionic acid from propionibacteria: current state, challenges and perspectives. Crit Rev Biotechnol 32:374–381. https://doi.org/10.3109/07388551.2011.651428

Loeker F, Marr PC, Howdle SM (2003) FTIR analysis of water in supercritical carbon dioxide microemulsions using monofunctional perfluoropolyether surfactants. Colloids Surf A 214:143–150. https://doi.org/10.1016/S0927-7757(02)00407-7

Longo S, Di Federico V (2015) Unsteady flow of shear-thinning fluids in porous media with pressure-dependent properties. Transport Porous Med 110:429–447. https://doi.org/10.1007/s11242-015-0565-y

Louvier-Hernandez JF, Luna-Barcenas G, Thakur R, Gupta RB (2005) Formation of chitin nanofibers by supercritical antisolvent. J Biomed Nanotechnol 1:109–114. https://doi.org/10.1166/jbn.2005.002

Lsaacs NS, Keating N (1992) The rates of a Diels-Alder reaction and supercritical carbon dioxide. J Chem Soc Chem Commun 12:876–877. https://doi.org/10.1039/C39920000876

Lu Y, Ao X, Tang J, Jia Y, Zhang X, Chen Y (2016) Swelling of shale in supercritical carbon dioxide. J Nat Gas Sci Eng 30:268–275. https://doi.org/10.1016/j.jngse.2016.02.011

Marathe RP, Mayadevi S, Pardhy SA, Sabne SM, Sivasanker S (2002) Alkylation of naphthalene with t-butanol in carbon dioxide. J Mol Catal A Chem 181:201–206. https://doi.org/10.1016/S1381-1169(01)00364-8

Martina M, Hutmacher DW (2007) Biodegradable polymers applied in tissue engineering research: a review. Polym Int 56:145–157. https://doi.org/10.1002/pi.2108

Masoudian MS (2016) Multiphysics of carbon dioxide sequestration in coalbeds: a review with a focus on geomechanical characteristics of coal. J Rock Mech Geotech Eng 8:93–112. https://doi.org/10.1016/j.jrmge.2015.08.002

Matuana LM, Park CB, Balatinecz JJ (1998) Structures and mechanical properties of microcellular foamed polyvinyl chloride. Cell Polym 17:1–16

Mayadevi S (2012) Reactions in supercritical carbon dioxide. Indian J Chem 51A:1298–1305. URL: http://nopr.niscair.res.in/handle/123456789/14665
McLeod MC, Gale WF, Roberts CB (2004) Metallic nanoparticle production utilizing a supercritical carbon dioxide flow process. Langmuir 20:7078–7082. https://doi.org/10.1021/la0493262
Merlet S, Marestin C, Schiets F, Romeyer O, Mercier R (2007) Preparation and characterization of nanocellular poly(phenylquinoxaline) foams. A new approach to nanoporous high-performance polymers. Macromolecules 40:2070–2078. https://doi.org/10.1021/ma062259t
Meziani MJ, Pathak P, Wang W, Desai T, Patil A, Sun Y-P (2005) Polymeric nanofibers from rapid expansion of supercritical solution. Ind Eng Chem Res 44:4594–4598. https://doi.org/10.1021/ie048796o
Middleton RS, Carey JW, Currier RP, Hyman JD, Kang Q, Karra S, Jiménez Martínez J, Porter ML, Viswanathan HS (2015) Shale gas and non-aqueous fracturing fluids: opportunities and challenges for supercritical CO_2. Appl Energy 147:500–509. https://doi.org/10.1016/j.apenergy.2015.03.023
Mohamed A, Trickett K, Chin SY, Cummings S, Sagisaka M, Hudson L, Nave S, Dyer R, Rogers SE, Heenan RK, Eastoe J (2010) Universal surfactant for water, oils, and CO_2. Langmuir 26:13861–13866. https://doi.org/10.1021/la102303q
Mohamed A, Sagisaka M, Guittard F, Cummings S, Paul A, Rogers SE, Heenan RK, Dyer R, Eastoe J (2011) Low Fluorine Content CO -philic Surfactants. Langmuir 27:10562–10569. https://doi.org/10.1021/la2021885
Mohamed A, Sagisaka M, Hollamby M, Rogers SE, Heenan RK, Dyer R, Eastoe J (2012) Hybrid CO_2-philic surfactants with low fluorine content. Langmuir 28:6299–6306. https://doi.org/10.1021/la3005322
Mori T, Hayashi H, Okamoto M, Yamasaki S, Hayami H (2009) Foam processing of polyethylene ionomers with supercritical CO_2. Compos Part A Appl Sci 40:1708–1716. https://doi.org/10.1016/j.compositesa.2009.08.018
Motiei M, Hacohen YR, Calderon-Moreno J, Gedanken A (2001) Preparing carbon nanotubes and nested fullerenes from supercritical CO(2)by a chemical reaction. J Am Chem Soc 123:8624–8625. https://doi.org/10.1021/ja015859a
Moutsatsou P, Coopman K, Smith MB, Georgiadou S (2015) Conductive PANI fibers and determining factors for the electrospinning window. Polymer 77:143–151. https://doi.org/10.1016/j.polymer.2015.08.039
Muljana H, Picchioni F, Knez Z, Heeres HZ, Janssen LPBM (2011) Insights in starch acetylation in sub- and supercritical CO_2. Carbohydr Res 346:1224–1232. https://doi.org/10.1016/j.carres.2011.04.002
Musie G, Wei M, Subramaniam B, Busch DH (2001) Catalytic oxidations in carbon dioxide-based reaction media, including novel CO_2-expanded phases. Coord Chem Rev 219-221:789–820. https://doi.org/10.1016/S0010-8545(01)00367-8
Myers T (2012) Potential contaminant pathways from hydraulically fractured shale to aquifers. Ground Water 50:872–882. https://doi.org/10.1111/j.1745-6584.2012.00933.x
Nagarajan R (1993) Polymer-induced structural transitions in microemulsions. Langmuir 9:369–375. https://doi.org/10.1021/la00026a002
Nalawade SP, Picchioni F, Janssen LPBM (2006) Supercritical carbon dioxide as a green solvent for processing polymer melts: processing aspects and applications. Prog Polym Sci 31:19–43. https://doi.org/10.1016/j.progpolymsci.2005.08.002
Niessen HG, Woelk K (2007) Investigations in supercritical fluids. Top Curr Chem 276:69–110. https://doi.org/10.1007/b135837
Nofar M, Park CB (2014) Poly (lactic acid) foaming. Prog Polym Sci 39:1721–1741. https://doi.org/10.1016/j.progpolymsci.2014.04.001
Nofar M, Ameli A, Park CB (2015) Development of polylactide bead foams with double crystal melting peaks. Polymer 69:83–94. https://doi.org/10.1016/j.polymer.2015.05.048
Noyori R (2005) Pursuing practical elegance in chemical synthesis. Chem Commun 14:1807–1811. https://doi.org/10.1039/B502713F

Oakes RS, Clifford AA, Rayner CM (2001) The use of supercritical fluids in synthetic organic chemistry. J Chem Soc Perkin Trans 1:917–941. https://doi.org/10.1039/B101219N

Odumabo SM, Karpyn ZT, Ayala HLF (2014) Investigation of gas flow hindrance due to fracturing fluid leakoff in low permeability sandstones. J Nat Gas Sci Eng 17:1–12. https://doi.org/10.1016/j.jngse.2013.12.002

Ohde H, Ye X-R, Wai CM, Rodriguez JM (2000) Synthesizing silver halide nanoparticles in supercritical carbon dioxide utilizing a water-in-CO_2 microemulsion. Chem Commun 23:2353–2354. https://doi.org/10.1039/B005924M

Ohde H, Hunt F, Wai CM (2001) Synthesis of silver and copper nanoparticles in a water-in-supercritical-carbon dioxide microemulsion. Chem Mater 13:4130–4135. https://doi.org/10.1021/cm010030g

Olsen GW, Church TR, Miller JP, Burris JM, Hanson KJ, Lundberg JK, Armitage JB, Herron RM, Medhdizadehkashi Z, Nobiletti JB, O'Neill EM, Mandel JH, Zobel LR (2003) Perfluorooctanesulfonate and other fluorochemicals in the serum of American Red Cross adult blood donors. Environ Health Perspect 111:1892–1901. https://doi.org/10.1289/ehp.6316

Orr FM Jr, Taber JJ (1984) Use of carbon dioxide in enhanced oil. Recov Sci 224:563–569. https://doi.org/10.1016/0196-8904(92)90059-6

Osborn SG, Vengosh A, Warner NR, Jackson RB (2011) Methane contamination of drinking water accompanying gas-well drilling and hydraulic fracturing. Proc Natl Acad Sci U S A 108:8172–8176. https://doi.org/10.1073/pnas.1100682108

Padrela L, Rodrigues MA, Duarte A, Dias AMA, Braga MEM, de Sousa HC (2018) Supercritical carbon dioxide-based technologies for the production of drug nanoparticles/nanocrystals—a compressive review. Adv Drug Deliv Rev 131:22–78. https://doi.org/10.1016/j.addr.2018.07.010

Pan Z, Connell LD, Camilleri M (2010) Laboratory characterisation of coal reservoir permeability for primary and enhanced coalbed methane recovery. Int J Coal Geol 82:252–261. https://doi.org/10.1016/j.coal.2009.10.019

Peach J, Eastoe J (2014) Supercritical carbon dioxide: a solvent like no other. Beilstein J Org Chem 10:1878–1895. https://doi.org/10.3762/bjoc.10.196

Pei P, Ling K, He J, Liu Z (2015) Shale gas reservoir treatment by a CO_2-based technology. J Nat Gas Sci Eng 26:1595–1606. https://doi.org/10.1016/j.jngse.2015.03.026

Perera MSA, Ranjith PG (2012) Carbon dioxide sequestration effects on coal's hydromechanical properties: a review. Int J Energy Res 36:1015–1031. https://doi.org/10.1002/er.2921

Perera MSA, Ranjitha PG, Choib SK, Airey D (2011a) The effects of sub-critical and super-critical carbon dioxide adsorption-induced coal matrix swelling on the permeability of naturally fractured black coal. Energy 36:6442–6450. https://doi.org/10.1016/j.energy.2011.09.023

Perera MSA, Ranjith PG, Airey DW, Choi SK (2011b) Sub- and super-critical carbon dioxide flow behavior in naturally fractured black coal: an experimental study. Fuel 90:3390–3397. https://doi.org/10.1016/j.fuel.2011.05.016

Pientka Z, Nemestothy N, Belafi-Bako K (2009) Application of polymeric foams for separation, storage and absorption of hydrogen. Desalination 241:106–110. https://doi.org/10.1016/j.desal.2008.01.060

Pintado-Sierra M, Delgado L, Aranaz I, Marcos-Fernández A, Reinecke H, Gallardo A, Zeugolis D, Elvira C (2014) Surface hierarchical porosity in poly(ε-caprolactone) membranes with potential applications in tissue engineering prepared by foaming in supercritical carbon dioxide. J Supercrit Fluids 95:273–284. https://doi.org/10.1016/j.supflu.2014.09.019

Pitt AR, Morley SD, Burbridge NJ, Quickenden EL (1996) The relationship between surfactant structure and limiting values of surface tension, in aqueous gelatin solution, with particular regard to multilayer coating. Colloids Surf A Physicochem Eng Asp 114:321–335. https://doi.org/10.1016/0927-7757(96)03593-5

Plasynski SI, Litynski JT, McIlvried HG, Srivastava RD (2009) Progress and new developments in carbon capture and storage. Crit Rev Plant Sci 28:123–138. https://doi.org/10.1080/07352680902776440

Potluri VK, Hamilton AD, Karanikas CF, Bane SE, Xu J, Beckman EJ, Enick RM (2003) The high CO_2-solubility of per-acetylated α-, β-, and γ-cyclodextrin. Fluid Phase Equilib 211:211–217. https://doi.org/10.1016/S0378-3812(03)00206-1

Rahmanian A, Ghaziaskar HS (2008) Selective extraction of maleic acid and phthalic acid by supercritical carbon dioxide saturated with trioctylamine. J Supercrit Fluids 46:118–122. https://doi.org/10.1016/j.supflu.2008.04.015

Ranathunga AS, Perera MSA, Ranjith PG, Zhang XG, Wu B (2017) Super-critical carbon dioxide flow behaviour in low rank coal: a meso-scale experimental study. J CO2 Util 20:1–13. https://doi.org/10.1016/j.jcou.2017.04.010

Randolph TW, Clark DS, Blanch HW, Prausnitz JM (1988) Enzymatic oxidation of cholesterol aggregates in supercritical carbon dioxide. Science 239:387–390. https://doi.org/10.1126/science.239.4838.387

Raveendran P, Wallen SL (2002) Sugar acetates as novel, renewable CO_2-philes. J Am Chem Soc 124:7274–7275. https://doi.org/10.1021/ja025508b

Reeves SR (2003) Enhanced CBM recovery, coalbed CO_2 sequestration assessed. Oil Gas J 101:49–53

Reilly JT, Bokis CP, Donohue MD (1995) An experimental investigation of Lewis acid-base interactions of liquid carbon dioxide using Fourier Transform Infrared (FT-IR) spectroscopy. Int J Thermophys 16:599–610. https://doi.org/10.1007/BF01438845

Reverchon E (2002) Supercritical-assisted atomization to produce micro and/or nanoparticles of controlled size and distribution. Ind Eng Chem Res 41:2405–2411. https://doi.org/10.1021/ie010943k

Reverchon E, Adami R (2006) Nanomaterials and supercritical fluids. J Supercrit Fluids 37:1–22. https://doi.org/10.1016/j.supflu.2005.08.003

Reverchon E, Caputo G, De Marco I (2003) Role of phase behavior and atomization in the supercritical antisolvent precipitation. Ind Eng Chem Res 42:6406–6414. https://doi.org/10.1021/ie0302138

Rezwan K, Chen QZ, Blaker JJ, Boccaccini AR (2006) Biodegradable and bioactive porous polymer/inorganic composite scaffolds for bone tissue engineering. Biomaterials 27:3413–3431. https://doi.org/10.1016/j.biomaterials.2006.01.039

Rindfleisch F, DiNoia TP, McHugh MA (1996) Solubility of polymers and copolymers in supercritical CO_2. J Phys Chem 100:15581–15587. https://doi.org/10.1021/jp9615823

Rizvi A, Park CB (2014) Dispersed polypropylene fibrils improve the foaming ability of a polyethylene matrix. Polymer 55:4199–4205. https://doi.org/10.1016/j.polymer.2014.06.014

Rodeheaver BA, Colton JS (2001) Open-celled microcellular thermoplastic foam. Polym Eng Sci 41:380–400. https://doi.org/10.1002/pen.10736

Rodrigues AR, Paiva A, da Silva MG, Simões P, Barreiros S (2011) Continuous enzymatic production of biodiesel from virgin and waste sunflower oil in supercritical carbon dioxide. J Supercrit Fluids 56(3):259–264. https://doi.org/10.1016/j.supflu.2010.10.031

Rogers RD, Seddon KR (2003) Ionic liquids--solvents of the future? Science 302:792–793. https://doi.org/10.1126/science.1090313

Roozbeh K, Karl-Heinz W, Hans B (2014) Sorption of CH_4 and CO_2 on carboniferous shale from Belgium using a manometric setup. Int J Coal Geol 128:153–161. https://doi.org/10.1016/j.coal.2014.04.014

Sagisaka M, Yoda S, Takebayashi Y, Otake K, Kitiyanan B, Kondo Y, Yoshino N, Takebayashi K, Sakai H, Abe M (2003) Preparation of a W/scCO_2 microemulsion using fluorinated surfactants. Langmuir 19:220–225. https://doi.org/10.1021/la020340t

Sagisaka M, Iwama S, Ono S, Yoshizawa A, Mohamed A, Cummings S, Yan C, James C, Rogers SE, Heenan RK, Eastoe J (2013) Nanostructures in water-in-CO_2 microemulsions stabilized by double-chain fluorocarbon solubilizers. Langmuir 29(25):7618–7628. https://doi.org/10.1021/la400376g

Sai H, Tan KW, Hur K, Asenath-Smith E, Hovden R, Jiang Y, Riccio M, Muller DA, Elser V, Estroff LA, Gruner SM, Wiesner U (2013) Hierarchical porous polymer scaffolds from block copolymers. Science 341:530–534. https://doi.org/10.1126/science.1238159

Sarbu T, Styranec T, Beckman EJ (2000) Non-fluorous polymers with very high solubility in supercritical CO_2 down to low pressures. Nature 405(6783):165–168. https://doi.org/10.1038/35012040

Scanlon BR, Reedy RC, Nicot JP (2014) Comparison of water use for hydraulic fracturing for unconventional oil and gas versus conventional oil. Environ Sci Technol 48:12386–12393. https://doi.org/10.1021/es502506v

Seki T, Grundwaldt D, Baiker A (2007) Continuous catalytic "one-pot" multi-step synthesis of 2-ethylhexanal from crotonaldehyde. Chem Commun 34:3562–3564. https://doi.org/10.1039/B710129E

Seki T, Grundwaldt D, van Vegten N, Baiker A (2008) Palladium supported on an acidic resin: a unique bifunctional catalyst for the continuous catalytic hydrogenation of organic compounds in supercritical carbon dioxide. Adv Synth Catal 350:691–705. https://doi.org/10.1002/adsc.200700532

Sellin MF, Webb PB, Cole-Hamilton DJ (2001) Continuous flow homogeneous catalysis: hydroformylation of alkenes in supercritical fluid–ionic liquid biphasic mixtures. Chem Commun 8:781–782. https://doi.org/10.1039/B101046H

Shah PS, Holmes JD, Doty RC, Johnston KP, Korgel BA (2000) Steric stabilization of nanocrystals in supercritical CO_2 using fluorinated ligands. J Am Chem Soc 122:4245–4246. https://doi.org/10.1021/ja9943748

Shah PS, Husain S, Johnston KP, Korgel BA (2002) Role of steric stabilization on the arrested growth of silver nanocrystals in supercritical carbon dioxide. J Phys Chem B 106:12178–12185. https://doi.org/10.1021/jp026180b

Shin S-K, Sim J-E, Kishimura H, Chun B-S (2012) Characteristics of menhaden oil ethanolysis by immobilized lipase in supercritical carbon dioxide. J Ind Eng Chem 18:546–550. https://doi.org/10.1016/j.jiec.2011.11.065

Soccol CR, Vandenberghe LPS, Rodrigues C, Pandey A (2006) New perspectives for citric acid production and application. Food Technol Biotechnol 44:141–149

Song H, Lee SY (2006) Production of succinic acid by bacterial fermentation. Enzym Microb Technol 39:352–361. https://doi.org/10.1016/j.enzmictec.2005.11.043

Song W, Ni H, Wang R, Shen Z, Zhao M (2016) Pressure controlling method for managed pressure drilling with supercritical carbon dioxide as the circulation fluid. Petrol Explor Dev 43(5):857–862. https://doi.org/10.1016/S1876-3804(16)30102-1

Span R, Wagner W (1996) A new equation of state for carbon dioxide covering the fluid region from the triple-point temperature to 1100 K at pressures up to 800 MPa. J Phys Chem Ref Data 25(6):1509–1596. https://doi.org/10.1063/1.555991

Stallings WE, Lamb HH (2003) Synthesis of nanostructured titania powders via hydrolysis of titanium isopropoxide in supercritical carbon dioxide. Langmuir 19:2989–2994. https://doi.org/10.1021/la020760i

Stevens JG, Bourne RA, Poliakoff M (2009) The continuous self aldol condensation of propionaldehyde in supercritical carbon dioxide: a highly selective catalytic route to 2-methylpentenal. Green Chem 11:409–416. https://doi.org/10.1039/B819687G

Stone MT, da Rocha SRP, Rossky PJ, Johnston KP (2003) Molecular differences between hydrocarbon and fluorocarbon surfactants at the CO_2/water interface. J Phys Chem B 107:10185–10192. https://doi.org/10.1021/jp035422k

Stone MT, Smith PG, Jr, da Rocha SRP, Rossky PJ, Johnston KP (2004) Low interfacial free volume of stubby surfactants stabilizes water-in-carbon Dioxide microemulsions. J Phys Chem B 108:1962-1966. https://doi.org/10.1021/jp036224w

Sun Y-P, Atorngitjawat P, Meziani MJ (2001) Preparation of silver nanoparticles via rapid expansion of water in carbon dioxide microemulsion into reductant solution. Langmuir 17:5707–5710. https://doi.org/10.1021/la0103057

Sun Y-P, Meziani MJ, Pathak P, Qu L (2005) Polymeric nanoparticles from rapid expansion of supercritical fluid solution. Chem Eur J 11:1366–1373. https://doi.org/10.1002/chem.200400422

Svetlana I, Gürcan G, John B, Scott WT (2015) Profitability of shale gas drilling: a case study of the Fayetteville shale play. Energy 81:382–393. https://doi.org/10.1016/j.energy.2014.12.051

Tadd AR, Marteel A, Mason MR, Davies JA, Abraham MA (2003) Hydroformylation of 1-hexene in supercritical carbon dioxide using a heterogeneous rhodium catalyst. 1. Effect of process parameters. J Supercritical Fluids 25(2):183–196. https://doi.org/10.1016/S0896-8446(02)00096-7

Tang M, Wang TC (2017) Foaming of poly(vinylidene fluoride-co hexafluoropropylene) using supercritical carbon dioxide. J Taiwan Inst Chem E 73:146–153. https://doi.org/10.1016/j.jtice.2016.09.038

Tapriyal D, Wang Y, Enick RM, Johnson JK, Crosthwaite J, Thies MC, Paik IH, Hamilton AD (2008) Poly(vinyl acetate), poly(1-*O*-(vinyloxy) ethyl-2,3,4,6-tetra-*O*-acetyl-β-D-glucopyranoside) and amorphous poly(lactic acid) are the most CO_2-soluble oxygenated hydrocarbon-based polymers. J Supercrit Fluids 46:252–257. https://doi.org/10.1016/j.supflu.2008.05.001

Tata M, John VT, Waguespack YY, McPherson GL (1994) Intercalation in novel organogels with a "Stacked" phenol microstructure. J Am Chem Soc 116:9464–9470. https://doi.org/10.1021/ja00100a008

Tezsevin I, Demirtas C, Onal I, Dilek C (2017) Density functional theory study of interactions between carbon dioxide and functionalized polyhedral oligomeric silsesquioxanes. Int J Quantum Chem 117:e25397. https://doi.org/10.1002/qua.25397

Tomasko DL, Li H, Liu D, Han X, Wingert MJ, Lee LJ, Koelling KW (2003) A review of CO_2 applications in the processing of polymers. Ind Eng Chem Res 42:6431–6456. https://doi.org/10.1021/ie030199z

Trickett K, Eastoe J (2008) Surfactant-based gels. Adv Colloid Interf Sci 144:66–74. https://doi.org/10.1016/j.cis.2008.08.009

Trickett K, Xing D, Enick R, Eastoe J, Hollamby MJ, Mutch KJ, Rogers SE, Heenan RK, Steytler DC (2010) Rod-like micelles thicken CO(2). Langmuir 26:83–88. https://doi.org/10.1021/la902128g

Turk M, Hils P, Helfgen B, Schaber K, Martin H-J, Wahl MA (2002) Micronization of pharmaceutical substances by the rapid expansion of supercritical solutions (RESS): a promising method to improve bioavailability of poorly soluble pharmaceutical agents. J Supercrit Fluids 22:75–84. https://doi.org/10.1016/S0896-8446(01)00109-7

Varma MN, Madras G (2007) Synthesis of biodiesel from castor oil and linseed oil in supercritical fluids. Ind Eng Chem Res 46:1–16. https://doi.org/10.1021/ie0607043

Varma MN, Deshpande PA, Madras G (2010) Synthesis of biodiesel in supercritical alcohols and supercritical carbon dioxide. Fuel 89:1641–1647. https://doi.org/10.1016/j.fuel.2009.08.012

Vasile C (2000) Handbook of polyolefins, 2nd edn. Marcel Dekker, New York. ISBN: 9780824786038

Vengosh A, Jackson RB, Warner N, Darrah TH, Kondash A (2014) A critical review of the risks to water resources from unconventional shale gas development and hydraulic fracturing in the United States. Environ Sci Technol 48:8334–8348. https://doi.org/10.1021/es405118y

Vishal V, Singh TN (2015) A laboratory investigation of permeability of coal to supercritical CO_2. Geotech Geol Eng 33:1009–1016. https://doi.org/10.1007/s10706-015-9882-8

Vishal V, Ranjith PG, Pradhan S, Singh T (2013) Permeability of sub-critical carbon dioxide in naturally fractured Indian bituminous coal at a range of down-hole stress conditions. Eng Geol 167:148–156. https://doi.org/10.1016/j.enggeo.2013.10.007

Wakayama H, Fukushima Y (1999) Porous platinum fibers synthesized using supercritical fluid. Chem Commun 4:391–392. https://doi.org/10.1039/A808255C

Wakayama H, Fukushima Y (2000a) Supercritical CO_2 as a solvent for synthesis of nanoporous materials. Ind Eng Chem Res 39:4641–4645. https://doi.org/10.1021/ie000148r

Wakayama H, Fukushima Y (2000b) Nanoporous silica prepared with activated carbon molds using supercritical CO_2. Chem Mater 12:756–761. https://doi.org/10.1021/cm990487e

Wakayama H, Itahara H, Tatsuda N, Inagaki S, Fukushima Y (2001) Nanoporous metal oxides synthesized by the nanoscale casting process using supercritical fluids. Chem Mater 13:2392–2396. https://doi.org/10.1021/cm001408y

Wakayama H, Inagaki S, Fukushima Y (2002) Nanoporous titania synthesized by a nanoscale casting process in supercritical carbon dioxide. J Am Ceram Soc 85:161–164. https://doi.org/10.1111/j.1151-2916.2002.tb00059.x

Wakayama H, Goto Y, Fukushima Y (2003) A novel method for tailoring porous structures of nanoporous materials using supercritical solvents. Phys Chem Chem Phys 5:3784–3788. https://doi.org/10.1039/B305653H

Wakayama H, Hatanaka T, Fukushima Y (2004) Synthesis of Pt-Ru nanoporous fibers by the nanoscale casting process using supercritical CO_2 for electrocatalytic applications. Chem Lett 33:658–659. https://doi.org/10.1021/ja002261e

Wang R, Ni H (2013) Wellbore heat transfer law of carbon dioxide coiled tubing drilling. J China Univ Petrol (Edition of Natural Science) 37(5):65–70. https://doi.org/10.3969/j.issn.1673-5005.2013.05.009

Wang Y, Dave RN, Pfeffer R (2004) Polymer coating/encapsulation of nanoparticles using a supercritical anti-solvent process. J Supercrit Fluids 28:85–99. https://doi.org/10.1016/S0896-8446(03)00011-1

Wang JS, Herman J, Enokida Y, Wai CM (2005) Dissolution of organic acids in supercritical CO_2 using a CO_2-philic Lewis base carrier. J Supercrit Fluids 36:98–105. https://doi.org/10.1016/j.supflu.2005.06.001

Wang Y, Hong L, Tapriyal D, Kim IC, Paik I-H, Crosthwaite JM, Hamilton AD, Thies MC, Beckman EJ, Enick RM, Johnson JK (2009) Design and evaluation of nonfluorous CO_2-soluble oligomers and polymers. J Phys Chem B 113:14971–14980. https://doi.org/10.1021/jp9073812

Wang J, Lessard BH, Maric M, Favis BD (2014) Hierarchically porous polymeric materials from ternary polymer blends. Polymer 55:3461–3467. https://doi.org/10.1016/j.polymer.2014.06.042

Wang J, Elsworth D, Wu Y, Liu J, Zhu W, Liu Y (2018) The influence of fracturing fluids on fracturing processes: a comparison between water, oil and SC-CO_2. Rock Mech Rock Eng 51:299–313. https://doi.org/10.1007/s00603-017-1326-8

Webb PB, Sellin MF, Kunene TE, Williamson S, Slawin AMZ, Cole-Hamilton DJ (2003) Continuous flow hydroformylation of alkenes in supercritical fluid-ionic liquid biphasic systems. J Am Chem Soc 125:15577–15588. https://doi.org/10.1021/ja035967s

West KN, Wheeler C, McCarney JP, Griffith KN, Bush D, Liotta CL, Eckert CA (2001) In situ formation of alkylcarbonic acids with CO_2. J Phys Chem A 105:3947–3948. https://doi.org/10.1021/jp003846y

Wu J, Mather PT (2009) POSS polymers: physical properties and biomaterials applications. Polym Rev 49:25–63. https://doi.org/10.1080/15583720802656237

Wu Y, Yang P (2001) Direct observation of vapor–liquid–solid nanowire growth. J Am Chem Soc 123:3165–3166. https://doi.org/10.1021/ja0059084

Xiangxin H, Mengting N, Xiumin J (2014) Combined fluidized bed retorting and circulating fluidized bed combustion system of oil shale: 2. Energy and economic analysis. Energy 74:788–794. https://doi.org/10.1016/j.energy.2014.07.050

Xing Z, Wu G, Huang S, Chen S, Zeng H (2008) Preparation of microcellular cross-linked polyethylene foams by a radiation and supercritical carbon dioxide approach. J Supercrit Fluids 47:281–289. https://doi.org/10.1016/j.supflu.2008.08.009

Xiong Y, Kiran E (1995) Miscibility, density and viscosity of poly(dimethylsiloxane) in supercritical carbon dioxide. Polymer 36:4817–4826. https://doi.org/10.1016/0032-3861(95)99298-9

Xu X, Ayyagari M, Tata M, John VT, McPherson GL (1993) Formation of novel organogels by the addition of phenols to AOT micelles in isooctane. J Phys Chem 97:11350–11353. https://doi.org/10.1021/j100145a038

Yang C, Zhe X, Zhang M, Wang M, Wu G (2017) Radiation effects on the foaming of atactic polypropylene with supercritical carbon dioxide. Radiat Phys Chem 131:35–40. https://doi.org/10.1016/j.radphyschem.2016.10.018

Yang ZZ, Yi LP, Li XG, Han XP, Li ZC (2019) Wellbore temperature and pressure calculation model for supercritical carbon dioxide jet fracturing. Energy Source Part A 41(2):185–200. https://doi.org/10.1080/15567036.2018.1511654

Ye X-R, Lin Y, Wang C, Wai CM (2003a) Supercritical fluid fabrication of metal nanowires and nanorods templated by multiwalled carbon nanotubes. Adv Mater 15:316–319. https://doi.org/10.1002/adma.200390077

Ye XR, Lin Y, Wai CM (2003b) Decorating catalytic palladium nanoparticles on carbon nanotubes in supercritical carbon dioxide. Chem Commun 7:642–643. https://doi.org/10.1039/B211350C

Ye XR, Wai CM, Lin Y, Young JS, Engelhard MH (2005) Supercritical fluid immersion deposition: a new process for selective deposition of metal films on silicon substrates. Surf Coat Technol 190:25–31. https://doi.org/10.1016/j.surfcoat.2004.06.023

Yuan C, Runze H, Robert JR, Eric M (2015) Life-cycle comparison of greenhouse gas emissions and water consumption for coal and shale gas fired power generation in China. Energy 86:335–343. https://doi.org/10.1016/j.energy.2015.04.034

Zhai WT, Kuboki T, Wang LL, Park CB, Lee EK, Naguib HE (2010) Cell structure evolution and the crystallization behavior of polypropylene/clay nanocomposites foams blown in continuous extrusion. Ind Eng Chem Res 49:9834–9845. https://doi.org/10.1021/ie101225f

Zhang J, Xiao M, Liu Z, Han B, Jiang T, He J, Yang G (2004a) Preparation of ZnS/CdS composite nanoparticles by coprecipitation from reverse micelles using CO_2 as antisolvent. J Colloid Interface Sci 273:160–164. https://doi.org/10.1016/j.jcis.2004.02.032

Zhang J, Liu Z, Han B, Liu D, Chen J, He J, Jiang T (2004b) A novel method to synthesize polystyrene nanospheres immobilized with silver nanoparticles by using compressed CO_2. Chem Eur J 10:3531–3536. https://doi.org/10.1002/chem.200305445

Zhang J, Liu Z, Han B, Jiang T, Wu W, Chen J, Li Z, Liu D (2004c) Preparation of polystyrene-encapsulated silver nanorods and nanofibers by combination of reverse micelles, gas antisolvent, and ultrasound techniques. J Phys Chem B 108:2200–2204. https://doi.org/10.1021/jp036408a

Zhang X, Lu Y, Tang J, Zhou Z, Liao Y (2016) Experimental study on fracture initiation and propagation in shale using supercritical carbon dioxide fracturing. Fuel 190:370–378. https://doi.org/10.1016/j.fuel.2016.10.120

Zhong X, Dehghani F (2012) Fabrication of biomimetic poly(propylene carbonate) scaffolds by using carbon dioxide as a solvent, monomer and foaming agent. Green Chem 14:2523–2533. https://doi.org/10.1039/C2GC35608B

Zosel K (1978) Separation with supercritical gases: practical applications. Angew Chem Int Ed Engl 17:702–709. https://doi.org/10.1002/anie.197807021

Zou Y, Li N, Ma X, Zhang S, Li S (2018) Experimental study on the growth behavior of supercritical CO_2-induced fractures in a layered tight sandstone formation. J Nat Gas Sci Eng 49:145–156. https://doi.org/10.1016/j.jngse.2017.11.005

Chapter 6
Extraction of Phenolic Compounds by Supercritical Fluid Extraction

Mudasir Yaqoob, Poonam Aggarwal, and Neha Purandare

Contents

M. Yaqoob (✉)
Department of Food Science and Technology, Punjab Agricultural University, Ludhiana, Punjab, India

Department of Food Science and Nutrition, Lovely Professional University, Phagwara, Punjab, India
e-mail: sachdev_poonam@pau.edu

P. Aggarwal
Department of Food Science and Technology, Punjab Agricultural University, Ludhiana, Punjab, India

N. Purandare
Department of Food Science and Nutrition, Lovely Professional University, Phagwara, Punjab, India

Inamuddin, A. M. Asiri (eds.), *Advanced Nanotechnology and Application of Supercritical Fluids*, Nanotechnology in the Life Sciences,
https://doi.org/10.1007/978-3-030-44984-1_6

Abbreviations

db	Dry basis
GAE	Gallic acid equivalent
MAE	Microwave-assisted extraction
MPa	Mega Pascals
PHWE	Pressurized hot water extraction
SC-CO_2	Supercritical carbon dioxide
SFE	Supercritical fluid extraction
SFE-GC	Supercritical fluid extraction-gas chromatography
SFE-HPLC-MS/MS	Supercritical fluid extraction-mass spectrometry liquid chromatography
SFE-SFC	Supercritical fluid extraction-supercritical fluid chromatography
SFE-US	Supercritical fluid extraction-ultrasound
UAE	Ultrasound-assisted extraction

6.1 Introduction

With the progress in nutrition and medical sciences, it has been perceived that not only the nutritive components of food are important but also the non-nutrient components such as phytochemicals. They play a very crucial role in keeping good health and reduce the incidence of common and chronic diseases (Sharma 2010; Sharma et al. 2012, 2016). Phenolic compounds are the plant nutrients that have a particular biological activity and constitute one of the most widely scattered groups of substances in the plant kingdom. Some studies have reported that the biological activity of various plant phytochemicals is effective than their synthetic counterparts with less or no side effects. Moreover the resistance of some microbial strains against drugs has also led to awareness among common masses against the use of synthetic products (Sundsford 2004; Edith et al. 2005; Hancock 2005; Isturiz 2008). This cause has forced scientists to develop improved methods of bioprocesses for the extraction and production of compounds from natural sources (Velioglu et al. 1998; Oreopoulou and Tzia 2007).

Extraction is a key step in phytochemical processing of plant materials for the determination of bioactive constituents (Dhanani et al. 2017). The efficacious extraction of phytochemicals could add value to the food processing industries (M'hiri et al. 2014). The richness of varied diversity of phytochemicals in naturally occurring foods like citrus fruits can be rendered possible by extraction. The rate and yield of extraction are linked to the solvent characteristics. It has been found that methanol is efficient in the extraction of lower weight polyphenols while aqueous acetone is suitable for the extraction of the higher molecular weight flavanols (Brglez Mojzer et al. 2016). Additional selection of suitable extraction technique and optimization of various factors are important for upscaling (bench scale to pilot plant level) (Babbar et al. 2011).

Commercially, different extraction techniques are used for the extraction of phytochemicals, but conventional extraction methods like Soxhlet extraction, maceration, and refluxing have certain limitations such as lower efficiency, high extraction temperature, low extraction yield, mass transfer resistance, use of large quantity of solvent and health hazards (Jadhav et al. 2009). Extraction yet again is a difficult step of valorization because of the sensitivity of the raw materials containing bioactives and the sensitivity of the molecules of interest. Therefore, novel extraction technique like supercritical fluid extraction can be an option to conventional extraction methods, having advantages with respect to extraction yields, solvent consumption, extraction time, and reproducibility (Safdar et al. 2017). SFE, an environment-friendly separation process using carbon dioxide gas, can be recycled and utilized again as part of the unit operation. It does not leave any chemical residue (Brunner 2005). It can also be used in conjunction with ultra-sonication and microwave for efficient extraction. However, these methods have not been used extensively and thus demand broad research to improve the understanding of design, extraction mechanism, and scale-up of these unique extraction systems for their industrial applications.

6.2 Phenolic Compounds

The phenolic compounds comprise a complex group of compounds varying from simple phenols to highly polymerized compounds such as tannins. These include hydroxycinnamic and hydroxybenzoic acids, terpenoids, stilbenes, acetophenones, flavonoids, and condensed tannins (Okwi and Emenike 2006) and these compounds have shown huge beneficial effects to the human health. Flavonoids are the nature's biological response modifiers, because they have the capability to adjust the body's reaction to allergens, carcinogens, and viruses. They show anti-inflammatory, anti-allergic, anti-microbial, and anti-cancer properties. Consumers and food industries have involved flavonoids for their medicinal properties, particularly their potential role in the prevention of cardiovascular disease and cancers. The composition of these phytochemicals in vegetables and fruits is variable according to the genotype and plant tissues (Lutz et al. 2015) and is usually affected by the pedoclimatic conditions, agricultural practices, processing techniques, and postharvest storage (Tomás-Barberá and Espín 2001). Nayak and Pereira (2006) reported that the therapeutic properties of plant bioactives have far more to offer in the future as a novel discovery. The utilization of phenolic compounds as chief bioactive component with multidimensional properties is achieving momentum and will be an efficient, inexpensive, and environment-friendly approach for the production of novel nutraceuticals and various other food products (Rafiq et al. 2018).

6.3 Supercritical Extraction of Phenolics

SFE is nowadays a widespread method for the extraction of many polyphenolic compounds due to its environment-friendly clean process. The thermal degradation of the heat liable valuable substances is prevented due to the reason that this process is conducted at low temperature. Also the absence of oxygen and light inhibits oxidation reactions. The effectiveness of SFE technique is affected by many factors, predominantly supercritical CO_2 flux, co-solvents, pressure, temperature, and extraction time (Diaaz-Reinoso et al. 2006; M'hiri et al. 2014). Camila et al. (2010) also suggested that supercritical fluid extraction can be used as a competitive alternate for the preparation of bioactive extracts. When both pressure and temperature of a fluid increase above the critical point values, it is said to be in a supercritical state. It therefore becomes necessary to explain these properties that determine the state of a substance and efficiency of extraction. However, in addition to critical constants, knowledge of certain important properties of the solvent and other compounds of interest such as density, compressibility, thermal conductivity, viscosity, diffusivity, dielectric constant, and heat capacity is necessary to maximize the output of the process. It implies that the solvent is the main part playing the main role in the extraction of different phenolic compounds by SFE. The extraction performed by solvents in conjugation with other procedures is discussed below.

6.3.1 Pure Carbon Dioxide

Utilization of pure carbon dioxide for the extraction of these bioactive compounds from natural sources has been done in previous studies. Pressure and temperature are the two major aspects regulating the process of supercritical fluid extraction using carbon dioxide (Khaw et al. 2017). Conducted studies from various experiments illustrated the use of pressure and temperature within range of 80–655 bar and 25–120 °C, respectively, for the extraction of phenolic compounds through SFE (Tyśkiewicz et al. 2018). Piantino et al. (2008) performed a study to extract phenolic compounds using pure CO_2 from *Baccharis dracunculifolia* at temperature of 60 °C and pressure of 400 bar. Results indicated two times elevated extraction yield in contrast to the yield obtained from ethanolic and methanolic extract.

Bimakr et al. (2011) used spearmint leaves to compare the extraction of supercritical fluid extraction (60 °C, 200 bar) and Soxhlet method (70% aqueous ethanol) which further exhibited comparable outcomes for both the methods. Various phenolic constituents such as epicatechin, catechin, naringenin, rutin, etc. were found in equivalent quantity in both cases.

In another study, Espinosa-Pardo et al. (2014) carried out the process of SFE using pure CO_2 in peaches. It was reported that optimal conditions for acquiring maximum yield of phenolic compounds were at temperature of 60 °C and pressure of 300 bar, whereas highest yield of carotenoids was attained at 40 °C and pressure of 200 bar. Pressure is another factor that influences the extraction of phenolic compounds in SFE, as pressure is positively correlated with solvent (CO_2) density (Norodin et al. 2016). Kryževičiute et al. (2016) demonstrated the effect of pressure on phenolic extraction yield of raspberry pomace using SFE at 60 °C (constant). It was reported 14 times increase in extraction yield was obtained when pressure was increased from 100 to 450 bar.

6.3.2 *Modified Carbon Dioxide*

Non-polar nature of carbon dioxide attributes to their incompetent dissolution of polar polyphenolic compounds (Farias-Campomanes et al. 2013). To overcome the abovementioned limitation of CO_2, polar co-solvents like water, methanol, or ethanol can be used which further increase the solubility of polyphenols in CO_2. Mechanism behind observed co-solvent effect is the formation of a charge-transfer complex amid co-solvent and solute.

Various physical interactions like dipole-dipole interaction or dipole-induced dipole interactions also get affected by addition of co-solvents. Consequently, these changes lead to density increase of the SFE solution (Ekart et al. 1992). Based on various types of modifiers used, SFE can be classified into five groups: (a) water; (b) pure and aqueous ethanol; (c) sequential extraction with ethanol, carbon dioxide, and water; (d) methanol; and (e) other solvents (ethyl acetate, hexane, ethyl lactate, and isopropyl alcohol) (Tyśkiewicz et al. 2018). Ethanol (aqueous or pure) is most commonly used as co-solvent with CO_2 to carry out the extractions (Manna et al. 2015).

Adil et al. (2007) applied pressure of 600 bar at 50 °C on extraction of apple and peach pomace. Highest yield of phenolic content was obtained with utilization of 20% (wt%) ethanol during supercritical fluid extraction (0.25 mg GAE/g for peach pomace and 0.427 mg GAE/g for apple pomace). Benelli et al. (2010) compared the extraction yield of orange pomace with pure CO_2 and modified CO_2 at pressure of 250 bar while using equal quality of ethanol. Results indicated two times improved extraction yield with modified CO_2. De Azevedo et al. (2008) used pure CO_2 and modified CO_2 (with 5% ethanol or 5% isopropyl alcohol) to extract green coffee beans. Extraction yield was found to be highest (17%) in case of extraction performed with CO_2 and 5% ethanol. Variation in selectivity and detection of specific compounds was noticed with the use of isopropyl alcohol and ethanol.

6.3.3 *Combined Supercritical Fluid Extraction*

Application of SFE with pure carbon and modified carbon to isolate phenolic compounds from natural sources is being widely practiced. Furthermore, various techniques of extraction and chromatography can be coupled with SFE to acquire polyphenol-rich constituents. Supercritical fluid extraction-gas chromatography (SFE-GC), supercritical fluid extraction-supercritical fluid chromatography (SFE-SFC), supercritical fluid extraction-mass spectrometry liquid chromatography (SFE-HPLC-MS/MS), supercritical fluid extraction-solid phase extraction, supercritical fluid extraction-Fourier transform infrared spectroscopy, ultrasound-assisted supercritical fluid extraction, and microwave-assisted supercritical fluid extraction supercritical fluid extraction-dispersive liquid-liquid micro-extraction are some of the combinations that are being employed (Tyśkiewicz et al. 2018).

Regarding SFE-SFC, mobile phase of SFC and solvent of SFE are integrated in analogous state to isolate, fractionate, measure, and detect the required components (Farias-Campomanes et al. 2013; Liu et al. (2018) reported utilization of SFE-SFC-MS/MS technique for the extraction of phenolic compounds from garlic at temperature of 50 °C for 9 min with addition of 30% methanol. Fractions were found to be loaded with different phenolic compounds such as *p*-coumaric acid, apigenin, ferulic acid, and naringenin. In another study by Hawthorne et al. (1989), technique of gas chromatography was utilized along with SFE (SFE-GC) at optimal conditions of 50 °C and 400 bar for the isolation and detection of phenolic compounds from hardwood smoke.

6.3.4 Supercritical Fluid Extraction Among Other Green Extraction Methods

SFE technique possesses several benefits that include safety, environment, and health maintenance. Nature of carbon dioxide makes it explicitly functional for the extraction of heat susceptible compounds (Chemat et al. 2012). SFE proficiently helps in curtailing the waste production along with checking the consumption of energy and solvent. These properties of SFE entitle it as an efficient waste management technique (Tyśkiewicz, et al. 2018).

Besides SFE, pressurized hot water extraction (PHWE), microwave-assisted extraction (MAE), pressurized liquid extraction (PLE), and ultrasound-assisted extraction (UAE) are some of the applicable modern green extraction techniques. A successful application of SFE coupled with US for extraction of phenols from pepper was documented by Dias et al. (2012). SFE-US technique was employed at 40 °C using pressure of 220 bar and power of 200 W. On further comparison of total phenolic content obtained with SFE and SFE-US extract, it was observed TPC value of SFE extract (0.23 mg gallic acid equivalent/g) (GAE/g) was three times lesser than SFE-US extract (0.62 mg GAE/g). Kraujalienė et al. (2017) compared the TPC value of goldenrod (*Solidago virgaurea*) extract obtained using SFE and SFE-PLE. For SFE-PLE, acetone with 70 or 140 °C (PLE) was combined with carbon dioxide at 70 °C. The results signified similar TPC values for both the cases (169–185 mg gallic acid equivalent/g).

6.4 Extraction of Phenolic Compounds from Various Plant Sources

6.4.1 Fruits

Phenolic compounds are extensively found in fruits, mostly represented by phenolic acids and flavonoids. The main phenolic subclasses in fruits are phenolic acids (hydroxycinnamic and hydroxybenzoic acids), flavonoids, coumarins, and hydrolyzable and condensed tannins. The various phenolic compounds are extracted from different fruits at specific critical pressure and temperature in supercritical fluid extraction (Table 6.1), such as in the case of grapes. Ghafoor et al. (2010) proposed that the optimum SFE conditions were at 40–43 °C temperature and pressure of 153–161 bar using ethanol as a modifier and CO_2 as a supercritical fluid which yielded total phenolic content of 2.156 mg GAE/100 mL. In mango, supercritical antisolvent extraction was used to obtain extract by the help of solid-liquid extraction with aqueous acetone by SC-CO_2 acting as an antisolvent keeping the temperature 35–45 °C and pressure 80–150 bar. According to Meneses et al. (2015), the main phenolic compounds obtained were quercetin, mangiferin, and kaempferol. Enzyme-assisted supercritical fluid extraction was developed by Mushtaq et al. (2015) for the extraction of phenolic compounds from pomegranate peel. The process was performed by supercritical carbon dioxide (SC-CO_2) combined with ethanol as co-solvent at 55 °C temperature and 300 bar pressure. Phenolic compounds, namely, vanillic, ferulic, and syringic acids, were obtained.

In apple and cherry, the phenolic compounds such as ferulic and coumaric acid, epicatechin, sakuranetin, and sakuranin were extracted at 50–60 °C temperature and a pressure of 250 bar (Serra et al. 2010) and 540–570 bar (Adil et al. 2007), respectively. In jamun, 162 bar pressure, 50 °C extraction temperature, and a co-solvent flow rate of 2.0 g/min were identified as optimum conditions for the extraction of phenolic compounds (Maran et al. 2014). Blueberry has the maximum antioxidant and polyphenol content in both the pulp and peel. According to Paes et al. (2014), for obtaining higher yields of extraction with pressurized liquids (PLE), ethanol, water, and acetone were used at different proportions. The temperature, pressure, and rate of the flow of solvent were kept constant at 40 °C, 20 MPa, and 10 mL/min. The highest yield of phytochemicals was extracted at 5% ethanol, 5% water, and 90% CO_2. During the process almost 16 anthocyanins were quantified and identified such as delphinidin 3-*O*-glucoside, delphinidin 3-*O*-galactoside, cyanidin 3-*O*-galactoside, cyanidin 3-*O*-glucoside, delphinidin 3-*O*-arabinoside, petunidin 3-*O*-galactoside, petunidin 3-*O*-glucoside, cyanidin 3-*O*-arabinoside, peonidin 3-*O*-galactoside, peonidin 3-*O*-glucoside, petunidin 3-*O*-arabinoside, malvidin 3-*O*-galactoside, malvidin 3-*O*-glucoside, malvidin 3-*O*-arabinoside, peonidin 3-*O*-arabinoside, and malvidin 3-*O*-xyloside. The phenolic compounds of tomato and watermelon extracted were gallic acid, quercetin, vanillic acid, sinapic acid, ferulic acid, naringenin, caffeic acid, rutin, and chlorogenic acid, respectively, using SFE with efficient yield, and the blueberry residues are rich in these phytochemicals (Martínez-Valverde et al. 2002; Kim et al. 2014).

According to Pereira et al. (2016), the extraction of myrtle leaves and berries was done by conventional as well as SFE method. SFE was done at 45 °C, 23 MPa, and a CO_2 flow rate of 0.3 kg h^{-1} by using ethanol as a co-solvent with a flow rate 0.09 kg h^{-1}. The antioxidant activity was determined by three methods like trolox equivalent antioxidant capacity (TEAC), oxygen radical absorbance capacity (ORAC), and Folin-Ciocalteu. SFE extracts showed higher yield and maximum antioxidant activity than the conventional methods.

Table 6.1 Extraction of phenolic compounds from different plants by SFE

Foods	Phenolic compounds extracted	Critical temperature (°C)	Critical pressure (bar)	Reference
Fruits				
Mango	Mangiferin, isomangiferin, quercetin, and kaempferol	35–45	80–150	Meneses et al. (2015)
Pomegranate	Vanillic, ferulic, syringic acids	55	300	Mushtaq et al. (2015)
Apple	Ferulic, coumaric acid, epicatechin	55.7–58.4	540–570	Adil et al. (2007)
Tomato	Quercetin, naringenin, caffeic acid, rutin, chlorogenic acid	40	–	Martínez-Valverde et al. (2002)
Cherry	Sakuranetin and sakuranin	50	250	Serra et al. (2010)
Watermelon	Gallic acid, quercetin, vanillic acid, sinapic acid, ferulic acid	100–300	–	Kim et al. (2014)
Jamun	Quinic acid, dihydroxybenzoic acid, caffeic acid, sinapic acid, kaempferol	50	162	Maran et al. (2014)
Blueberry	Phenolic acids, flavonoids	40	200	Paes et al. (2014)
Orange	Hydroxycinnamic acids, flavonoid hesperidin	40–60	160	Espinosa-Pardo et al. (2017)
Vegetables				
Potato	Gallic, chlorogenic, caffeic, protocatechuic, syringic, and ferulic acids	180	60	Singh and Saldaña (2011)
Broccoli	Quercetin, kaempferol	35	150	Arnáiz et al. (2016)
Cabbage	Quercetin, kaempferol, *p*-coumaric, sinapic, and ferulic acids	40–80	50–100	Santos et al. (2012)
Onion	Quercetin	165	90–131	Min-Jung Ko et al. (2011)
Carrot	Chlorogenic acid	59	306	de Andrade Lima et al. (2018)
Cauliflower	Kaempferol glycosides	35	–	Huynh et al. (2014)
Spices				
Clove	Eugenol	50	300	Oliveira et al. (2016)
Coriander	Gallic, chlorogenic, ferulic, and *p*-coumaric acid, quercetin, rutin, luteolin, apigenin, and kaempferol	100–200	30–90	Zeković et al. (2016)
Saffron	Phenolic acids, flavonoids	62	164	Ahmadian-Kouchaksaraie et al. (2017)
Pepper	Thymol, transcaryophyllene, germacrene D, and carvacrol	60	400	Garmus et al. (2015)
Herbs				
Origanum majorana L. (Marjoram)	Ursolic acid, carnosic acid, carnosol	60	400	Vagi et al. (2005)
Mentha spicata L. (Spearmint)	Catechin, rutin, luteolin, naringenin, epicatechin	60	200	Bimakr et al. (2011)
Oregano	Dihydroquercetin, eriodictyol, dihydrokaempferol, naringenin	40–60	150–350	Cavero et al. (2006)
Lamiaceae (rosemary, sage, thyme, hyssop)	Carnosol, carnosic acid	100	350	Babovic et al. (2010)

6.4.2 Vegetables

Various phenolic compounds from vegetables have been extracted by this green technology at various processing variables such as critical pressure and temperature, percentage of ethanol and dynamic extraction time, and the use of SC-CO_2. The specificity of various phenolic compounds is extracted at particular critical pressure and temperature depending on what conditions the particular compounds get diffused into the solvent. The amount of secondary metabolites (phenolic compounds) vary from part to part of the plant, as these compounds provide a defense mechanism to the plant against various pathogenic organisms and thus against various diseases. It therefore becomes important on which part of the vegetable we have to extract these valuable compounds. Sometimes the peels which are discarded are rich sources of phenolic compounds and need to extract from the same.

Singh and Saldaña (2011) studied the extraction of various polyphenolic compounds in potato peel, and it was found that the peel contained eight phenolic compounds: caffeic acid, gallic acid, syringic acid, protocatechuic acid, ferulic acid, *p*-hydroxyl benzoic acid, coumaric acid, and chlorogenic acid using subcritical water. In cabbage, a study was performed with ethanol as a solvent and extraction temperature (313–393 K), pressure (5–10 MPa), and static extraction time (3–15 min) (Santos et al. 2012). The optimal conditions for obtaining maximum mass yield (5.31%, db) in carrot was found at 306 bar, 58.5 °C, and 14.3% of co-solvent (ethanol), and at 15.5% ethanol, temperature of 59.0 °C, and a pressure of 349 bar, phenolic percentage was high (86.1%) (de Andrade Lima et al. 2018). Min-Jung Ko et al. (2011) utilized subcritical water extraction for quercetin (flavonoid) from onion skin by varying the temperature (100–190 °C) and extraction time (5–30 min) to get the maximum yield from the peel. The extraction of different phenolic compounds from other vegetables is shown in Table 6.1.

6.4.3 Spices

A variety of phenolic compounds have been extracted from spices such as phenolic acids, phenolic diterpenes, volatile oils, and flavonoids. Phenolic substances and the antioxidant activity are in co-relation in these plant materials. Mostly spices are the rich sources of essential oils, SFE-CO_2 being better option to extract these non-polar compounds. Even the utilization of less temperature and a closed system would prevent their oxidation and thus their activity also than conventional distillation method.

In clove, essential oils were obtained with SFE-CO_2 in the experimental conditions of critical pressures of 100, 200, and 300 bar and temperature 40 and 50 °C. The identification of oils was evaluated by gas chromatography-mass spectrometer (GCMS), and eugenol was known to be the major component (Oliveira et al. 2016).

In coriander, the seeds were extracted by subcritical water extraction at pressure 30–90 bar and temperature 100–200 °C, and 3,4-dimethoxycinnamic acid was found to be the major compound (Zeković et al. 2016). In saffron, the extraction temperature (40–80 °C), extraction pressure (90–180 bar), and extraction time (20–60 min) were optimized (Ahmadian-Kouchaksaraie et al. 2017) for the extraction of phenolic compounds for maximum yield. In pepper, Garmus et al. (2015) conducted the sequential extraction of bioactive compounds by carbon dioxide, ethanol, and water as solvents. They found thymol, transcaryophyllene, germacrene D, and carvacrol as the major components of bioactive compounds.

According to Guan et al. (2007), the compounds of clove bud were extracted by four methods—SFE, steam distillation, Soxhlet extraction, and hydro-distillation. A total of 23 compounds were identified, but the concentration of compounds was relatively different from each method. The high qualities of compounds were obtained by the method of supercritical fluid extraction.

6.4.4 Flowers

Spilanthes acmella var. *oleracea*, commonly known as jambú, is a source of secondary metabolites, for example, alkylamides. These are responsible for the antiseptic, anti-inflammatory, and anesthetic bioactivities of the plant. The flower bud consists of a grassy taste continued by a tingling sensation and creates more salivation, with an experience of cooling in the throat. They are used as flavoring in chewing tobacco. The extracts were extracted from jambú leaves, stems, and flowers by a fractionated procedure that involved a SFE using carbon dioxide as solvent, with an enhanced solvent extraction (ESE) step by using CO_2 with ethanol, water, and their mixtures as solvent enhancers. The flowers were found to be rich in spilanthol, which gives the highest antioxidant/total phenolic ratio (Dias et al. 2012).

Flowers of Macela (*Achyrocline satureioides*) were extracted for their phenolic compounds by SFE with ethanol as co-solvent. Several experiments were conducted at pressure of 20 and 30 MPa and temperature of 303.15 and 318.15 K. These experiments were carried out with various compositions and mass flow rate of supercritical solvents in 420 min. The extracts were analyzed by spectrophotometry and gas chromatography. The quantity of quercetin and gallic acid and the constituents of the essential oil were obtained (Hatami et al. 2012).

Bergeron et al. (2005) carried out comparison of the chemical compounds by conducting the extraction from *Scutellaria lateriflora* using SFE and accelerated solvent extraction (ASE) versus 70% ethanol extraction or standard hot water. The flavonoids were compared with conventional extraction methods through hot water extraction and the use of 70% ethanol. ASE at the temperature of 85 °C with water acting as a solvent gives the best results for flavonoids, whereas SFE gives higher results for aglycone flavonoid.

6.5 Conclusion

Plants containing the secondary metabolites (phenolic compounds) are important for various diseases' prevention and also in terms of their natural antioxidant activity. Conventional extraction processes are inefficient for their extraction and also have certain disadvantages related to their biological activity. Therefore, novel and greener extraction technique (SFE) has been developed which utilizes less solvent and less time and delivers efficient yields. Supercritical carbon dioxide is a safe solvent for the purpose of extraction, and the use of co-solvents like ethanol enhances its efficiency. Therefore, SFE is the efficient non-conventional technique for the extraction of phenolic compounds. Yet it still becomes hard to decide the most appropriate critical pressure and temperature for the extraction of particular phenolics for different plant materials.

References

Adil IH, Cetin HI, Yener ME, Bayindirli A (2007) Subcritical (carbon dioxide + ethanol) extraction of polyphenols from apple and peach pomaces, and determination of the antioxidant activities of the extracts. J Supercrit Fluids 43:55–63. https://doi.org/10.1016/j.supflu.2007.04.012

Ahmadian-Kouchaksaraie Z, & Niazmand, R (2017) Supercritical carbon dioxide extraction of antioxidants from Crocus sativus petals of saffron industry residues: Optimization using response surface methodology. The Journal of Supercritical Fluids, 121:19–31

Arnáiz E, Bernal J, Martín MT, Diego JC, Bernal JL, Recio LT (2016) Optimisation of the supercritical fluid extraction of antioxidants from broccoli leaves. Food Anal Methods 9(8):2174–2181

Babbar N, Oberoi HS, Uppal DS, Patil RT (2011) Total phenolic content and antioxidant capacity of extracts obtained from six important fruit residues. Food Res Int 44:391–396. https://doi.org/10.1016/j.foodres.2010.10.001

Babovic N, Djilas S, Jadranin M, Vajs V, Ivanovic J, Petrovic S, Zizovic I (2010) Supercritical carbon dioxide extraction of antioxidant fractions from selected Lamiaceae herbs and their antioxidant capacity. Innovative Food Sci Emerg Technol 11(1):98–107. https://doi.org/10.1016/j.ifset.2009.08.013

Benelli P, Riehl CAS, Smania A Jr, Smania EFA, Ferreira SRS (2010) Bioactive extracts of orange (Citrus sinensis L. Osbeck) pomace obtained by SFE and low pressure techniques: mathematical modeling and extract composition. J Supercrit Fluids 55:132–141. https://doi.org/10.1016/j.supflu.2010.08.015

Bergeron C, Gafner S, Clausen E, Carrier DJ (2005) Comparison of the chemical composition of extracts from Scutellaria lateriflora using accelerated solvent extraction and supercritical fluid extraction versus standard hot water or 70% ethanol extraction. J Agric Food Chem 53(8):3076–3080. https://doi.org/10.1021/jf048408t

Bimakr M, Rahman RA, Taip FS, Ganjloo A, Salleh LM, Selamat J, Hamid A, Zaidul ISM (2011) Comparison of different extraction methods for the extraction of major bioactive flavonoid compounds from spearmint (Mentha spicata L.) leaves. Food Bioprod Proc 89:67–72. https://doi.org/10.1016/j.fbp.2010.03.002

Brglez Mojzer E, Knez Hrnčič M, Škerget M, Knez Ž, Bren U (2016) Polyphenols extraction methods, antioxidative action, bioavailability and anticarcinogenic effects. Molecules 21:901–908. https://doi.org/10.3390/molecules21070901

Brunner G (2005) Supercritical fluids technology and application to food processing. J Food Eng 67:21–33. https://doi.org/10.1016/j.jfoodeng.2004.05.060
Camila G, Pereira M, Angela A, Meireles B (2010) Supercritical fluid extraction of bioactive compounds: fundamentals, applications and economic perspectives. Food Bioprocess Technol 3:340–372. https://doi.org/10.1007/s11947-009-0263-2
Cavero S, García-Risco MR, Marín FR, Jaime L, Santoyo S, Senoráns FJ et al (2006) Supercritical fluid extraction of antioxidant compounds from oregano: chemical and functional characterization via LC–MS and in vitro assays. J Supercrit Fluids 38(1):62–69. https://doi.org/10.1016/j.supflu.2005.01.003
Chemat F, Vian MA, Cravotto G (2012) Green extraction of natural products: concept and principles. Int J Mol Sci 13(7):8615–8627
de Andrade Lima M, Charalampopoulos D, Chatzifragkou A (2018) Optimisation and modelling of supercritical CO2 extraction process of carotenoids from carrot peels. J Supercrit Fluids 133:94–102
De Azevedo ABA, Mazzafera P, Mohamed RS, de Melo SAB, Kieckbusch TG (2008) Extraction of caffeine, chlorogenic acids and lipids from green coffee beans using supercritical carbon dioxide and co-solvents. Braz J Chem Eng 825:543–552. https://doi.org/10.1590/s0104-66322008000300012
Dhanani T, Shah S, Gajbhiye NA, Kumar S (2017) Effect of extraction methods on yield, phytochemical constituents and antioxidant activity of Withania somnifera. Arab J Chem 10:1193–1199. https://doi.org/10.1016/j.arabjc.2013.02.015
Diaaz-Reinoso B, Moure S, Domianguez H, Parajoa JC (2006) Supercritical CO_2 extraction and purification of compounds with antioxidant activity. J Agric Food Chem 54:2441–2469. https://doi.org/10.1021/jf052858j
Dias AMA, Santos P, Seabra IJ, Júnior RNC, Braga MEM, De Sousa HC (2012) Spilanthol from Spilanthes acmella flowers, leaves and stems obtained by selective supercritical carbon dioxide extraction. J Supercrit Fluids 61:62–70. https://doi.org/10.1016/j.supflu.2011.09.020
Edith A, Mofolusho F, Omonike O, Larry A, Dora A (2005) In vivo antimalarial and cytotoxic properties of Annona senegalensis extract. Afr J Tradit Complement Altern Med 3:138–139. https://doi.org/10.4314/ajtcam.v12i6.1s
Ekart MP, Bennett KL, Ekart SM, Gurdial GS, Liotta CL, Eckert CA (1992) Cosolvent interactions in supercritical fluids solutions. AIChE J 39:235–348. https://doi.org/10.1002/aic.690390206
Espinosa-Pardo FA, Martinez J, Martinez-Correa HA (2014) Extraction of bioactive compounds from peach palm pulp (Bactris gasipaes) using supercritical CO_2. J Supercrit Fluid 93:2–6. https://doi.org/10.1016/j.supflu.2014.05.010
Espinosa-Pardo FA, Nakajima VM, Macedo GA, Macedo JA, Martínez J (2017) Extraction of phenolic compounds from dry and fermented orange pomace using supercritical CO2 and cosolvents. Food Bioprod Process 101:1–10
Farias-Campomanes AM, Rostagno MA, Meireles MAA (2013) Production of polyphenol extracts from grape bagasse using supercritical fluid: yield, extract composition and economic evaluation. J Supercrit Fluids 77:70–78
Garmus TT, Paviani LC, Queiroga CL, Cabral FA (2015) Extraction of phenolic compounds from pepper-rosmarin (Lippia sidoides Cham.) leaves by sequential extraction in fixed bed extractor using supercritical CO2, ethanol and water as solvents. J Supercrit Fluids 99:68–75. https://doi.org/10.1016/j.supflu.2015.01.016
Ghafoor K, Park J, Choi YH (2010) Optimization of supercritical fluid extraction of bioactive compounds from grape (Vitis labrusca B.) peel by using response surface methodology. Innovative Food Sci Emerg Technol 11(3):485–490. https://doi.org/10.1016/j.ifset.2010.01.013
Guan W, Li S, Yan R, Tang S, Quan C (2007) Comparison of essential oils of clove buds extracted with supercritical carbon dioxide and other three traditional extraction methods. Food Chem 101(4):1558–1564. https://doi.org/10.1016/j.foodchem.2006.04.009
Hancock EW (2005) Mechanisms of newer antibiotics for gram-positive pathogens. Lancet Infect Dis 5:209–218. https://doi.org/10.1016/s1473-3099(05)70051-7

Hatami T, Cavalcanti RN, Takeuchi TM, Meireles MAA (2012) Supercritical fluid extraction of bioactive compounds from Macela (Achyrocline satureioides) flowers: kinetic, experiments and modeling. J Supercrit Fluids 65:71–77. https://doi.org/10.1016/j.supflu.2012.03.001

Hawthorne SB, Krieger MS, Miller DJ (1989) Supercritical carbon dioxide extraction of polychlorinated biphenyls, polycyclic aromatic hydrocarbons, heteroatom-containing polycyclic aromatic hydrocarbons, and n-alkanes from polyurethane foam sorbents. Anal Chem 61(7):736–740

Huynh NT, Smagghe G, Gonzales GB, Van Camp J, Raes K (2014) Enzyme-assisted extraction enhancing the phenolic release from cauliflower (Brassica oleracea L. var. botrytis) outer leaves. J Agric Food Chem 62(30):7468–7476. https://doi.org/10.1021/jf502543c

Isturiz R (2008) Global resistance trends and the potential impact on empirical therapy. Int J Antimicrob Agents 32:201–206. https://doi.org/10.1016/s0924-8579(09)70003-2

Jadhav D, Rekha BN, Gogate PR, Rathod VK (2009) Extraction of vanillin from vanilla pods: a comparison study of conventional soxhlet and ultrasound assisted extraction. J Food Eng 93:421–426. https://doi.org/10.1016/j.jfoodeng.2009.02.007

Khaw KY, Parat MO, Shaw PN, Falconer JR (2017) Solvent supercritical fluid technologies to extract bioactive compounds from natural sources: a review. Molecules 22(7):1186. https://doi.org/10.3390/molecules22071186

Kim SJ, Matsushita Y, Fukushima K, Aoki D, Yagami S, Yuk HG, Lee SC (2014) Antioxidant activity of a hydrothermal extract from watermelons. LWT-Food Sci Technol 59(1):361–368. https://doi.org/10.1016/j.lwt.2014.04.041

Ko MJ, Cheigh CI, Cho SW, Chung MS (2011) Subcritical water extraction of flavonol quercetin from onion skin. J Food Eng 102(4):327–333. https://doi.org/10.1016/j.jfoodeng.2010.09.008

Kraujalienė V, Pukalskas A, Venskutonis PR (2017) Biorefining of goldenrod (Solidago virgaurea L.) leaves by supercritical fluid and pressurized liquid extraction and evaluation of antioxidant properties and main phytochemicals in the fractions and plant material. J Funct Foods 37:200–208

Kryževičiute N, Kraujalis P, Venskutonis PR (2016) Optimization of high pressure extraction processes for the separation of raspberry pomace into lipophilic and hydrophilic fractions. J Supercrit Fluids 108:61–68. https://doi.org/10.1016/j.supflu.2015.10.025

Liu J, Ji F, Chen F, Guo W, Yang M, Huang S et al (2018) Determination of garlic phenolic compounds using supercritical fluid extraction coupled to supercritical fluid chromatography/tandem mass spectrometry. J Pharm Biomed Anal 159:513–523

Lutz M, Hernández J, Henríquez C (2015) Phenolic content and antioxidant capacity in fresh and dry fruits and vegetables grown in Chile. CYTA-J Food 13:541–547

M'hiri N, Ioannou I, Boudhrioua NM, Ghoul M (2014) Effect of different operating conditions on the extraction of phenolic compounds in orange peel. Food Bioprod Proc 96:161–170. https://doi.org/10.1016/j.fbp.2015.07.010

Manna L, Bugnone C, Banchero M (2015) Valorization of hazelnut, coffee and grape wastes through supercritical fluid extraction of triglycerides and polyphenols. J Supercrit Fluids 104:204–211. https://doi.org/10.1016/j.supflu.2015.06.012

Maran JP, Priya B, Manikandan S (2014) Modeling and optimization of supercritical fluid extraction of anthocyanin and phenolic compounds from Syzygium cumini fruit pulp. J Food Sci Technol 51(9):1938–1946. https://doi.org/10.1007/s13197-013-1237-y

Martínez-Valverde I, Periago MJ, Provan G, Chesson A (2002) Phenolic compounds, lycopene and antioxidant activity in commercial varieties of tomato (Lycopersicum esculentum). J Sci Food Agric 82(3):323–330

Meneses MA, Caputo G, Scognamiglio M, Reverchon E, Adami R (2015) Antioxidant phenolic compounds recovery from Mangifera indica L. by-products by supercritical antisolvent extraction. J Food Eng 163:45–53. https://doi.org/10.1016/j.jfoodeng.2015.04.025

Mushtaq M, Sultana B, Anwar F, Adnan A, Rizvi SS (2015) Enzyme-assisted supercritical fluid extraction of phenolic antioxidants from pomegranate peel. J Supercrit Fluids 104:122–131. https://doi.org/10.1016/j.supflu.2015.05.020

Nayak BS, Pereira LMP (2006) Catharanthus roseus flower extract has wound-healing activity in Sprague Dawley rats. BMC Complement Altern Med 6(41):1–6

Norodin NSM, Salleh LM, Mustafa NM (2016). Supercritical carbon dioxide (SC-CO2) extraction of essential oil from Swietenia mahagoni seeds. In: Materials science and engineering conference series, vol 162, no. 1, p 012030. https://doi.org/10.1088/1757-899x/162/1/012030,

Oliveira MS, da Costa WA, Pereira DS, Botelho JRS, de Alencar Menezes TO, de Aguiar Andrade EH et al (2016) Chemical composition and phytotoxic activity of clove (Syzygium aromaticum) essential oil obtained with supercritical CO2. J Supercrit Fluids 118:185–193. https://doi.org/10.1016/j.supflu.2016.08.010

Okwi DE, Emenike IN (2006) Evaluation of the phytonutrients and vitamins contents of citrus fruits. Int J Mol Med 2:233–239. https://doi.org/10.1002/9781118253649.ch5

Oreopoulou CV, Tzia C (2007) Utilization of by-products and treatment of waste in the food. In: Oreopoulou CV, Russ W (eds) . Springer, New York, pp 209–232. https://doi.org/10.1007/978-0-387-35766-9_11

Paes J, Dotta R, Barbero GF, Martínez J (2014) Extraction of phenolic compounds and anthocyanins from blueberry (Vaccinium myrtillus L.) residues using supercritical CO2 and pressurized liquids. J Supercrit Fluids 95:8–16. https://doi.org/10.1016/j.supflu.2014.07.025

Pereira P, Cebola MJ, Oliveira MC, Bernardo-Gil MG (2016) Supercritical fluid extraction vs conventional extraction of myrtle leaves and berries: comparison of antioxidant activity and identification of bioactive compounds. J Supercrit Fluids 113:1–9. https://doi.org/10.1016/j.supflu.2015.09.006

Piantino CR, Aquino FWB, Follegatti-Romero LA, Cabral FA (2008) Supercritical CO_2 extraction of phenolic compounds from Baccharis dracunculifolia. J Supercrit Fluids 47:209–214. https://doi.org/10.1016/j.supflu.2008.07.012

Rafiq S, Kaul R, Sofi SA, Bashir N, Nazir F, Nayik GA (2018) Citrus peel as a source of functional ingredient: a review. J Saudi Soc Agric Sci 17:351–358. https://doi.org/10.1016/j.jssas.2016.07.006

Santos DT, Veggi PC, Meireles MAA (2012) Optimization and economic evaluation of pressurized liquid extraction of phenolic compounds from jabuticaba skins. J Food Eng 108(3):444–452. https://doi.org/10.1016/j.jfoodeng.2011.08.022

Safdar MN, Kausar T, Jabbar S, Mumtaz A, Ahad K, Saddozai A (2017) Extraction and quantification of polyphenols from kinnow (Citrus reticulate L.) peel using ultrasound and maceration techniques. J Food Drug Anal 25:488–500

Serra AT, Seabra IJ, Braga ME, Bronze MR, de Sousa HC, Duarte CM (2010) Processing cherries (Prunus avium) using supercritical fluid technology. Part 1: recovery of extract fractions rich in bioactive compounds. J Supercrit Fluids 55(1):184–191. https://doi.org/10.1016/j.supflu.2010.06.005

Sharma SK (2010) Functional foods and nutraceuticals. In: Postharvest management and processing of fruits and vegetables – instant notes. New India Publishing Agency, New Delhi, pp 283–288. https://doi.org/10.1016/b978-0-12-812698-1.00001-7

Sharma SK, Mangal M, Bansal S (2012) Dietary fibers in health foods. In: Gupta RK, Bansal S, Mangal M (eds) Health foods recent trends and technological interventions, New Delhi, pp 123–146

Sharma SK, Bansal S, Mangal M, Dixit AK, Gupta RK, Mangal AK (2016) Utilization of food processing by-products as dietary, functional, and novel fiber. Crit Rev Food Scinutr 56:1647–1661. https://doi.org/10.1080/10408398.2013.794327

Singh PP, Saldaña MD (2011) Subcritical water extraction of phenolic compounds from potato peel. Food Res Int 44(8):2452–2458. https://doi.org/10.1016/j.foodres.2011.02.006

Sundsford A, Simonsen GS, Haldorsen BC, Haaheim H, Hjelmevoll SO, Littauer P, Dahl KH (2004) Genetic methods for detection of antimicrobial resistance. APMIS 112(11-12):815–837. https://doi.org/10.1111/j.1600-0463.2004.apm11211-1208.x

Tomás-Barberán F, Espín J (2001) Phenolic compounds and related enzymes as determinants of quality in fruits and vegetables. J Sci Food Agri 81:853–876

Tyśkiewicz K, Konkol M, Rój E (2018) The application of supercritical fluid extraction in phenolic compounds isolation from natural plant materials. Molecules 23(10):2625. https://doi.org/10.3390/molecules23102625

Vagi E, Rapavi E, Hadolin M, Vasarhelyine Peredi K, Balazs A, Blazovics A, Simandi B (2005) Phenolic and triterpenoid antioxidants from Origanum majorana L. herb and extracts obtained with different solvents. J Agric Food Chem 53(1):17–21. https://doi.org/10.1021/jf048777p

Velioglu YS, Mazza G, Gao L, Oomah BD (1998) Antioxidant activity and total phenolics in selected fruits, vegetables, and grain products. J Agric Food Chem 46:4113–4117. https://doi.org/10.1021/jf9801973

Zeković Z, Kaplan M, Pavlić B, Olgun EO, Vladić J, Canlı O, Vidović S (2016) Chemical characterization of polyphenols and volatile fraction of coriander (Coriandrum sativum L.) extracts obtained by subcritical water extraction. Ind Crop Prod 87:54–63. https://doi.org/10.1016/j.indcrop.2016.04.024

Chapter 7
The Application of Supercritical Carbon Dioxide in the Extraction of Biomolecules

Natascha Cheikhyoussef

Contents

Abbreviations

BIC	Broken and intact cell
GRAS	"Generally recognized green solvent"
HBD	Hot ball diffusion
LM	Logistic model
ORAC	Oxygen radical absorbance capacity
RSM	Response surface methodology
SC	Shrinking core
SC-CO_2	Supercritical carbon dioxide
SF	Supercritical fluid

N. Cheikhyoussef (✉)
Research and Innovation, Ministry of Higher Education, Technology and Innovation, Windhoek, Namibia
e-mail: natascha.cheikhyoussef@mheti.gov.na

Inamuddin, A. M. Asiri (eds.), *Advanced Nanotechnology and Application of Supercritical Fluids*, Nanotechnology in the Life Sciences,
https://doi.org/10.1007/978-3-030-44984-1_7

7.1 Introduction

The development of green extraction processes is aimed at reducing the use of organic solvents and the energy consumption through reduced reaction and processing times that will produce safe products of good quality and purity (Chemat et al. 2019). Chemat et al. (2019) have described the six principles for identifying the green extraction process (Chemat et al. 2012) and developed accordingly the good practices guidelines for each of these principles. The principles described by Chemat et al. (2012, 2019) are (1) "innovation by selection of varieties and use of renewable plant resources", (2) "use of alternative solvents and principally water or agro-solvents", (3) "reduction of energy consumption by energy recovery and using innovative technologies", (4) "production of co-products instead of waste towards bio-refinery concepts", (5) "reduction of unit operations number and development of safe, robust and controlled processes" and (6) "aim for green extract with green values and non-denatured and biodegradable extract without contaminants". The application of supercritical carbon dioxide (SC-CO_2), a "generally recognized green solvent" (GRAS) (Rovetto and Aieta 2017; Chemat et al. 2019), for the use of the extraction of biomolecules fulfils the requirements under principle 2 and is the most common used green solvent for supercritical fluid (SF) extractions (Stuart et al. 1996; Lang and Wai 2001; Rovetto and Aieta 2017). This is also evident from the numerous review articles published in the past 20 years on the use of SFs, which have provided numerous reports on the SC-CO_2 extraction applications (Table 7.1). This phenomena is attributed to the properties of the carbon dioxide as being non-flammable, non-toxic, cheap and non-corrosive (Lang and Wai 2001; Huang et al. 2012) with the SC-CO_2 extraction process being highly selective, obtaining solvent free products with no development of co-products (Knez et al. 2019). Extracts obtained from a SF extraction have been commonly found to be of greater quality when compared to other methods (Fornari et al. 2012a). A wide array of published reviews discuss the supercritical extraction of bioactives (Table 7.1), which are essentially biomolecules possessing specific biological activities or functions (Fig. 7.1) with a great range of applications in industry (da Silva et al. 2016). De Melo et al. (2014) have described reported SF extractions from vegetable matrices of about 600 essays for the period 2000–2013, including modelling, operating conditions, scale-up and an economic assessment. A review by Khaw et al. published in 2017 focusses on SF extraction of bioactives from different natural sources. The review, including other green extraction methods, provides insights on the SC-CO_2 extraction of bioactives and operating conditions from about 40 plant species. A summary of the reported SC-CO_2 extraction parameters of bioactives, among other newer methods such as "subcritical water extraction", "ultrasound-assisted extraction" and "microwave-assisted extraction", specifically from marine macroalgae was reported by Cikoš et al. (2018). Gallego et al. (2019) have described reported bioactives extracted by subcritical and supercritical fluid extraction from various plant sources, seaweeds, microalgae and food by-products for the period from 2015 to 2019. Previously, Herrero et al. (2015) published a review on the same theme for the period 2006–2014. The bioactives extracted from these sources using

Table 7.1 Published reviews on the use of supercritical fluids for the extraction of biomolecules

Title of the review	References
Analytical supercritical fluid extraction of natural products	Modey et al. (1996)
Compounds of agricultural significance using environmental analytical supercritical fluid extraction	Stuart et al. (1996)
Natural extracts using supercritical carbon dioxide	Mukhopadhyay (2000)
Supercritical fluid extraction in herbal and natural product studies	Lang and Wai (2001)
Supercritical extraction from solid: Process design data (2001–2003)	Meireles (2003)
Supercritical fluid extraction and fractionation of natural matter	Reverchon and De Marco (2006)
Comparative assessment of technologies for extraction of artemisinin	Lapkin et al. (2006)
Supercritical fluid extraction in plant essential and volatile oil analysis	Pourmortazavi and Hajimirsadeghi (2007)
Supercritical fluid extraction of bioactive compounds: Fundamentals, applications and economic perspectives	Pereira and Meireles (2010)
Isolation of essential oil from different plants and herbs by supercritical fluid extraction	Fornari et al. (2012a)
Principles of supercritical fluid extraction and applications in the food, beverage and nutraceutical industries	Knez et al. (2013)
Supercritical fluid extraction of vegetable matrices: Applications, trends and future perspectives of a convincing green technology	De Melo et al. (2014)
Essential oils: Extraction, bioactivities, and their uses for food preservation	Tongnuanchan and Benjakul (2014)
Supercritical carbon dioxide extraction of carotenoids from pumpkin (*Cucurbita* spp.)	Durante et al. (2014)
Extraction of triacylglycerols and fatty acids using supercritical fluids—Review	Martínez and de Aguiar (2014)
Plants, seaweeds, microalgae and food by-products as natural sources of functional ingredients obtained using pressurized liquid extraction and supercritical fluid extraction	Herrero et al. (2015)
Supercritical fluid extraction of bioactive compounds	Da Silva et al. (2016)
Supercritical fluid extraction of bioactive compounds from plant materials	Wrona et al. (2017)
Plant growth biostimulants, dietary feed supplements and cosmetics formulated with supercritical CO_2 algal extracts	Michalak et al. (2017)
Solvent supercritical fluid technologies to extract bioactive compounds from natural sources: A review	Khaw et al. (2017)
Overview on the application of modern methods for the extraction of bioactive compounds from marine macroalgae	Cikoš et al. (2018)
Recent applications of on-line supercritical fluid extraction coupled to advanced analytical techniques for compounds extraction and identification	Sánchez-Camargo et al. (2019)
Are supercritical fluids solvents for the future?	Knez et al. (2019)
Supercritical fluid extraction of essential oils	Yousefi et al. (2019)
Sub- and supercritical fluid extraction of bioactive compounds from plants, food-by-products, seaweeds and microalgae—An update	Gallego et al. (2019)
Green extraction techniques in green analytical chemistry	Armenta et al. (2019)

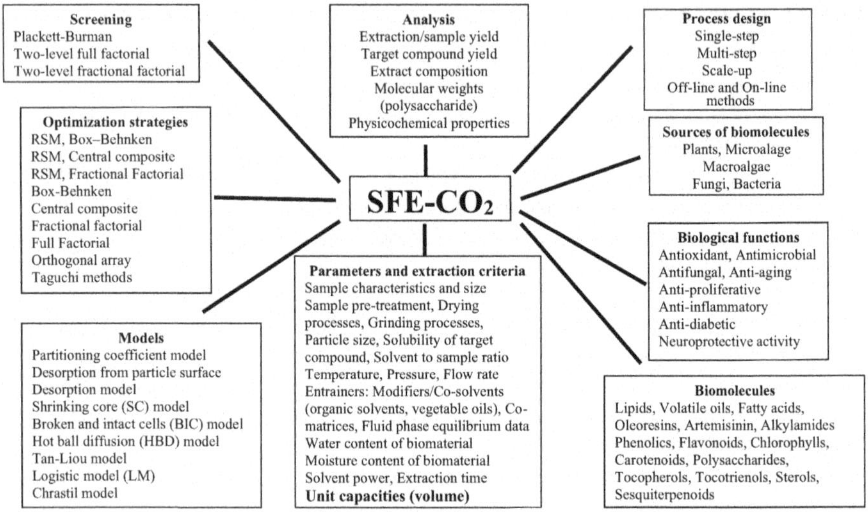

Fig. 7.1 Overview of the supercritical carbon dioxide (SF-CO_2) extraction process of biomolecules (Grosso et al. 2010; Sovová 2012; Huang et al. 2012; da Silva et al. 2016; Shukla et al. 2019; Yousefi et al. 2019)

specifically SC-CO_2, with or without co-solvents, include carotenoids, anthocyanins, phenolic compounds, polyphenolics, sesquiterpenes, antioxidant compounds, polyphenols, monoterpenes, vitamin E, cannabinoids, non-polar flavonoids, lycopene, piperine, tetrahydrocannabinol, colchicine, tocols, tocopherols, polar and non-polar lipids, oils, essential oils, oleanolic acids, ursolic acids, fatty acids, chlorophyll A, ergosterol, fucosterol, fucoxanthin and triacylglycerides (Gallego et al. 2019). Michalak et al. (2017) have presented a summary of reported biologically active compounds (fucoxanthin, beta-carotene, carotenoids, astaxanthin, canthaxanthin, chlorophyll, polyphenols, fatty acids, lipids, oil, auxins, cytokinins, micro- and macro-elements) that have been obtained via SC-CO_2 extraction from algal biomass. Algae extracts obtained by this method are proposed by Michalak et al. (2017) to be used in cosmetics and dietary feeds and as growth stimulants due to their components being solvent-free. Table 7.3 provides a summary of reported SC-CO_2 extractions of biomolecules discussed in this chapter. The chapter intends to present recent applications of SC-CO_2 extraction of selected biomolecules with potential uses in industry and provides an overview of published reports on this critical area of Green Chemistry.

7.2 Design and Optimization of the Supercritical Carbon Dioxide Extraction Process

In order to ensure high extraction efficiencies and to obtain high quality and purified extractions (Chemat et al. 2019; Yousefi et al. 2019), the design and the optimization strategies are critical. In particular, with SF extractions, the aim is on the reduction of extraction time, amount of solvent, energy usage, costs, waste produced and the environmental impact (Chemat et al. 2019; Yousefi et al. 2019). Figure 7.1 presents a summary of the various components of the SC-CO_2 extraction process of biomolecules. Understanding the mass transfer mechanisms of the extraction process is imperative for starting the design of the extraction process (Huang et al. 2012) and in choosing the appropriate mathematical model to be applied (Kumhom et al. 2011). The development of mathematical models for the SC-CO_2 extraction process was reported by various research groups (Table 7.2). These models are generally described as the models that are based on "heat transfer analogies", models based on "differential mass balances", empirical models and the shrinking core model

Table 7.2 Reported theoretical models for supercritical fluid extraction

Title of the publication	References
Mathematical modeling of sunflower seed extraction by supercritical CO_2	Perrut et al. (1997)
Mass transfer modelling of apricot kernel oil extraction with supercritical carbon dioxide	Özkal et al. (2005)
Mathematical modelling of supercritical CO_2 extraction of volatile oils from aromatic plants	Grosso et al. (2010)
Prediction of isoflavone extraction from soybean meal using supercritical carbon dioxide with cosolvents	Kumhom et al. 2011
Theoretical models for supercritical fluid extraction	Huang et al. (2012)
Kinetic study of the supercritical CO_2 extraction of different plants from the *Lamiaceae* family	Fornari et al. (2012b)
Modeling the supercritical fluid extraction of essential oils from plant materials	Sovová (2012)
Effects of high water content and drying pre-treatment on supercritical CO_2 extraction from *Dunaliella salina* microalgae: Experiments and modelling	Mouahid et al. (2016)
Modeling of the kinetics of supercritical fluid extraction of lipids from microalgae with emphasis on extract desorption	Sovová et al. (2016)
Supercritical carbon dioxide extraction of *Calendula officinalis*: Kinetic modeling and scaling up study	López-Padilla et al. (2017)
Modeling of the kinetics of supercritical fluid extraction of lipids from microalgae with emphasis on extract desorption	Sovová et al. (2016)
Broken-and-intact cell model for supercritical fluid extraction: Its origin and limits	Sovová (2017)
Supercritical carbon dioxide extraction of pomegranate (*Punica granatum* L.) seed oil: Kinetic modelling and solubility evaluation	Natolino and Da Porto (2019)
New developments in the modelling of carotenoids extraction from microalgae with supercritical CO_2	Sovová and Stateva (2019)
Extraction of vetiver (*Chrysopogon zizanioides*) root oil by supercritical CO_2, pressurized-liquid, and ultrasound-assisted methods and modelling of supercritical extraction kinetics	Santos et al. (2019)
Evaluation of the effects of temperature and pressure on the extraction of eugenol from clove (*Syzygium aromaticum*) leaves using supercritical CO_2	Frohlich et al. (2019)

Table 7.3 Reported publications on the supercritical carbon dioxide extraction of biomolecules

Primary source for biomolecule(s)	Target biomolecule(s)	References
Chlorella vulgaris	Carotenoids, lipids	Mendes et al. (1995)
Artemisia annua L.	Artemisinin, artemisinic acid	Kohler et al. (1997)
Sargassum hemiphyllum (turn.) C. Ag. J.	Lipids	Cheung et al. (1998)
Red seaweed	*n*-3 fatty acids	Cheung (1999)
Cunninghamella echinulata	Fungal oil containing γ-linolenic acid	Certik and Horenitzky (1999)
Echinacea angustifolia	Alkylamides	Sun et al. (2002)
Botryococcus braunii *Chlorella vulgaris* *Dunaliella salina* *Arthrospira maxima*	Alkadienes Carotenoids β-Carotene γ-Linolenic acid	Mendes et al. (2003)
Foeniculum vulgare	Volatile oil	Coelho et al. (2003)
Artemisia annua L. leaves	Artemisinin	Quispe-Condori et al. (2005)
Artemisia annua L.	Artemisinin	Lin et al. (2006)
Nannochloropsis sp.	Lipids	Andrich et al. (2005)
Artemisia annua L.	Scopoletin, artemisinin	Tzeng et al. (2007)
Blakeslea trispora NRRL 2895 and 2896	Lycopene	Choudhari, Singhal (2008)
Haematococcus pluvialis	Astaxanthin	Krichnavaruk et al. (2008)
Italian coriander seeds	Volatile oil	Grosso et al. (2008)
Santolina chamaecyparissus	Volatile oil	Grosso et al. (2009a)
Satureja montana	Volatile oil (thymoquinone)	Grosso et al. (2009b)
Ampelopsis grossedentata stems	Bioactive compounds	Wang et al. (2011)
Soybean meal	Isoflavone	Kumhom et al. (2011)
Chlorella vulgaris microalgae	Lipids	Dejoye et al. (2011)
Citrus grandis (L.) Osbeck (pomelo) peel	Flavonoids	He et al. (2012)
Origanum vulgare, Thymus zygis, Salvia officinalis, Rosmarinus officinalis	Essential oil	Fornari et al. (2012b)
Tiger nut (*Cyperus esculentus* L.)	Oil	Lasekan and Abdulkarim (2012)
Cannabis sativa L. (hemp) seed	Seed oil, oxidative stability	Da Porto et al. (2012a)
Cannabis sativa L. (hemp) seed	Seed oil, oxidative stability	Da Porto et al. (2012b)
Grape	Seed oil, polyphenol co-extraction	Rombaut et al. (2014)
Moringa oleifera	Seed oil	Ruttarattanamongkol et al. (2014)
Prunus persica seeds	Oil, phytosterols	Ekinci and Gürü (2014)
Sasa palmata (bamboo) leaves	Phenolics	Zulkafli et al. (2014)
Tetraselmis sp. (green algae)	Lipids	Li et al. (2014)
Sasa palmata (bamboo) leaves	Phenolic compounds	Zulkafli et al. 2014
Artemisia sphaerocephala Krasch seeds	Polysaccharides	Chen et al. (2014)
Brazilian plants	Phenolics	Veggi et al. (2014a)
Hymenaea courbaril L. (jatoba)	Phenolic compounds	Veggi et al. (2014b)
Cannabis sativa L. (hemp)	Seed oil	Aladić et al. (2015)
Tiger nuts	Oil, phenolic compounds	Koubaa et al. (2015)
Euterpe oleracea (açaí)	Berry oil, fatty acids, phenolics, anthocyanins	De Cássia et al. (2016)
Dunaliella salina microalgae	Carotenoids	Mouahid et al. (2016)

(continued)

Table 7.3 (continued)

Primary source for biomolecule(s)	Target biomolecule(s)	References
Sunflower seed	Oil	Rai et al. (2016)
Diospyros kaki L. (persimmon)	Carotenoids	Zaghdoudi et al. (2016)
Artemisia annua L.	Artemisinin	Baldino et al. (2017)
Eremanthus erythroppapus (Candeia)	Oil	Santos et al. (2017)
Cannabis sativa L. (hemp)	Cannabinoids δ^9-Tetrahydrocannabinolic acid, δ^9-tetrahydrocannabinol	Rovetto and Aieta (2017)
Artemisia annua L. leaves	Bioactive extracts	Martinez-Correa et al. (2017)
Ocimum sanctum Linn.	Eugenol	Chatterjee et al. (2017)
Calendula officinalis	Oleoresin	López-Padilla et al. (2017)
Spinach by-products	Lutein and chlorophyll	Derrien et al. (2018)
Artemisia annua L.	Artemisinin	Ciftzi et al. (2018)
Phyllostachys heterocycle (bamboo)	Polysaccharide	Zou et al. (2018)
Oats	Phenolic acids, avenanthramides and antioxidant activity	Walters et al. (2018)
Craterellus tubaeformis (Finnish wild mushrooms)	Volatile compounds (aroma compounds)	Chen et al. (2018)
Algerian *Thymus munbyanus*	Essential oils	Bendif et al. (2018)
Avena sativa L. (oats)	Polyphenols	Escobedo-Flores et al. (2018)
Cacao pod husk	Phenolic compounds	Valadez-Carmona et al. (2018)
Garlic	Phenolic compounds	Liu et al. (2018)
Radish leaves	Bioactive compounds	Goyeneche et al. (2018)
Oenocarpus bacaba (bacaba)	Oil	Pinto et al. (2018)
Avena sativa L. (oats)	Oil, main fatty acids, polyphenols	Fernández-Acosta et al. (2019)
Soybean residue	Phytochemicals	Alvarez et al. (2019)
Parboiled rice bran	Rice bran oil	Juchen et al. (2019)
Solidago gigantea Ait.	Lipids	Wrona et al. (2019)
Euterpe oleracea Mart. (lyophilized açaí)	Pulp oil	Silva et al. (2019)
Artemisia annua L.	Artemisinin	Rodrigues et al. (2019)
Ginger rhizomes	Volatile oil and gingerols enriched oleoresin	Shukla et al. (2019)
Chrysopogon zizanioides	Vetiver essential oil	Santos et al. (2019)
"Horchata" by-products	Oils-phenolic profile	Roselló-Soto et al. (2019a)
"Horchata" by-products	Fatty acid profile, α-tocopherol, phenolic compounds and lipid oxidation parameters	Roselló-Soto et al. (2019b)
Punica granatum L. (pomegranate)	Seed oil	Natolino and Da Porto (2019)
Curcuma longa, *Curcuma amada*	Oleoresin	Nagavekar and Singhal (2019)
Origanum vulgare L.	Oil	García-Pérez et al. (2019)
Syzygium aromaticum (clove) leaves	Eugenol	Frohlich et al. (2019)
Oenocarpus distichus Mart. (bacaba-de-leque)	Oil	Cunha et al. (2019)
Virola surinamensis (ucuúba)	Seed oil	Cordeiro et al. (2019)

(Özkal et al. 2005). Huang et al. (2012) have described several models such as the "broken and intact cell" model (BIC), hot ball diffusion (HBD) model, shrinking core (SC) model, Tan-Liou model, partitioning coefficient model and the logistic model (LM) for different SC-CO_2 extraction systems (Fig. 7.1). The most widely used model for obtaining extracts via SC-CO_2 extraction from plant sources is the BIC model (Huang et al. 2012). The BIC model has been found suitable for the extraction of oleoresin from marigold (*Calendula officinalis*) plants (López-Padilla et al. 2017). Natolino and Da Porto (2019) have applied kinetic (BIC model) and solubility (Chrastil model) modelling for the SC-CO_2 extraction of pomegranate seed oil from *Punica granatum* L. The "shrinking core model" was applied for the isoflavone extraction (SC-CO_2 and methanol) from soybean meal by Kumhom et al. (2011) and investigated the axial dispersion coefficient, effective diffusivity, solubility and the film "mass transfer coefficient", of which the film "mass transfer coefficient" and the solubility were found to be the most significant. Sovová (2012, 2017), Sovová et al. (2016) and Sovová and Stateva (2019) have described the modelling of the SF extraction process for essential oils, lipids and carotenoids, which have been readily adapted to represent extraction curves.

Optimization targets (Fig. 7.1) are based on varying mainly the pressure, temperature and the flow rate of the SC-CO_2. Co-solvents such as water, methanol, diethyl ether, ethanol, acetone, acetonitrile or dichloromethane are added at varying concentrations to observe the effect on the extraction yield and the extract composition, as these improve the solvating power of the SC-CO_2 (Michalak et al. 2017; Rovetto and Aieta 2017). This is necessary for the extraction of polar compounds since SC-CO_2 is a non-polar solvent and its polarity can be influenced with the use of polar modifiers (Nagavekar and Singhal 2019). Additional parameters (Fig. 7.1) that are targeted for optimization are the sample's particle size, extraction time and solvent power (Ekinci and Gürü 2014). The effects of the solvent power of SC-CO_2 are critical during any extraction process, whereby the effects on solvent power and extraction selectivity are discussed by Ekinci and Gürü (2014). Solvent density is significantly affected by changes in pressure (Derrien et al. 2018). Rovetto and Aieta (2017) investigated the effect of different pressures, flow rates and co-solvent (ethanol) on the yield of cannabinoids from *Cannabis sativa* L., whereby pressure and plant material had notable effects on the extraction yield.

7.3 Lipids, Volatile Oils and Oleoresins

Lipids have been extracted from various macroalgae (MA) and microalgae (MI) species using SC-CO_2 such as *Chlorella vulgaris* (MI) (Mendes et al. 1995; Dejoye et al. 2011), *Tetraselmis* sp. (MI) (Li et al. 2014), *Chaetomorpha linum* (MA) (Aresta et al. 2005), *Sargassum hemiphyllum* (MA) (Cheung et al. 1998) and *Hypnea charoides* (MA) (Cheung 1999). Fatty acids from *Arthrospira maxima* (*Spirulina maxima*) (MI) have been extracted with SC-CO_2 and with 10% ethanol (Mendes et al. 2003). Michalak et al. (2017) reported a comparative summary of the

SF extraction methods and parameters used for the extraction of algae bioactives. Mainly, SC-CO_2 extraction alone has been applied with some reported to have added organic solvents, water or vegetable oils to the extraction process. Biomolecules that are oxidized easily and that are heat labile are suitably extracted with SC-CO_2 due to its non-oxidant nature (Michalak et al. 2017). SC-CO_2 has been identified as an appropriate method to extract lipids (Li et al. 2014). A low temperature of 30 °C and a pressure of 35 MPa were found to be optimum for the SC-CO_2 extraction of *Moringa oleifera* seed oil (75.27%) and in having the greatest solvation power (Ruttarattanamongkol et al. 2014). The solvent CO_2 showed higher selectivity at the low pressure of 15 MPa towards the extraction of sterols, tocopherols and fatty acids (Ruttarattanamongkol et al. 2014). At conditions of 15 MPa and 35 °C, biomolecules β-sitosterol, campesterol, γ-tocopherol and α-tocopherol were found to be at highest concentrations of 2310.9, 1179.2, 106.8 and 230.3 mg/kg, respectively (Ruttarattanamongkol et al. 2014). SC-CO_2 extraction of hemp seed oil from *Cannabis sativa* L. has been previously reported with oil yields of 21.50% w/w (40 °C, 300 bar, particle size of 0.71 mm) (Da Porto et al. 2012a) and 22% (300 bar and 40 °C and at 400 bar and 80 °C) obtained (Da Porto et al. 2012b). Hemp (*Cannabis sativa* L.) seed oil was extracted at varying temperature of 40 and 60 °C and at a constant pressure of 300 bar with a CO_2 flow rate of 1.94 kg/h (Aladić et al. 2015). The yield of seed oil was not affected by changes in temperature but increased with increasing pressure (Aladić et al. 2015). The fatty acid concentration was affected by pressure, whilst extraction time did not affect the content (Aladić et al. 2015). Temperature had a varying effect on the content of the fatty acids (Aladić et al. 2015). The common vegetable oil, sunflower oil (54.37 wt%), has been extracted with SC-CO_2 from the sunflower seed (particle size, 0.75 mm) with a flow rate of 10 g/min, 5% co-solvent at 400 bar and 80 °C (Rai et al. 2016). Oil from *Eremanthus erythropappus* (candeia wood) has been optimally extracted at 70 °C and 24 MPa with 2 ml/min flow rate and ethanol (1.3% v/v) and ethyl acetate (5% v/v) as co-solvents, yielding 2.35 wt% oil (Santos et al. 2017). The highest concentration of α-bisabolol (16.53 g/kg), a naturally occurring sesquiterpene alcohol, was obtained with 5% ethanol (Santos et al. 2017). A review on the properties (pharmacological), mechanisms of action and the applications of α-bisabolol and oils rich in α-bisabolol has been published by Kamatou and Viljoen (2010). The yield of total lipids extracted from *Solidago gigantea* Ait. (goldenrod) has been evaluated using a "Box-Behnken design with three variables" studied, temperature (40–80 °C), pressure (20–80 MPa) and the flow rate of CO_2 (3–7 kg/h) (Wrona et al. 2019). The optimum conditions reported are temperature at 313.95 K, pressure at 68.07 MPa and CO_2 flow rate of 3.18 kg/h yielding 203.32 mg stearic acid equivalent/g dry mass, with temperature having a negligible effect on the content of total lipids (Wrona et al. 2019).

Shukla et al. (2019) have applied SC-CO_2 (single-step) extraction and fractionation process to obtain oleoresin enriched with gingerols and essential oil from dried ginger rhizomes. Optimum conditions reported by Shukla et al. (2019) for obtaining 28.3 wt% volatile oil and 37.97 wt% major actives were pressure at 276 bar, temperature at 40 °C and flow rate of 30 g/min for 153 min. Shukla et al.

(2019) have presented a summary of reported literature on obtaining ginger extracts using SC-CO_2 extraction processes. Oregano oil (*Origanum vulgare* L.) extracted with SC-CO_2 (100 bar, 40 °C, 8 g/min ethanol) with a highest yield of 13.40% showed high antimicrobial and antioxidant activity (García-Pérez et al. 2019). The volatile compound, carvacrol (29.99%), and the fatty acids (70.9–76.8%), α-linolenic (C18:3ω3, 20.55–24.66%), palmitic (C16:0, 22.76–23.65%), oleic (C18:1ω9c, 15.19–16.63%) and linoleic acids (C18:2ω6c, 12.16–13.35%), were found at highest concentrations in the oil (García-Pérez et al. 2019). Oil has been extracted from parboiled rice bran with SC-CO_2 and ethanol, and the effects of varying pressures (100, 150, 200 bar) and temperatures (40, 60, 80 °C) and ethanol to rice bran (0:1, 0.5:1, 1:1, 2:1) were analysed (Juchen et al. 2019). Conditions for one experimental run for 250 min at 200 bar, 40 °C, 45.94 g CO_2/g bran and 1:1 ethanol to rice bran yielded 25.48 wt% of rice bran oil. Two sequential extractions at the same conditions yielded 26.32 wt% (Juchen et al. 2019). Roselló-Soto et al. (2019a) have compared the SC-CO_2 (10–40 MPa, 40 °C) extraction with the conventional extraction (modified Folch et al. 1957) of oil extracted from the "horchata" by-products with the SC-CO_2 extraction obtaining higher amounts of α-tocopherol and total phenolic compounds. The α-tocopherol concentration decreased, whilst total phenolic compounds increased with increasing pressure, respectively. The highest oil yield was obtained with conventional extraction (14.85%). Oil yield and pressure during the SC-CO_2 extraction were shown to have a linear relationship (10 MPa = 0.61%; 40 MPa = 7.36%), which has been reported by several other researchers (Lasekan and Abdulkarim (2012); Rombaut et al. 2014; Koubaa et al. 2015). Pinto et al. (2018) have obtained a yield of 60.39% after SC-CO_2 extraction of "bacaba oil" from *Oenocarpus bacaba* at temperature of 60 °C and pressure of 420 bar. A good quality bacaba-de-leque pulp oil from *Oenocarpus distichus* Mart. has been extracted using SC-CO_2 extraction with a yield of 46% at a pressure of 270 bar and temperature of 60 °C (Cunha et al. 2019). SC-CO_2 extraction (320 bar and 60 °C) has been used in the extraction of pomegranate (*Punica granatum* L.) seed oil that contained a higher punicic acid concentration and with a higher oxidation stability as compared to the Soxhlet extraction method (Natolino and Da Porto 2019). The extraction times with SC-CO_2 (2 h) were much shorter than the period needed for Soxhlet extraction (8 h) to reach the asymptotic yield. A CO_2 flow rate of 8.0 kg/h at 320 bar and 60 °C resulted in the highest yield of the seed oil (Natolino and Da Porto 2019). Fernández-Acosta et al. (2019) recently studied the effect of chemical pre-treatment, pressure, temperature, dynamic and static time and particle size on the SC-CO_2 extraction of oil *Avena sativa* L. (oats). The oat oil yield was significantly influenced by pressure and the size of particles. The pre-treatment, temperature and particle size significantly influenced the fatty acid composition and oxygen radical absorbance capacity (ORAC) antioxidant activity. Polyphenol concentration and total phenolic content were affected by pre-treatment and temperature (Fernández-Acosta et al. 2019). Cordeiro et al. (2019) recently investigated the SC-CO_2 extraction of ucuúba oil from the seeds of *Virola surinamensis*, a tree growing in the Amazon, with antimicrobial activity against *Staphylococcus aureus*. Extraction conditions applied were pressure (350 bar), temperature (40, 60 or 80 °C) and a CO_2 mass flow of 7.9 × 10 kg/s (Cordeiro et al. 2019). De Cássia et al. (2016) have extracted the oil from the popular fruit (berries) açaí (*Euterpe oleracea* Mart.) obtained from Pará, Brazil, under varying temperature and pressure conditions. Highest oil yield was obtained at 70 °C and 490 bar. The fatty composition was affected by the varying operating conditions with significant effects reported on type and concentration of fatty acid detected (De Cássia et al. 2016). Silva et al. (2019) have evaluated the effects of pressure (350, 420, 490 bar), temperature (50, 60 or 70 °C) and geographical location on the extraction of the oil from açaí in lyophilized form. Oil yields between 49.28 (location, Anajás, 60 °C, 420 bar) and 57.06% (location, Chaves, 70 °C, 490 bar) were obtained (Silva et al. 2019) with the operating conditions not affecting the fatty acid composition within each study area. Operating conditions of 60 °C and 420 bar and 70 °C and 490 bar resulted in optimum results of antioxidant capacity, total anthocyanins and total phenolic compounds in the oil from Chaves (Silva et al. 2019). Kerrihard and Pegg (2015) reported on the suitability of the application of SC-CO_2 extraction of oils containing higher concentrations of γ-linolenic acid (GLA; 18:3*n*-6), which have as a result higher anti-inflammatory functionalities. Scale-up experiments and validation processes for the SC-CO_2 extraction of lipids and volatile oils have been investigated by several researchers with promising results of potential extraction at a commercial scale (Shukla et al. 2019; Wrona et al. 2019).

Essential oils, also referred to as volatile oils, are commonly used in traditional medicine and aromatherapy and as natural additives due to their reported antimicrobial activities (Chávez-González et al. 2016) in the food and cosmetics industry (Fornari et al. 2012a). Grosso et al. (2010) have tested five mathematical models for the modelling of the extraction of aromatic plants, fennel, coriander, cotton lavender, savoury, winter savoury and thyme by which extraction was impacted by particle size, internal mass transfer coefficient, internal diffusion, and pressure and temperature changes.

Sovová (2012) has described various published mathematical models for the extraction of essential oils from plants for a 15-year period up to 2012. Tongnuanchan and Benjakul (2014) have discussed the extraction methods, including aspects of SC-CO_2 extraction among others, and the uses and bioactivities of essential oils. Yousefi et al. (2019) have compiled a review that investigates the SF extraction of essential oils from plants. The review describes conventional extraction methods as compared to the SF extraction, optimization and modelling techniques commonly applied and the effects of the various operating parameters on the supercritical extraction process (Yousefi et al. 2019). SC-CO_2 has been investigated for the extraction of the volatile oil from the flower heads of *Santolina chamaecyparissus* L. under different operating conditions. Notably, pressure increase to 9 MPa enriched the content of sesquiterpene in the extracted *Santolina chamaecyparissus* essential oil (Grosso et al. 2009a). The volatile oil from *Satureja montana* L. extracted by SC-CO_2 was reported by Grosso (2009b) to contain higher concentrations of thymoquinone (1.6–3.0%) as compared to the hydrodistillation, a conventional extraction method. Essential oils produced via SC-CO_2 extraction from various herbs and plants have been reported by Fornari et al. (2012a). A cooled mill was used to crush the dried leaves (*O. vulgare*, *Thymus vulgaris*, *Salvia officinalis*, *Rosmarinus officinalis*, *Ocimum basilicum*, *Origanum majorana*) and petals (*Calendula officinalis*) and then sieved (200–600 μm). Essential oils were then extracted at 30 MPa and 40 °C with a flow rate of 60 g/min for 5 h (Fornari et al. 2012a). The economically important vetiver essential oil was extracted with SC-CO_2 among other methods ("pressurized liquid and ultrasound-assisted methods") from *Chrysopogon zizanioides* (L.) Roberty, syn. *Vetiveria zizanioides* (L.) Nash root with the highest yield (2.23% m/m) obtained at 20 MPa and 60 °C (Santos et al. 2019). A higher yield (2.66% (m/m)) was obtained after the addition of 5% (v/v) of the co-solvent, ethanol (Santos et al. 2019). The dominant compounds were isovalencenol (9.04% SC-CO_2; 8.70% SC-CO_2+ 5% EtOH 5%), khusimol (30.49% SC-CO_2; 31.33% SC-CO_2+ 5% EtOH 5%), zizanoic acid (8.33% SC-CO_2; 6.82% SC-CO_2+ 5% EtOH 5%) and α-vetivone (6.42% SC-CO_2; 6.61% SC-CO_2+ 5% EtOH 5%). Leaf powder (20 g, dp = 0.42 mm) of *Ocimum sanctum* Linn. has been used for the extraction of eugenol (2.96 mg/g dry leaves) with SC-CO_2 at a flow rate of 2.5 L/min, 200 bar and 50 °C for 90 min (Chatterjee et al. 2017). Eugenol yield was affected only by pressure changes, and the kinetics of the extraction were reported to be first-order kinetics (Higuchi model) (Chatterjee et al. 2017). Eugenol (29.84%) has been extracted from clove at 40 °C and 220 bar with highest antioxidant activity of the extract found after extraction at 40 °C and 150 bar (Frohlich et al. 2019). In comparison to the Soxhlet extraction, it was revealed that the SC-CO_2 was more efficient in terms of yield, antioxidant activity, sample clean-up, reaction time and temperature (Frohlich et al. 2019).

Oleoresin has been optimally extracted from *Curcuma longa* (conventional turmeric) and *Curcuma amada* (mango ginger) at 65 °C and 350 bar for 150 min and at 40 °C and 300 bar for 30 min, respectively (Nagavekar and Singhal 2019). Modifier (30% ethanol) and pre-treatment with Stargen®002 enzyme significantly improved the yield (Nagavekar and Singhal 2019).

7.4 Artemisinin

The active pharmaceutical ingredient, artemisinin, is extracted from the herbaceous plant *Artemisia annua* L. (sweet wormwood) and is used in pharmaceutical applications for the treatment of malaria and cancer (Rodrigues et al. 2019). Artemisinin is a sesquiterpene that is highly oxygenated with a 1,2,4-trioxane ring structure (Brown 2010). Faurant (2011) have provided a historic background on the discovery of artemisinin and market-related developments, with Brown (2010) having provided a review on the photochemistry of the plant *A. annua* L. and the biosynthesis of the compound artemisinin. The SC-CO_2 extraction has commonly been employed in the high yield extraction of artemisinin from *A. annua* L. yielding high purity and clean extracts (Kohler et al. 1997; Quispe-Condori et al. 2005; Lin et al. 2006; Tzeng et al. 2007; Baldino et al. 2017; Martinez-Correa et al. 2017; Ciftzi et al. 2018; Rodrigues et al. 2019). Lapkin et al. (2006) have published a comparative analysis on existing conventional and green technologies such as SC-CO_2 for the extraction of artemisinin.

Pure, non-degraded extracts of artemisinin have been obtained with a flow rate of 2 ml/min of SC-CO_2/3% methanol at 50 °C and 15 MPa within 20 min (Kohler et al. 1997). Artemisinin yields of 0.62% and 0.70% have been achieved at conditions of 150 bar and 30 °C and 300 bar and 50 °C, respectively (Quispe-Condori et al. 2005). Artemisinin of high purity within a short reaction time was yielded at optimized conditions of 33 °C, 18.72 MPa, SC-CO_2/16.25 wt% of *n*-hexane after an extraction time of 1.5 h (Lin et al. 2006). The highest purity was obtained at 60 °C and 17.34 MPa and proved to be far better than when Soxhlet extraction (hexane) was applied (Lin et al. 2006). Martinez-Correa et al. (2017) reported the suitability of a two-step extraction procedure, starting with SC-CO_2 extraction (60 °C, 40 MPa), followed by further extractions with either ethanol (25 °C) or water (60 °C). Baldino et al. (2017) on the other hand reported the optimum procedure to be a one-step extraction at 40 °C and 100 bar for 600 min. Ciftzi et al. (2018) reported optimum conditions to be 33 °C and 30 MPa for yielding 1.09% (predicted yield). The cost-effectiveness of the SC-CO_2 extraction process of artemisinin as compared to the conventional extraction (ethanol) was evaluated by Rodrigues et al. (2019), which determined that this process is more attractive economically and viability could be achieved if reduced cost for the raw material could be achieved. An artemisinin content of 23.4% was reached at 50 °C and 200 bar for 60 min (Rodrigues et al. 2019). A reaction time of 180 min was needed to obtain 6% of artemisinin using conventional extraction with ethanol (Rodrigues et al. 2019).

7.5 Alkylamides

Alkylamides from the dried roots of *Echinacea angustifolia* were extracted by Sun et al. (2002) at highest concentration as compared to the fresh roots, by which the yield was positively affected by temperature and pressure.

7.6 Phenolics, Flavonoids, Chlorophylls and Carotenoids

Phenolics and flavonoids extracted from plants via the SC-CO_2 process have been reported by several researchers (Table 7.3). *Ampelopsis grossedentata* stems were used to optimally extract flavonoids and phenolics at 40 °C and 250 bar for 50 min with 1:3 v/v methanol/ethanol and 1:1 v/v methanol/ethanol, respectively (Wang et al. 2011). Pomelo peel was used for the extraction of flavonoids (2.37%) at 39 MPa, 80 °C, 85% ethanol for 49 min (He et al. 2012). SC-CO_2 extracted flavonoids had higher scavenging activities as compared to the conventional process (He et al. 2012). Increased concentration of total phenolic content and antioxidant activity were obtained with the addition of co-solvents (ethanol and ethyl acetate) to the

$SC\text{-}CO_2$ extraction process of the oil from *Eremanthus erythropappus* (candeia wood) (Santos et al. 2017). Total phenolics from radishes (*Raphanus sativus* L.) were reported at concentrations of 1375 mg GAE/100 g and 1455 mg GAE/100 g at operating conditions of 400 bar at 35 °C and 40 °C, respectively (Goyeneche et al. 2018). Escobedo-Flores et al. (2018) extracted polyphenols from *Avena sativa* L. (oats) obtaining maximum yields at 55 °C and 38 MPa. An optimum concentration of 1437.57 mg/g was predicted with the generated quadratic models (Escobedo-Flores et al. 2018). The effects of particle size on the concentrations of avenanthramides and phenolics extracted from medium oat bran, whole flour (WF), low bran and fine bran have been reported by Walters et al. (2018), with larger sizes presenting a limiting factor during $SC\text{-}CO_2$ extractions. Higher radical scavenging activities were observed with the extracts via the $SC\text{-}CO_2$ extraction of the defatted fraction of the fine particles (Walters et al. 2018). Avenanthramides have been reported to possess strong anti-inflammatory properties (Sur et al. 2008). The yield of total phenolics and total chlorophylls extracted from *Solidago gigantea* Ait. (goldenrod), a medicinal plant, has been evaluated using a "Box-Behnken design with three variables" studied, temperature (40–80 °C), pressure (20–80 MPa) and the flow rate of CO_2 (3–7 kg/h) (Wrona et al. 2019). The three variables studied had an effect on the yield of total phenolics and total chlorophylls with optimum conditions reported as temperature of 313.59 K and 352.22 K, pressure of 79.14 MPa and 74.59 MPa and CO_2 flow rate of 3.25 kg/h and 3.00 kg/h, respectively (Wrona et al. 2019). Increased temperature resulted in a decrease in the total phenolic content, but increased when pressure was increased (Wrona et al. 2019). Alvarez et al. (2019) determined that conditions of 40 MPa and 35 °C with the co-solvent ethanol were optimum or the extraction of polyphenols and flavonoids with highest antioxidant activity. Roselló-Soto et al. (2019a) affirmed the suitability of applying $SC\text{-}CO_2$ extraction of lipophilic phenolic compounds when compared to the conventional extraction. Isohydroxymatairesinol was extracted at highest concentrations at pressures of 30 MPa (756.22 ppb) and 40 MPa (1331.45 ppb). Increasing pressures improved the extraction of the phenolic compounds, including the antioxidant activity (Roselló-Soto et al. 2019a). A $SC\text{-}CO_2$ extraction process to obtain an extract rich in phenolic compounds from cacao (*Theobroma cacao*) pod husk was developed by Valadez-Carmona et al. (2018). The yield was influenced by pressure and co-solvent percentage, and a yield of 0.52% was obtained at optimum conditions of 299 bar, 60 °C and 13.7% ethanol with high selectivity towards antioxidants (Valadez-Carmona et al. 2018).

Carotenoids are pigments and the secondary metabolites of plants and some microorganisms (Zaghdoudi et al. 2016) and can be successfully extracted using $SC\text{-}CO_2$. Sovová et al. (2001) have published data on the effects of different temperatures and pressures with and without ethanol and vegetable oil on the solubility of β-carotene in $SC\text{-}CO_2$. Knowledge of the solubility of biomolecules such as the carotenoids in $SC\text{-}CO_2$ is necessary to develop appropriate $SC\text{-}CO_2$ extraction processes. De la Fuente et al. (2006) have determined the solubility of lycopene and astaxanthin at different temperatures and pressures. Similar to the behaviour of β-carotene in SC-CO_2, lycopene and astaxanthin solubility was greater with increased temperature (313–333 K) and constant pressure (30 MPa) as compared to increased pressure (30–50 MPa) at constant temperature (313 K) (de la Fuente et al. 2006). A review on the extraction of carotenoids from pumpkin (*Cucurbita* spp.) via the SC-CO_2 process and the influence temperature and pressure, pre-treatment effects, entrainers (modifier or co-solvents) and co-matrices on total carotenoid yield and carotenoid composition has been published by Durante et al. (2014). Choudhari and Singhal (2008) have extracted lycopene, a red-coloured tetraterpenic C40 carotenoid, from *Blakeslea trispora*, a zygomycete, at optimized conditions of 349 bar and 52 °C for 1.1 h and using an entrainer such as acetone, yielding 92%. Astaxanthin has been extracted from the microalgae *Haematococcus pluvialis* with the use of olive oil and soybean oil as co-solvents achieving a yield of 51.03% and 36.36%, respectively (Krichnavaruk et al. 2008). The microalgae *Haematococcus pluvialis* has been reported to be one of the greatest sources of the natural occurring astaxanthin, a carotenoid with potent antioxidant activity (Shah et al. 2016).

Xanthophylls (all-*trans*-lutein (15.46 μg/g), all-*trans*-zeaxanthin (16.81 μg/g) and all-*trans*-β-cryptoxanthin (33.23 μg/g)) have been optimally extracted from persimmon fruits (*Diospyros kaki* L.) at a flow rate of 3 ml/min, 300 bars, 60 °C and 25% (w/w) ethanol for 30 min obtaining higher yields as compared to the Soxhlet extraction method (Zaghdoudi et al. 2016). Conditions of flow rate of 1 ml/min, 100 bars, 40 °C and 25% (w/w) ethanol for 30 min were better suited for the extraction of 11.19 μg/g all-*trans*-β-carotene (Zaghdoudi et al. 2016). Spinach by-products have been used by Derrien et al. (2018) for the optimization of SC-CO_2 extraction of chlorophyll and lutein. Optimized conditions that resulted in a 72% and 50% yield of lutein and chlorophyll, respectively, were reported to be 39 MPa, 56 °C with a co-solvent of 10% ethanol for 3.6 h (Derrien et al. 2018). The SC-CO_2 extraction process of carotenoids from microalgae with the use of published data was described and modelled by Sovová and Stateva (2019). The model confirmed that higher temperatures and pressures increased yield of carotenoid in oil and extraction rate due to increased solubility of carotenoid in the supercritical fluid and the reduced capacity of adsorption of the microalga (Sovová and Stateva 2019). The phase equilibrium was found to be responsible in controlling the extraction process (Sovová and Stateva 2019).

7.7 Polysaccharides

Polysaccharides have been extracted via the SC-CO_2 process from the seeds of *Artemisia sphaerocephala* Krasch. at optimum conditions of temperature (extraction, 45 °C; separation, 56 °C), pressure (extraction, 45 MPa; separation, 10 MPa), a flow rate of 20 L/h for 2 h resulted in a yield of 18.59% (w/w) (Chen et al. 2014).

The 551.3 kDa polysaccharide was composed of the monosaccharides, mannose (10.8 mg/g), rhamnose (8.78 mg/g), galactose (9.86 mg/g), glucose (16.2 mg/g), arabinose (8.48 mg/g), xylose (38.48 mg/g) and fucose (10.09 mg/g) (Chen et al. 2014). The polysaccharide extracted from *Artemisia sphaerocephala* Krasch. has been reported to have medicinal applications (Xing et al. 2009). A polysaccharide (2.47%) from the leaves of bamboo (*Phyllostachys heterocycla*) has been extracted with SC-CO_2/ethanol modifier (30 ml) at optimized parameters of 50 °C and 40 MPa with a 2-h reaction time (Zou et al. 2018).

7.8 Tocopherols and Sterols

Plant cells synthesize α-, β-, γ- and δ-tocopherol, which are then stored in their seeds and leaves (Bendif et al. 2018). Tocopherols have strong antioxidant properties (Bendif et al. 2018). The content of tocopherols in extracted *Cannabis sativa* L. seed oil is significantly affected by temperature and pressure with higher temperatures and pressures resulting in a negative response (Aladić et al. 2015). Potential sources of tocopherol were reported by Bendif et al. (2018) to be contained in the SC-CO_2 extracts obtained from *Thymus munbyanus* subsp. *coloratus* (α-tocopherol [1580 μg/g], β-tocopherol [170 μg/g], γ-tocopherol [220 μg/g], δ-tocopherol [160 μg/g]) and *Thymus munbyanus* subsp. *munbyanus* (α-tocopherol [780 μg/g], β-tocopherol [140 μg/g], γ-tocopherol [120 μg/g], δ-tocopherol [130 μg/g]). The *Thymus* extracts were obtained with SC-CO_2 extraction with a flow rate of 2 L/min at 70 °C and 45 MPa for 210 min (Bendif et al. 2018). Sitosterol (1220 mg/kg seed) has been found to be contained in peach oil (extracted yield, 35.3 g/100 g seed) after SC-CO_2 extraction from the *Prunus persica* seeds (0.3 mm) at optimum conditions of 200 bar, flow rate of 7 ml/min and at 40 °C after 3 h (Ekinci and Gürü 2014).

7.9 Conclusions

A wide variety of biomolecules can be efficiently extracted with the SC-CO_2 extraction process. Optimization strategies are targeting to achieve parameters that can achieve high yield products with high purity and quality at less costs and environmental impacts. Research on the cost-effectiveness, economics and possible facility designs for the scaling up of the SC-CO_2 extraction process is underway with some successful examples already in place. The advent of the Green Chemistry era has made it possible to explore more opportunities in the development of innovative extraction and processing technologies.

References

Aladić K, Jarni K, Barbir T, Vidović S, Vladić J, Bilić M, Jokić S (2015) Supercritical CO_2 extraction of hemp (*Cannabis sativa* L.) seed oil. Ind Crops Prod 76:472–478. https://doi.org/10.1016/j.indcrop.2015.07.016

Alvarez MV, Cabred S, Ramirez CL, Fanovich MA (2019) Valorization of an agroindustrial soybean residue by supercritical fluid extraction of phytochemical compounds. J Supercrit Fluids 143:90–96. https://doi.org/10.1016/j.supflu.2018.07.012

Andrich G, Nesti U, Venturi F, Zinnai A, Fiorentini R (2005) Supercritical fluid extraction of bioactive lipids from the microalga *Nannochloropsis* sp. Eur J Lipid Sci Technol 107:381–386. https://doi.org/10.1002/ejlt.200501130

Aresta M, Dibenedetto A, Carone M, Colonna T, Fragale C (2005) Production of biodiesel from macroalgae by supercritical CO_2 extraction and thermochemical liquefaction. Environ Chem Lett 3(3):136–139. https://doi.org/10.1007/s10311-005-0020-3

Armenta S, Garrigues S, Esteve-Turrillas FA, de la Guardia M (2019) Green extraction techniques in green analytical chemistry. Trends Anal Chem 116:248–253. https://doi.org/10.1016/j.trac.2019.03.016

Baldino L, Reverchon E, Della Porta G (2017) An optimized process for SC-CO_2 extraction of antimalarial compounds from *Artemisia annua* L. J Supercrit Fluids 128:89–93. https://doi.org/10.1016/j.supflu.2017.05.018

Bendif H, Adouni K, Miara MD, Baranauskienė R, Kraujalis P, Venskutonis PR, Nabavi SM, Maggi F (2018) Essential oils (EOs), pressurized liquid extracts (PLE) and carbon dioxide supercritical fluid extracts (SFE-CO_2) from Algerian *Thymus munbyanus* as valuable sources of antioxidants to be used on an industrial level. Food Chem 260:289–298. https://doi.org/10.1016/j.foodchem.2018.03.108

Brown GD (2010) The biosynthesis of artemisinin (Qinghaosu) and the phytochemistry of *Artemisia annua* L. (Qinghao). Molecules (Basel, Switzerland) 15(11):7603–7698. https://doi.org/10.3390/molecules15117603

Certik M, Horenitzky R (1999) Supercritical CO_2 extraction of fungal oil containing γ-linolenic acid. Biotechnol Tech 13(1):11–15. https://doi.org/10.1023/A:1008853214591

Chatterjee D, Ghosh PK, Ghosh S, Bhattacharjee P (2017) Supercritical carbon dioxide extraction of eugenol from tulsi leaves: process optimization and packed bed characterization. Chem Eng Res Des 118:94–102. https://doi.org/10.1016/j.cherd.2016.11.025

Chávez-González ML, Rodríguez-Herrera R, Aguilar CN (2016) Essential oils: A natural alternative to combat antibiotics resistance. In: Kon K, Rai M (eds) Antibiotic resistance. Academic, Cambridge, pp 227–237. https://doi.org/10.1016/B978-0-12-803642-6.00011-3

Chemat F, Abert-Vian A, Cravotto G (2012) Green extraction of natural products: concept and principles. Int J Mol Sci 13:8615–8627. https://doi.org/10.3390/ijms13078615

Chemat F, Abert-Vian M, Fabiano-Tixier AS, Strube J, Uhlenbrock L, Gunjevic V, Cravotto G (2019) Green extraction of natural products. Origins, current status, and future challenges. Trends Anal Chem 118:248–263. https://doi.org/10.1016/j.trac.2019.05.037

Chen J, Li J, Sun AD, Zhang B-L, Qin S-G, Zhang Y-Q (2014) Supercritical CO_2 extraction and pre-column derivatization of polysaccharides from *Artemisia sphaerocephala* Krasch. Seeds via gas chromatography. Ind Crop Prod 60:138–143. https://doi.org/10.1016/j.indcrop.2014.06.013

Chen X, Huynh N, Cui H, Zhou P, Zhang X, Yang B (2018) Correlating supercritical fluid extraction parameters with volatile compounds from Finnish wild mushrooms (*Craterellus tubaeformis*) and yield prediction by partial least squares regression analysis. RSC Adv 8:5233. https://doi.org/10.1039/c7ra12472d

Cheung PCK (1999) Temperature and pressure effects on supercritical carbon dioxide extraction of n-3 fatty acids from red seaweed. Food Chem 65(3):399–403. https://doi.org/10.1016/S0308-8146(98)00210-6

Cheung PCK, Leung AYH, Ang PO Jr (1998) Comparison of supercritical carbon dioxide and Soxhlet extraction of lipids from a brown seaweed, *Sargassum hemiphyllum* (Turn.) C. Ag. J. Agric Food Chem 46:4228–4232. https://doi.org/10.1021/jf980346h

Choudhari SM, Singhal RS (2008) Supercritical carbon dioxide extraction of lycopene from mated cultures of *Blakeslea trispora* NRRL 2895 and 2896. J Food Eng 89(3):349–354. https://doi.org/10.1016/j.jfoodeng.2008.05.016

Ciftzi ON, Cahyadi J, Guigard SE, Saldaña MDA (2018) Optimization of artemisinin extraction from *Artemisia annua* L. with supercritical carbon dioxide + ethanol using response surface methodology. Electrophoresis 39:926–1933. https://doi.org/10.1002/elps.201800084

Cikoš A-M, Jokić S, Šubarić D, Jerković I (2018) Overview on the application of modern methods for the extraction of bioactive compounds from marine macroalgae. Mar Drugs 16:348. https://doi.org/10.3390/md16100348

Coelho JAP, Pereira AP, Mendes RL, Palavra AMF (2003) Supercritical carbon dioxide extraction of *Foeniculum vulgare* volatile oil. Flavour Fragr J 18:316–319. https://doi.org/10.1002/ffj.1223

Cordeiro RM, de S e Silva AP, RHH P, da Costa WA, da Silva SHM, de Souza Pinheiro WB, MSP A, Carvalho RN Jr (2019) Supercritical CO_2 extraction of ucuúba (*Virola surinamensis*) seed oil: global yield, kinetic data, fatty acid profile, and antimicrobial activities. Chem Eng Commun 206(1):86–97. https://doi.org/10.1080/00986445.2018.1474741

Cunha VMB, da Silva MP, de Sousa SHB, do Nascimento Bezerra P, EGO M, da Silva NJN, da Silva Banna DAD, Araújo ME, de Carvalho RN Jr (2019) Bacaba-de-leque (*Oenocarpus distichus* Mart.) oil extraction using supercritical CO_2 and bioactive compounds determination in the residual pulp. J Supercrit Fluids 144:81–90. https://doi.org/10.1016/j.supflu.2018.10.010

Da Porto C, Voinovich D, Decorti D, Natolino A (2012a) Response surface optimization of hemp seed (*Cannabis sativa* L.) oil yield and oxidation stability by supercritical carbon dioxide extraction. J Supercrit Fluids 68:45–51. https://doi.org/10.1016/j.supflu.2012.04.008

Da Porto C, Decorti D, Tubaro F (2012b) Fatty acid composition and oxidation stability of hemp (*Cannabis sativa* L.) seed oil extracted by supercritical carbon dioxide. Ind Crop Prod 36:401–404. https://doi.org/10.1016/j.indcrop.2011.09.015

Da Silva RPFF, Rocha-Santos TAP, Duarte AC (2016) Supercritical fluid extraction of bioactive compounds. Trends Analyt Chem 76:40–51. https://doi.org/10.1016/j.trac.2015.11.013

De Cássia C, Batista R, de Oliveira MS, Araújo ME, Rodrigues AMC, Botelho JRS, da Silva Souza Filho AP, Machado NT, Carvalho RN Jr (2016) Supercritical CO_2 extraction of açaí (*Euterpe oleracea*) berry oil: global yield, fatty acids, allelopathic activities, and determination of phenolic and anthocyanins total compounds in the residual pulp. J Supercrit Fluids 107:364–369. https://doi.org/10.1016/j.supflu.2015.10.006

De la Fuente JC, Oyarzún B, Quezada N, del Valle JM (2006) Solubility of carotenoid pigments (lycopene and astaxanthin) in supercritical carbon dioxide. Fluid Phase Equilib 247(1–2):90–95. https://doi.org/10.1016/j.fluid.2006.05.031

De Melo MMR, Silvestre AJD, Silva CM (2014) Supercritical fluid extraction of vegetable matrices: applications, trends and future perspectives of a convincing green technology. J Supercrit Fluids 92:115–176. https://doi.org/10.1016/j.supflu.2014.04.007

Dejoye C, Abert Vian M, Lumia G, Bouscarle C, Charton F, Chemat F (2011) Combined extraction processes of lipid from *Chlorella vulgaris* microalgae: microwave prior to supercritical carbon dioxide extraction. Int J Mol Sci 12:9332–9341. https://doi.org/10.3390/ijms12129332

Derrien M, Aghabararnejad M, Gosselin A, Desjardins Y, Angers P, Boumghar Y (2018) Optimization of supercritical carbon dioxide extraction of lutein and chlorophyll from spinach by-products using response surface methodology. LWT Food Sci Technol 93:79–87. https://doi.org/10.1016/j.lwt.2018.03.016

Durante M, Salvatore Lenucci M, Mita G (2014) Supercritical carbon dioxide extraction of carotenoids from pumpkin (*Cucurbita* spp.): a review. Int J Mol Sci 15:6725–6740. https://doi.org/10.3390/ijms15046725

Ekinci MS, Gürü M (2014) Extraction of oil and β-sitosterol from peach (*Prunus persica*) seeds using supercritical carbon dioxide. J Supercrit Fluids 92:319–323. https://doi.org/10.1016/j.supflu.2014.06.004

Escobedo-Flores Y, Chavez-Flores D, Salmeron I, Molina-Guerrero C, Perez-Vega S (2018) Optimization of supercritical fluid extraction of polyphenols from oats (*Avena sativa* L.) and their antioxidant activities. J Cereal Sci 80:198–204. https://doi.org/10.1016/j.jcs.2018.03.002

Faurant C (2011) From bark to weed: the history of artemisinin. Parasite (Paris, France) 18(3):215–218. https://doi.org/10.1051/parasite/2011183215

Fernández-Acosta K, Salmeron I, Chavez-Flores D, Perez-Reyes I, Ramos V, Ngadi M, Kwofie EM, Perez-Vega S (2019) Evaluation of different variables on the supercritical CO_2 extraction of oat (*Avena sativa* L.) oil; main fatty acids, polyphenols, and antioxidant content. J Cereal Sci 88:118–124. https://doi.org/10.1016/j.jcs.2019.05.017

Folch J, Lees M, Sloane Stanley GH (1957) A simple method for the isolation and purification of total lipides from animal tissues. J Biol Chem 226(1):497–509. http://www.jbc.org/content/226/1/497.long

Fornari T, Vicente G, Vázquez E, García-Risco MR, Reglero G (2012a) Isolation of essential oil from different plants and herbs by supercritical fluid extraction. J Chromatogr A 1250:34–48. https://doi.org/10.1016/j.chroma.2012.04.051

Fornari T, Ruiz-Rodriguez A, Vicente G, Vázquez E, García-Risco MR, Reglero G (2012b) Kinetic study of the supercritical CO_2 extraction of different plants from the *Lamiaceae* family. J Supercrit Fluids 64:1–8. https://doi.org/10.1016/j.supflu.2012.01.006

Frohlich PC, Santos KA, Palu F, Cardozo-Filho L, da Silva C, da Silva EA (2019) Evaluation of the effects of temperature and pressure on the extraction of eugenol from clove (*Syzygium aromaticum*) leaves using supercritical CO_2. J Supercrit Fluids 143:313–320. https://doi.org/10.1016/j.supflu.2018.09.009

Gallego R, Bueno M, Herrero M (2019) Sub- and supercritical fluid extraction of bioactive compounds from plants, food-by-products, seaweeds and microalgae-an update. Trends Anal Chem 116:198–213. https://doi.org/10.1016/j.trac.2019.04.030

García-Pérez JS, Cuéllar-Bermúdez SP, de la Cruz-Quiroz R, Arévalo-Gallegos A, Esquivel-Hernandez DA, Rodríguez-Rodríguez J, García-García R, Iqbal HMN, Parra-Saldivar R (2019) Supercritical CO_2-based tailor made valorization of *Origanum vulgare* L. extracts: a green approach to extract high-value compounds with applied perspectives. J Environ Manag 232:796–802. https://doi.org/10.1016/j.jenvman.2018.11.117

Goyeneche R, Fanovichc A, Rodrigues CR, Nicolao MC, Di Scala K (2018) Supercritical CO_2 extraction of bioactive compounds from radish leaves: yield, antioxidant capacity and cytotoxicity. J Supercrit Fluids 135:78–83. https://doi.org/10.1016/j.supflu.2018.01.004

Grosso C, Ferraro V, Figueiredo AC, Barroso JG, Coelho JA, Palavra AM (2008) Supercritical carbon dioxide extraction of volatile oil from Italian coriander seeds. Food Chem 111:197–203. https://doi.org/10.1016/j.foodchem.2008.03.031

Grosso C, Figueiredo AC, Burillo J, Mainar AM, Urieta JS, Barroso JG, Coelho JA, Palavra AM (2009a) Supercritical fluid extraction of the volatile oil from *Santolina chamaecyparissus*. J Sep Sci 32:3215–3222. (errata in J Separat Sci 32:3365–3366). https://doi.org/10.1002/jssc.200900350

Grosso C, Figueiredo AC, Burillo J, Mainar AM, Urieta JS, Barroso JG, Coelho JA, Palavra AMF (2009b) Enrichment of the thymoquinone content in volatile oil from *Satureja montana* using supercritical fluid extraction. J Sep Sci 32:328–334. https://doi.org/10.1002/jssc.200800490

Grosso C, Coelho JP, Pessoa FLP, Fareleira JMNA, Barroso JG, Urieta JS, Palavra AF, Sovova H (2010) Mathematical modelling of supercritical CO_2 extraction of volatile oils from aromatic plants. Chem Eng Sci 65:3579–3590. https://doi.org/10.1016/j.ces.2010.02.046

He J-Z, Shao P, Liu J-H, Ru Q-M (2012) Supercritical carbon dioxide extraction of flavonoids from pomelo (*Citrus grandis* (L.) Osbeck) peel and their antioxidant activity. Int J Mol Sci 13:13065–13078. https://doi.org/10.3390/ijms131013065

Herrero M, del Pilar S-CA, Cifuentes A, Ibáñez E (2015) Plants, seaweeds, microalgae and food by-products as natural sources of functional ingredients obtained using pressurized liquid extraction and supercritical fluid extraction. Trends Anal Chem 71:26–38. https://doi.org/10.1016/j.trac.2015.01.018

Huang Z, Shi X-H, Jiang W-J (2012) Theoretical models for supercritical fluid extraction. J Chromatogr A 1250:2–26. https://doi.org/10.1016/j.chroma.2012.04.032

Juchen PT, Araujo MN, Hamerski F, Corazza ML, Voll FAP (2019) Extraction of parboiled rice bran oil with supercritical CO_2 and ethanol as CO-solvent: kinetics and characterization. Ind Crops Prod 139:111506. https://doi.org/10.1016/j.indcrop.2019.111506

Kamatou GPP, Viljoen AM (2010) Review of the application and pharmacological properties of α-bisabolol and α-bisabolol-rich oils. J Am Oil Chem Soc 87(1):1–7. https://doi.org/10.1007/s11746-009-1483-3

Kerrihard AL, Pegg RB (2015) Utilizing the bioactive contents of specialty oils and fats. In: Talbot G (ed) Food science, technology and nutrition, specialty oils and fats in food and nutrition. Woodhead Publishing, Sawston, pp 317–348. https://doi.org/10.1016/B978-1-78242-376-8.00013-2

Khaw K-Y, Parat M-O, Shaw PN, Falconer JR (2017) Solvent supercritical fluid technologies to extract bioactive compounds from natural sources: a review. Molec 22:1186. https://doi.org/10.3390/molecules22071186

Knez Ž, Škerget M, KnezHrnčič M (2013) Principles of supercritical fluid extraction and applications in the food, beverage and nutraceutical industries. In: Rizvi SSH (ed) Food science, technology and nutrition, separation, extraction and concentration processes in the food, beverage and nutraceutical industries. Woodhead Publishing, Sawston, pp 3–38. https://doi.org/10.1533/9780857090751.1.3

Knez Z, Pantić M, Cör D, Novak Z, Knez Hrnčič M (2019) Are supercritical fluids solvents for the future? Chem Eng Process Process Intensification 141:107532. https://doi.org/10.1016/j.cep.2019.107532

Kohler M, Haerdi W, Christen P, Veuthey J-L (1997) Extraction of artemisinin and artemisinic acid from *Artemisia annua* L. using supercritical carbon dioxide. J Chromatogr A 785(1–2):353–360. https://doi.org/10.1016/S0021-9673(97)00403-2

Koubaa M, Barba FJ, Mhemdi H, Grimi N, Koubaa W, Vorobiev E (2015) Gas assisted mechanical expression (GAME) as a promising technology for oil and phenolic compound recovery from tiger nuts. Innov Food Sci Emerg Technol 32:172–180. https://doi.org/10.1016/j.ifset.2015.09.019

Krichnavaruk S, Shotipruk A, Goto M, Pavasant P (2008) Supercritical carbon dioxide extraction of astaxanthin from *Haematococcus pluvialis* with vegetable oils as cosolvent. Bioresour Technol 99(13):5556–5560. https://doi.org/10.1016/j.biortech.2007.10.049

Kumhom T, Elkame A, Douglas PL, Douglas S, Pongamphai S, Teppaitoon W (2011) Prediction of isoflavone extraction from soybean meal using supercritical carbon dioxide with cosolvents. Chem Eng J 172:1023–1032. https://doi.org/10.1016/j.cej.2011.07.019

Lang Q, Wai CM (2001) Supercritical fluid extraction in herbal and natural product studies-a practical review. Talanta 53:771–782. https://doi.org/10.1016/s0039-9140(00)00557-9

Lapkin AA, Plucinski PK, Cutler M (2006) Comparative assessment of technologies for extraction of artemisinin. J Nat Prod 69:1653–1664. https://doi.org/10.1021/np060375j

Lasekan O, Abdulkarim SM (2012) Extraction of oil from tiger nut (*Cyperus esculentus* L.) with supercritical carbon dioxide (SC-CO_2). LWT Food Sci Technol 47(2):287–292. https://doi.org/10.1016/j.lwt.2012.01.021

Li Y, Naghdi FG, Garg S, Adarme-Vega TC, Thurecht KJ, Ghafor WA, Tannock S, Schenk PM (2014) A comparative study: the impact of different lipid extraction methods on current microalgal lipid research microbial cell factories. Microb Cell Factories 13:1–9. https://doi.org/10.1186/1475-2859-13-14

Lin YL, Yang CC, Hsu HK, Hsu SL, Chang CMJ (2006) Response surface methodology to supercritical fluids extraction of artemisinin and the effects on rat hepatic stellate cell in vitro. J Supercrit Fluids 39:48–53. https://doi.org/10.1016/j.supflu.2006.02.012

Liu J, Ji F, Chen F, Guo W, Yang M, Huang S, Zhang F, Liu Y (2018) Determination of garlic phenolic compounds using supercritical fluid extraction coupled to supercritical fluid chromatography/tandem mass spectrometry. J Pharm Biomed Anal 159:513–523. https://doi.org/10.1016/j.jpba.2018.07.020

López-Padilla A, Ruiz-Rodriguez A, Reglero G, Fornari T (2017) Supercritical carbon dioxide extraction of *Calendula officinalis*: kinetic modeling and scaling up study. J Supercrit Fluids 130:292–300. https://doi.org/10.1016/j.supflu.2017.03.033

Martínez J, de Aguiar AC (2014) Extraction of triacylglycerols and fatty acids using supercritical fluids - review. Curr Anal Chem 10:67–77. https://doi.org/10.2174/1573411011410010006

Martinez-Correa HA, Bitencourt RG, Kayano ACAV, Magalhães PM, Costa FTM, Cabral FA (2017) Integrated extraction process to obtain bioactive extracts of *Artemisia annua* L. leaves using supercritical CO_2, ethanol and water. Ind Crop Prod 95:535–542. https://doi.org/10.1016/j.indcrop.2016.11.007

Meireles MAA (2003) Supercritical extraction from solid: process design data (2001–2003). Curr Opin Solid State Mater Sci 7(4–5):321–330. https://doi.org/10.1016/j.cossms.2003.10.008

Mendes RL, Fernandes HL, Coelho JP, Reis EC, Cabral JMS, Novais JM, Palavra AF (1995) Supercritical CO_2 extraction of carotenoids and other lipids from *Chlorella vulgaris*. Food Chem 53:99–103. https://doi.org/10.1016/0308-8146(95)95794-7

Mendes RL, Nobre BP, Cardoso MT, Pereira AP, Palavra AF (2003) Supercritical carbon dioxide extraction of compounds with pharmaceutical importance from microalgae. Inorg Chim Acta 356:328–334. https://doi.org/10.1016/S0020-1693(03)00363-3

Michalak I, Chojnacka K, Saeid A (2017) Plant growth biostimulants, dietary feed supplements and cosmetics formulated with supercritical CO_2 algal extracts. Molecules 22:66. https://doi.org/10.3390/molecules22010066

Modey WK, Mulholland DA, Raynor MW (1996) Analytical supercritical fluid extraction of natural products. Phytochem Anal 7:1–15. https://doi.org/10.1002/(SICI)1099-1565(199601)7:1<1::AID-PCA275>3.0.CO;2-U

Mouahid A, Crampon C, Toudji S-A A, Badens E (2016) Effects of high water content and drying pre-treatment on supercritical CO_2 extraction from *Dunaliella salina* microalgae: experiments and modelling. J Supercrit Fluids 116:271–280. https://doi.org/10.1016/j.supflu.2016.06.007

Mukhopadhyay M (2000) Natural extracts using supercritical carbon dioxide. CRC Press, Boca Raton. https://doi.org/10.1201/9781420041699

Nagavekar N, Singhal RS (2019) Supercritical fluid extraction of *Curcuma longa* and *Curcuma amada* oleoresin: optimization of extraction conditions, extract profiling, and comparison of bioactivities. Ind Crops Prod 134:134–145. https://doi.org/10.1016/j.indcrop.2019.03.061

Natolino A, Da Porto C (2019) Supercritical carbon dioxide extraction of pomegranate (*Punica granatum* L.) seed oil: kinetic modelling and solubility evaluation. J Supercrit Fluids 151:30–39. https://doi.org/10.1016/j.supflu.2019.05.002

Özkal SG, Yener ME, Bayindirli L (2005) Mass transfer modelling of apricot kernel oil extraction with supercritical carbon dioxide. J Supercrit Fluids 35:119–127. https://doi.org/10.1016/j.supflu.2004.12.011

Pereira CG, Meireles MAA (2010) Supercritical fluid extraction of bioactive compounds: fundamentals, applications and economic perspectives. Food Bioprocess Technol 3:340–372. https://doi.org/10.1007/s11947-009-0263-2

Perrut M, Clavier JY, Poletto M, Reverchon E (1997) Mathematical modeling of sunflower seed extraction by supercritical CO_2. Ind Eng Chem Res 36:430–435. https://doi.org/10.1021/ie960354s

Pinto RHH, Sena C, Santos OV, Costa WA, Rodrigues AMC, Carvalho RN Jr (2018) Extraction of bacaba (*Oenocarpus bacaba*) oil with supercritical CO_2: global yield isotherms, fatty acid composition, functional quality, oxidative stability, spectroscopic profile and antioxidant activity. Grasas Aceites 69(2):1–8. https://doi.org/10.3989/gya.0883171

Pourmortazavi SM, Hajimirsadeghi SS (2007) Supercritical fluid extraction in plant essential and volatile oil analysis. J Chromatogr A 1163(2):2–24. https://doi.org/10.1016/j.chroma.2007.06.021

Quispe-Condori S, Sánchez D, Foglio MA, Rosa PTV, Zetzl C, Brunner G, Meireles MAA (2005) Global yield isotherms and kinetic of artemisinin extraction from *Artemisia annua* L. leaves using supercritical carbon dioxide. J Supercrit Fluids 36:40–48. https://doi.org/10.1016/j.supflu.2005.03.003

Rai A, Mohanty B, Bhargava R (2016) Supercritical extraction of sunflower oil: a central composite design for extraction variables. Food Chem 192:647–659. https://doi.org/10.1016/j.foodchem.2015.07.070

Reverchon E, De Marco I (2006) Supercritical fluid extraction and fractionation of natural matter. J Supercrit Fluids 38:146–166. https://doi.org/10.1016/j.supflu.2006.03.020

Rodrigues MFF, Sousa IMO, Vardanega R, Nogueira GC, Meireles MAA, Foglio MA, Marchese JA (2019) Techno-economic evaluation of artemisinin extraction from *Artemisia annua* L. using supercritical carbon dioxide. Ind Crops Prod 132:336–343. https://doi.org/10.1016/j.indcrop.2019.02.049

Rombaut N, Savoire R, Thomasset B, Bélliard T, Castello J, Van Hecke É, Lanoisellé J-L (2014) Grape seed oil extraction: interest of supercritical fluid extraction and gas-assisted mechanical extraction for enhancing polyphenol co-extraction in oil. C R Chim 17(3):284–292. https://doi.org/10.1016/j.crci.2013.11.014

Roselló-Soto E, Barba FJ, Lorenzo JM, Dominguez R, Pateiro M, Mañes J, Moltó JC (2019a) Evaluating the impact of supercritical-CO_2 pressure on the recovery and quality of oil from "horchata" by-products: fatty acid profile, α-tocopherol, phenolic compounds and lipid oxidation parameters. Food Res Int 120:888–894. https://doi.org/10.1016/j.foodres.2018.11.054

Roselló-Soto E, Barba FJ, Lorenzo JM, Munekata PES, Gómez B, Moltó JC (2019b) Phenolic profile of oils obtained from "horchata" by-products assisted by supercritical-CO_2 and its relationship with antioxidant and lipid oxidation parameters: triple TOF-LC-MS-MS characterization. Food Chem 274:865–871. https://doi.org/10.1016/j.foodchem.2018.09.055

Rovetto LJ, Aieta NV (2017) Supercritical carbon dioxide extraction of cannabinoids from *Cannabis sativa* L. J Supercrit Fluids 129:16–27. https://doi.org/10.1016/j.supflu.2017.03.014

Ruttarattanamongkol K, Siebenhandl-Ehn S, Schreiner M, Petrasch AM (2014) Pilot-scale supercritical carbon dioxide extraction, physico-chemical properties and profile characterization of *Moringa oleifera* seed oil in comparison with conventional extraction methods. Ind Crops Prod 58:68–77. https://doi.org/10.1016/j.indcrop.2014.03.020

Sánchez-Camargo AP, Parada-Alonso F, Ibáñez E, Cifuentes A (2019) Recent applications of on-line supercritical fluid extraction coupled to advanced analytical techniques for compounds extraction and identification. J Sep Sci 42:243–257. https://doi.org/10.1002/jssc.201800729

Santos KA, Frohlich PC, Hoscheid J, Tiuman TS, Gonçalves JE, Cardozo-Filho L, da Silva EA (2017) Candeia (*Eremanthus erythroppapus*) oil extraction using supercritical CO_2 with ethanol and ethyl acetate cosolvents. J Supercrit Fluids 128:323–330. https://doi.org/10.1016/j.supflu.2017.03.029

Santos KA, Klein EJ, da Silva C, da Silva EA, Cardozo-Filho L (2019) Extraction of vetiver (*Chrysopogon zizanioides*) root oil by supercritical CO_2, pressurized-liquid, and ultrasound-assisted methods and modelling of supercritical extraction kinetics. J Supercrit Fluids 150:30–39. https://doi.org/10.1016/j.supflu.2019.04.005

Shah MM, Liang Y, Cheng JJ, Daroch M (2016) Astaxanthin-producing green microalga *Haematococcus pluvialis*: from single cell to high value commercial products. Front Plant Sci 7:531. https://doi.org/10.3389/fpls.2016.00531

Shukla A, Naik SN, Goud VV, Das C (2019) Supercritical CO_2 extraction and online fractionation of dry ginger for production of high-quality volatile oil and gingerols enriched oleoresin. Ind Crops Prod 130:352–362. https://doi.org/10.1016/j.indcrop.2019.01.005

Silva MP, Cunha VMB, Sousa SHB, Menezes EGO, do Nascimento Bezerra P, de Farias Neto JT, GNR F, Araújo ME, de Carvalho RN Jr (2019) Supercritical CO_2 extraction of lyophilized Açaí

(*Euterpe oleracea* Mart.) pulp oil from three municipalities in the state of Pará, Brazil. J CO2 Utilization 31:226–234. https://doi.org/10.1016/j.jcou.2019.03.019
Sovová H (2012) Modeling the supercritical fluid extraction of essential oils from plant materials. J Chromatogr A 1250:27–33. https://doi.org/10.1016/j.chroma.2012.05.014
Sovová H (2017) Broken-and-intact cell model for supercritical fluid extraction: its origin and limits. J Supercrit Fluids 129:3–8. https://doi.org/10.1016/j.supflu.2017.02.014
Sovová H, Stateva RP (2019) New developments in the modelling of carotenoids extraction from microalgae with supercritical CO_2. J Supercrit Fluids 148:93–103. https://doi.org/10.1016/j.supflu.2019.03.002
Sovová H, Stateva RP, Galushko AA (2001) Solubility of beta-carotene in supercritical CO_2 and the effect of entrainers. J Supercrit Fluids 21(3):195–203. https://doi.org/10.1016/S0896-8446(01)00101-2
Sovová H, Nobre BP, Palavra A (2016) Modeling of the kinetics of supercritical fluid extraction of lipids from microalgae with emphasis on extract desorption. Materials 9(6):423–441. https://doi.org/10.3390/ma9060423
Stuart IA, MacLachlan J, McNaughtan A (1996) Compounds of agricultural significance using environmental analytical supercritical fluid extraction. A review. Analyst 121 (IIR-28R):11R. https://doi.org/10.1039/AN996210011R
Sun L, Rezaei KA, Temelli F, Ooraikul B (2002) Supercritical fluid extraction of alkylamides from *Echinacea angustifolia*. J Agric Food Chem 50:3947–3953. https://doi.org/10.1021/jf0200265
Sur R, Nigam A, Grote D, Liebel F, Southall MD (2008) Avenanthramides, polyphenols from oats, exhibit anti-inflammatory and anti-itch activity. Arch Dermatol Res 300(10):569–574. https://doi.org/10.1007/s00403-008-0858-x
Tongnuanchan P, Benjakul S (2014) Essential oils: extraction, bioactivities, and their uses for food preservation. J Food Sci 79(7):R1231–R1249. https://doi.org/10.1111/1750-3841.12492
Tzeng TC, Lin YL, Jong TT, Chang CMJ (2007) Ethanol modified supercritical fluids extraction of scopoletin and artemisinin from *Artemisia annua* L. Sep Purif Technol 56:18–24. https://doi.org/10.1016/j.seppur.2007.01.010
Valadez-Carmona L, Ortiz-Moreno A, Ceballos-Reyes G, Mendiola JA, Ibáñez E (2018) Valorization of cacao pod husk through supercritical fluid extraction of phenolic compounds. J Supercrit Fluids 131:99–105. https://doi.org/10.1016/j.supflu.2017.09.011
Veggi PC, Cavalcanti RN, Meireles MAA (2014a) Production of phenolic-rich extracts from Brazilian plants using supercritical and subcritical fluid extraction: experimental data and economic evaluation. J Food Eng 131:96–109. https://doi.org/10.1016/j.jfoodeng.2014.01.027
Veggi PC, Prado JM, Bataglion GA, Eberlin MN, Meireles MAA (2014b) Obtaining phenolic compounds from jatoba (*Hymenaea courbaril* L.) bark by supercritical fluid extraction. J Supercrit Fluid 89:68–77. https://doi.org/10.1016/j.supflu.2014.02.016
Walters M, Ribeiro APL, Hosseinian F, Tsopmo A (2018) Phenolic acids, avenanthramides, and antioxidant activity of oats defatted with hexane or supercritical fluid. J Cereal Sci 79:21–26. https://doi.org/10.1016/j.jcs.2017.09.010
Wang Y, Ying L, Sun D, Zhang S, Zhu Y, Xu P (2011) Supercritical carbon dioxide extraction of bioactive compounds from *Ampelopsis grossedentata* stems: process optimization and antioxidant activity. Int J Mol Sci 12:6856–6870. https://doi.org/10.3390/ijms12106856
Wrona O, Rafińska K, Możeński C, Buszewski B (2017) Supercritical fluid extraction of bioactive compounds from plant materials. J AOAC Int 100:1624–1635. https://doi.org/10.5740/jaoacint.17.0232
Wrona O, Rafińska K, Możeński C, Buszewski B (2019) Supercritical carbon dioxide extraction of *Solidago gigantea* Ait.: optimization at quarter technical scale and scale up the process to half technical plant. Ind Crops Prod 130:316–324. https://doi.org/10.1016/j.indcrop.2018.12.050
Xing XH, Zhang ZM, Hu XZ, Wu RQ, Xu C (2009) Antidiabetic effects of *Artemisia sphaerocephala* Krasch gum, a novel food additive in China, on streptozotocin-induced type 2 diabetic rats. J Ethnopharmacol 125(3):410–416. https://doi.org/10.1016/j.jep.2009.07.021

Yousefi M, Rahimi-Nasrabadi M, Pourmortazavi SM, Wysokowski M, Jesionowski T, Ehrlich H, Mirsadeghi S (2019) Supercritical fluid extraction of essential oils. Trends Anal Chem 118:182–193. https://doi.org/10.1016/j.trac.2019.05.038

Zaghdoudi K, Framboisier X, Frochot C, Vanderesse R, Barth D, Kalthoum-Cherif J, Blanchard F, Guiavarćh Y (2016) Response surface methodology applied to supercritical fluid extraction (SFE) of carotenoids from persimmon (*Diospyros kaki* L.). Food Chem 208:209–219. https://doi.org/10.1016/j.foodchem.2016.03.104

Zou X, Liu Y, Tao C, Liu Y, Liu M, Wu J, Lv Z (2018) CO_2 supercritical fluid extraction and characterization of polysaccharide from bamboo (*Phyllostachys heterocycla*) leaves. J Food Meas Charact 12(1):35–44. https://doi.org/10.1007/s11694-017-9614-2

Zulkafli ZD, Wang H, Miyashita F, Utsumi N, Tamura K (2014) Cosolvent-modified supercritical carbon dioxide extraction of phenolic compounds from bamboo leaves (*Sasa palmata*). J Supercrit Fluids 94:123–129. https://doi.org/10.1016/j.supflu.2014.07.008

Chapter 8
Chemistry of Ionic Liquid, Switchable Solvents, Supercritical Carbon Dioxide and Sub/Supercritical Water

Ederson R. Abaide, Cristiano C. Muller, Crisleine P. Draszewski, Marcus V. Tres, Marcio A. Mazutti, and Giovani L. Zabot

Contents

8.1 Introduction

The concern of industry and researchers with the environment is crescent, mainly due to the alarming quantity of greenhouse gas emissions. Global greenhouse gas emissions in 2017 reached 49.2 billion tons of carbon dioxide (CO_2) equivalent, which is the highest value recorded since the initial evaluation (Kuramochi et al.

E. R. Abaide · M. V. Tres · G. L. Zabot (✉)
Laboratory of Agroindustrial Processes Engineering (LAPE), Federal University of Santa Maria (UFSM), Cachoeira do Sul, RS, Brazil
e-mail: giovani.zabot@ufsm.br

C. C. Muller
Prati Donaduzzi & Cia. Ltda., Pharmaceutical Industry, Toledo, PR, Brazil

C. P. Draszewski · M. A. Mazutti
Department of Chemical Engineering, Federal University of Santa Maria (UFSM), Santa Maria, RS, Brazil

Inamuddin, A. M. Asiri (eds.), *Advanced Nanotechnology and Application of Supercritical Fluids*, Nanotechnology in the Life Sciences,
https://doi.org/10.1007/978-3-030-44984-1_8

2018). Carbon dioxide emissions can be considered the worse problem, with its emission corresponding to approximately 75% of greenhouse gas emissions (Ritchie and Roser 2017).

Many industries are using biofuels and other alternatives for replacing the current fuels. The use of biofuels instead of hydrocarbons is not enough for satisfying a completely sustainable process because most of these processes require raw material pretreatment. Among the various stages of a process, some pretreatments generate more polluting agents, since they require the use of large volumes of organic solvents or inorganic acids and a high energy consumption (Tobiszewski et al. 2009). Many pretreatments are used in biorefineries, which cost up to 40% of the total cost of biofuels and are responsible for the higher energy expenditure of the process, further increasing the challenge of using green solvents that must be energetically efficient without increasing the financial value invested in the process (Sindhu et al. 2016).

Besides working as a reaction medium in the production of biofuels, most of the solvents are used in extraction and separation processes. The processes produce considerable amounts of wastes when they are conducted at industrial scale, thus representing a matter of concern. Also, several conventional solvents cause problems to human health and the environment, contributing to the pollution of land, water, and air (Herrero et al. 2015).

Extraction processes of compounds from natural sources are fulfilled through an exhaustive extraction, which requires chemical purification steps to separate the desired molecule. Conventional extraction processes are long and use a lot of solvent volumes, which may cause a total or partial degradation of the target molecules and loss of some volatiles. Despite the high amount of energy and solvents used, the yield is often low (Chemat et al. 2012).

The petroleum industry is one of the most energy-consuming chemical industries, and it consumes approximately 50% of the energy in the separation stage. This consumption occurs mainly in the fractionation of oil that takes place in distillation equipment. Although effective, distillations have high energy consumption. Therefore, solvent-based separations can be an option to replace the distillation columns (Kiss et al. 2016).

All processes that use solvents are based on the yield increase, number of cycles, and reduction of extraction time and solvents' consumption. The process responses depend on operational conditions and properties of solvents (Pollet et al. 2014). Therefore, green solvents used in sustainable processes should be evaluated by a large extent as an alternative to conventional solvents aiming at optimizing operational conditions, reducing time, increasing yield, and minimizing energy consumption. Green solvents which will be highlighted in this book chapter are ionic liquid, switchable solvents, sub/supercritical water, and supercritical carbon dioxide.

Based on this context, green solvents can be considered as a sustainable option to replace commonly used industrial solvents that are harmful to human health and the environment. However, the use of green solvents depends on the desired properties within an operational range at an industrial scale. Furthermore, depending on the technology, the energetic and economic responses are different. Therefore, a reasonable evaluation based on theoretical and experimental data must be done to choose the best green solvents. In this way, this book chapter presents a critical and constructive overview of using green solvents in sustainable processes, highlighting the fundamentals of green solvents and recent studies that have used environmentally friendly solvents.

8.2 Fundamentals

The wider use of environmentally friendly technologies demonstrates that many industries are concerned about the negative environmental impact their activities may have on the environment. There is a need to design and apply more sustainable industrial operations using green solvents (Sunol and Sunol 2014; Zabot et al. 2018). The green solvents, as switchable solvents, ionic liquids, water in sub/supercritical state, and carbon dioxide in the supercritical state have been extensively used in chemical researches, such as energy storage, carbon capture, and reaction and extraction processes, among others. Therefore, knowing the fundamentals of these green solvents is interesting in order to use their potentialities in different applications.

8.2.1 Ionic Liquid

In a gross term, ionic liquids can be described as low-temperature molten salts, usually composed of asymmetric organic cations and inorganic anions that are fluid at relatively low temperature, from room temperature, near to 25 °C, to 100 °C (Amarasekara 2016). Ionic liquids are generally low-melting organic salts, in some cases below 25 °C, quite different from commonly known inorganic salts. Several applications are utilizing ionic liquids such as in energy production and storage (MacFarlane et al. 2016), for lignocellulosic biomass pretreatment (Halder et al. 2019), organic synthesis and catalysis (Vekariya 2017), and extraction and separation processes. The cations and anions, which form the ionic liquids, can be combined in several ways, and this can enhance their physical-chemical properties (Fig. 8.1). Therefore, it is possible to change the properties of these compounds to attain the desired application (Sunol and Sunol 2014). The properties of such solvents may be altered according to the ions and structure.

Components mainly based on chloroaluminate (III) or halide anions were the first generation of ionic liquid reported in 1982. However, no further attention was given to them due to their instability. The second generation of ionic liquids appeared in 1992, using the tetrafluoroborate ion as an anion (Sunol and Sunol 2014). In the early 2000s, it was introduced the third generation of ionic liquids (Davis 2004).

Ionic liquids have some special properties, such as high viscosity, low volatility, high conductivity, and long thermal stability. The designable properties of ionic

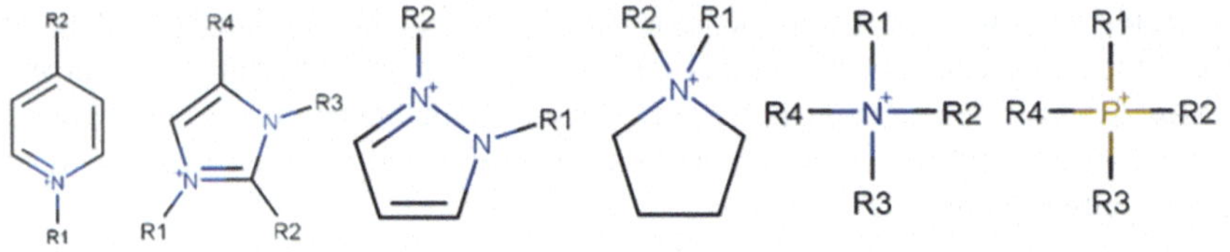

Pyridinium Imidazolium Pyrazolium Pyrrolidinium Ammonium Phosphonium

Cl^-, Br^-, I^- $Al_2Cl_7^-$, $Al_2Cl_{10}^-$ NO_3^-, PO_4^-, HSO_4^-, SO_4^- $(CF_3SO_2)_2N^-$, $N(CN)_2^-$, $(CF_3SO_2)_3C^-$

$Sb_2F_{11}^-$, $Fe_2Cl_7^-$, $Zn_2Cl_5^-$, $Zn_3Cl_7^-$ PF_6^-, SbF_6^-, BF_4^- BR_4^-, $RCB_{11}H_{11}^-$

$CuCl_2^-$, $SnCl_2^-$ $CF_3SO_3^-$, $ROSO_3^-$, $CF_3CO_2^-$, $C_6H_5SO_3^-$

Fig. 8.1 Ionic liquid molecules with organic cationic structures and possible counterions

liquids have promoted them to be an alternative to conventional solvents as a medium in reactions and green separation processes. For the first generation of ionic liquids, the chloroaluminate is highlighted. The percentage of aluminum (III) chloride can determine the physicochemical properties of ionic liquids. The fluids that have aluminum (III) chloride higher than 50 mol% are referred to as acidic, while the others are referred to as basic. The exactly 50 mol% composition is referred to as neutral (Kenneth 1997).

The second generation of ionic liquids consists of 1-ethyl-3-methylimidazolium as cation and tetrafluoroborate and hexafluorophosphate as anions, with the advantage that they could be prepared in an inert atmosphere. These ionic liquids, when dried under vacuum at temperatures in the order of 150 °C, can have water content lower than 1 mg/L (Endres and Frank 2006).

The task-specific ionic liquids, termed as third-generation ionic liquids, are fluids that incorporate functional groups designed to confer properties or reactivities. The polarity of a compound can be increased by the inclusion of certain functional groups, as with 1-methyl-3-propylimidazolium bis(triflyl)-imide when an ether or alcohol group is included in its structure. Catalytic activation or increased covalent binding capacity of salt may be altered due to changes in functional groups. Depending on specific uses, the functional groups can vary, including urea, thiourea, and a metal complex (Davis 2004).

Another important property that can be observed is the viscosity since modifications of functional groups in an ionic liquid can drastically change its value. At room temperature, approximately 25 °C, ionic liquids present viscosities from 10 to 40,000 mPa s (Joglekar et al. 2006). This information is essential because it is possible to calculate the fluid flow through the pumping and mixing systems.

Density is another important property that is particularly used in the design and volumetric sizing of equipment and for flow calculations. In general, the density of ionic liquids is high due to the presence of heavy anions. For pure ionic liquids, the densities vary depending on the choice of anion and cation. Typically, the values range from 1.01 to 1.57 g/cm^3 at 90 °C (Sunol and Sunol 2014).

8.2.2 Switchable Solvents

According to Jessop et al. (2005), the first switchable solvents were reported in the early 2000s. Thereafter, promising applications have been developed. Switchable solvents can be classified as switchable polarity solvents, switchable hydrophilicity solvents, and switchable water (Jessop et al. 2012). Therefore, this section presents an overview of switchable solvents in order to expose the fundamentals of switchable polarity solvents, switchable hydrophilicity solvents, and switchable water.

8.2.2.1 Switchable Polarity Solvents

Mixing an amine and alcohol base can form switchable polarity solvents. Such mixing can suitably occur by the addition of carbon dioxide in the gaseous form at pressure and temperature of approximately 0.1 MPa and 25 °C, forming alkyl mono-carbonate, which instantly produces a salt (Fig. 8.2). If nitrogen gas is subsequently added, the amine-alcohol mixture is obtained again (Jessop 2015).

The different properties obtained in systems using switchable polarity solvents can be a great advantage. An example is an amine-alcohol mixture, which is miscible in many conventional organic solvents. When added to switchable polarity solvents, it has properties that provide heterogeneous phase separation systems. This is an interesting feature, especially for extraction processes. Switchable polarity solvents have been employed for carbon dioxide capture and process separation and as reaction medium (Marıa 2014).

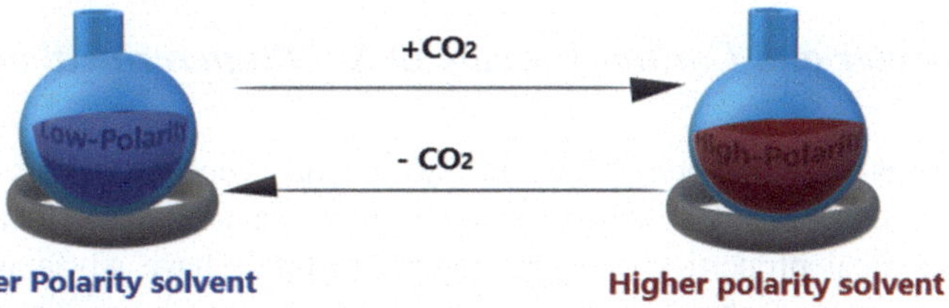

Fig. 8.2 Representative scheme of polarity change by the addition or removal of carbon dioxide, CO_2, in switchable polarity solvents

8.2.2.2 Switchable Hydrophilicity Solvents

Concerning switchable hydrophilicity solvent liquids, it is interesting to point out that they can alternate between hydrophilic and hydrophobic states, only with the addition of water, forming a biphasic mixture. An example of switchable hydrophilicity is *N,N,N*-tributylpentanamidine which is an amidine with several tertiary amines (Jessop et al. 2011).

Switchable hydrophilicity solvents can be used in situations involving the separation of the desired compound from the reaction medium. Normally, distillation, which requires much energy, increases the amount of solvents to be separated afterward. However, if the solvent is a hydrophobic switchable hydrophilicity solvent, it can be separated from the desired compound using simple extraction with water and dissolved carbon dioxide, which is then separated from water by removal of carbon dioxide, creating a biphasic mixture (Jessop et al. 2012).

8.2.2.3 Switchable Water

Switchable water has the ability to modify its ionic strength by selecting the conditions. More precisely, switchable water can be defined as an aqueous solution that under a stimulus can change its ionic strength. A solution containing amine or polyamine, for example, has its high ionic strength when in contact with carbon dioxide at 101.325 kPa (Jessop et al. 2012).

Problems of organic synthesis and catalysis can be solved using switchable water. In homogeneous catalysis, mixtures of biphasic solvents as organic and aqueous phases require the separation of product and the reaction medium. A reaction is difficult when reagents are poorly soluble in aqueous phase. This can be avoided by performing a reaction in a homogeneous medium and subsequently using a trigger to cause phase separation (Mercer et al. 2010).

8.2.3 Water and Carbon Dioxide in Sub/Supercritical State

New products with high purity and more sustainable processes can be obtained and designed by using technologies that involve subcritical and supercritical fluids. The physicochemical properties of sub/supercritical fluids have advantages for their wide use in various technological processes. Substances with pressure and temperature above their critical values are defined as supercritical fluids, whereas those above saturation pressure but below their critical values are defined as subcritical fluids (Knez et al. 2019).

Fluid in the supercritical state has no phase/liquid/gas distinction and is able to easily alternate its properties as a function of changes in pressure or temperature. Therefore, properties such as viscosity, diffusivity, and electrical constant can be intentionally modified according to the need of the process (Montoya et al. 2019).

Sub/supercritical water and carbon dioxide are widely used solvents, as well as having their properties easily alternated, do not harm human health, do not burn, and are easy to thermodynamically operate due to their high stability (Amaral et al. 2017).

Water in different thermodynamic states, when analyzed for temperature, can be classified as high-temperature water from room temperature, approximately 25 to 100 °C, subcritical water from 100 to 374 °C, and supercritical water, above approximately 374 °C and 22 MPa (Hartonen and Riekkola 2017). For carbon dioxide, the critical point is defined at 31.1 °C and 7.39 MPa (Amaral et al. 2017).

Using NIST (2018) database, Table 8.1 was constructed, which presents some properties of water and carbon dioxide at different thermodynamic states. The dielectric constant, viscosity, and density in water and carbon dioxide decrease with increasing pressure and temperature, except for the viscosity of carbon dioxide at 0.1 MPa, which demonstrates a singular behavior when compared to other solvents (Knez et al. 2019).

For carbon dioxide, there are no major changes in dielectric constant values (Table 8.1). However, there is a drastic reduction in this property for water, profoundly changing the value of the ionic product. This explains why subcritical water is more dissociated than supercritical water. The dielectric constant is reduced and supercritical water behaves as a nonpolar solvent (Guzonas et al. 2018). A high ionic product favors ionic reactions. For example, the concentration of hydronium and oxydryl at 22 MPa and 300 °C is 3×10^{-6} mol/L. At 400 °C, in the same pressure, the concentration is 3×10^{-10} mol/L. Therefore, water in the supercritical state has a lower concentration of hydronium and oxydryl ions when compared to subcritical water (Cocero et al. 2018).

When there is a change from subcritical to supercritical state, drastic changes in water properties occur. Thus, knowing the value of the ionic product at different temperatures and pressures is of enormous importance for the process where it is desired to dissociate the lignocellulosic biomass to obtaining sugars and bioproducts (Abaide et al. 2019a).

Water can be widely used for ionic reactions at 25 MPa and at temperatures below 300 °C since under these conditions its ionic product remains close to 12–13 (Akiya and Savage 2002). However, it is interesting to emphasize that the ionic product can rise to around 20 when the pressure is kept constant and the temperature is increased to 400 °C. This change in the ionic product affects the kinetics of glucose and fructose. This behavior affects reaction kinetics, favoring radical reactions (Promdej and Matsumura 2011).

Regarding the viscosity values, presented in Table 8.1, a high difference can be evidenced when comparing the viscosity values for the water and the carbon dioxide. This is a parameter that justifies the use of carbon dioxide in extraction processes, low viscosity in general. Low viscosity fluids diffuse more easily into the solid matrix of a sample and still penetrate more rapidly, which increases yield and extraction efficiency (Pouliot et al. 2014).

The viscosity of water is reduced when the temperature is equal or higher than 300 °C (Table 8.1). In fact, generally, water is not used in the extraction of compounds with biological activity under these conditions, since many compounds are

Table 8.1 Properties of water and carbon dioxide at different temperatures and pressures

Temperature [°C]	Density [kg/m^3]	Viscosity [μPa s]	Heat capacity [J/g K]	Dielectric constant
Water at 0.1 MPa				
0	999.84	1790.9	4.21	69.96
50	988.03	546.76	4.18	1.00
100	958.63/0.58[a]	282.9/12.25[a]	4.21/2.07[a]	1.00
150	0.51	14.18	1.98	1.00
200	0.46	16.17	1.97	1.00
250	0.41	18.22	1.98	1.00
300	0.37	20.29	2.01	1.00
350	0.34	22.37	2.03	1.00
400	0.32	24.45	2.07	1.00
Water at 10 MPa				
0	1004.8	1767.3	4.17	88.37
50	992.3	548.53	4.15	70.31
100	962.93	284.36	4.19	55.90
150	922.31	184.84	4.27	44.41
200	870.92	136.40	4.44	35.13
250	805.69	107.77	4.79	27.36
300	715.26	86.45	5.68	20.29
350	44.56	20.26	4.01	1.31
400	37.82	22.15	3.09	1.23
Water at 50 MPa				
0	1023.8	1696	4.02	89.98
50	1008.7	577.09	4.08	71.66
100	980.26	295.05	4.11	57.21
150	942.69	194.54	4.17	45.78
200	896.96	145.60	4.28	36.65
250	842.40	117.42	4.46	29.23
300	776.46	98.47	4.78	22.96
350	693.23	83.23	5.37	17.39
400	577.76	67.98	6.79	11.99
Carbon dioxide at 0.1 MPa				
0	1.95	13.71	0.82	1.00
50	1.64	16.13	0.87	1.00
100	1.42	18.47	0.91	1.00
150	1.25	20.72	0.96	1.00
200	1.11	22.89	0.99	1.00
250	1.01	24.96	1.03	1.00
300	0.92	26.96	1.06	1.00
350	0.84	28.87	1.08	1.00
400	0.78	30.71	1.11	1.00

(continued)

Table 8.1 (continued)

Temperature [°C]	Density [kg/m³]	Viscosity [μPa s]	Heat capacity [J/g K]	Dielectric constant
Carbon dioxide at 10 MPa				
0	974.05	113.92	2.17	1.60
50	384.33	28.36	5.80	1.21
100	188.56	21.79	1.52	1.09
150	145.56	22.84	1.25	1.07
200	122.13	24.45	1.17	1.06
250	106.49	26.21	1.15	1.05
300	94.99	27.99	1.15	1.04
350	86.05	29.75	1.15	1.04
400	78.83	31.45	1.17	1.04
Carbon dioxide at 50 MPa				
0	1104.3	172.62	1.74	1.70
50	962.45	108.28	1.75	1.59
100	818.74	76.05	1.73	1.49
150	686.94	59.01	1.67	1.39
200	580.03	49.9	1.56	1.32
250	500.31	45.14	1.47	1.27
300	440.68	42.77	1.40	1.23
350	395.24	41.75	1.35	1.21
400	359.51	41.48	1.33	1.19

[a]Phase equilibrium

degraded at high temperatures, which would make the operation unfeasible (Montoya et al. 2019).

Some properties of carbon dioxide in supercritical state are relevant in the extraction process like the ability to dissolve nonpolar compounds. Other properties are the high affinity with oxygenated organic compounds and the ability to avoid the solubilization of proteins, polysaccharides, sugars, and minerals (Brunner 2005).

8.3 Recent Studies Using Green Solvents

Based on the properties and versatility of green solvents, new applications for ionic liquid, switchable solvents, sub/supercritical water, and supercritical carbon dioxide in green processes are increasingly being studied. In this section, an overview of the different approaches is presented, which is illustrated by selected examples and recent studies.

8.3.1 Ionic Liquids

Studies related to separation and extraction processes (Shukla et al. 2018), carbon capture (Wai et al. 2016), energy storage (MacFarlane et al. 2016), and uses as a natural dissolution medium for polymers such as biomass fractionation (Halder et al. 2019) have been using ionic liquids as efficient agents. Thus, ionic liquids are helping in the development of many fields of study, which can be observed by the growing number of academic publications.

It is relevant to observe with critical viewpoint the last 5 years of scientific production, until July 2019, that studied the application of ionic liquids. Figure 8.3a shows the number of articles find in the Science Direct database that employed ionic liquids for the most observed applications. The search was done by analyzing the production of research articles with the following keywords: "ionic liquid" and

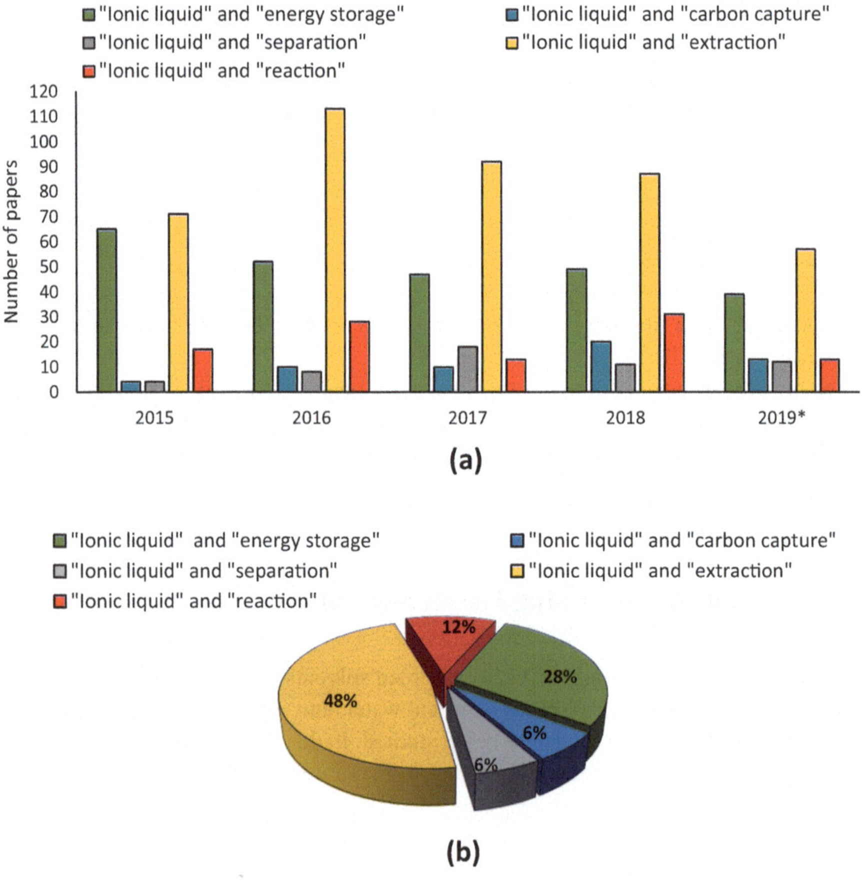

Fig. 8.3 Scientific production in the last 5 years related to studies with ionic liquids; asterisk—data gathered until July 2019

"energy storage"; "ionic liquid" and "carbon capture"; "ionic liquid" and "separation"; "ionic liquid" and "extraction"; and "ionic liquid" and "reaction."

There is a higher production of scientific articles related to using ionic liquids in extraction operation, followed by research that studied this green solvent in energy storage (Fig. 8.3a). In the last 4 years, especially in 2016, the largest amount of research was observed using ionic liquids in extraction processes. However, there is a decrease in the number of articles related to this topic from 2016 to 2019. For another field of study, this decrease in the number of scientific productions was not observed, since the number of articles produced remained practically constant, oscillating around the main value.

A total of 884 scientific articles from 2015 to July 2019 studied the application of ionic liquids related to the keywords mentioned in this document. Some different areas have been intensively studied (Fig. 8.3b). Researches that used ionic liquids in the extraction process represented approximately 48% of the surveys evaluated in the last years. The second and third most studied topics were related to energy storage (28%) and reactions (12%).

Most of the studies reported that used ionic liquids in extraction processes in the last 5 years, until July 2019, have focused on metals, aromatics, and oil. Regarding the application of ionic liquids for energy storage, it is possible to highlight works that use these solvents as electrolytes in batteries and fabrication of supercapacitors. Still, ionic liquids were used as catalysts in esterification reactions to study their effects. Some of these studies are presented in Table 8.2.

The bis(1-ethyl-3-methylimidazolium) tetrathiocyanatocobaltate and bis(1-butyl-3-methylimidazolium) tetrathiocyanatocobaltate ionic liquids have been employed in the extraction of aromatic hydrocarbons from pyrolysis gasoline (Larriba et al. 2017). Firstly, the authors evaluated the extraction of toluene from heptane solvent using the tetrathiocyanatocobaltate as an ionic liquid. Afterward, the extraction of toluene and *p*-xylene from the pyrolysis gasoline using tetrathiocyanatocobaltate was done. The authors observed a reduction from 66.1 to 0.09% in the content of aromatic compounds in pyrolysis gasoline.

Ionic liquids are also used to extract lipids from microalgae (Pan et al. 2016). The 1-butyl-3-methylimidazolium hydrogen sulfate was used as an ionic liquid, and the process was assisted by microwave irradiation. The reactor tube (100 mL) was heated to 120 °C. The extraction with ionic liquid assisted by microwave showed higher efficiency (370%) than using organic solvents, being 15 times more quickly.

An ionic liquid was used for energy storage (Deyab 2018; Thangavel et al. 2018). The performance of a lead-acid battery was tested in order to inhibit the corrosion of lead electrode and hydrogen gas evolution using 1-ethyl-3-methylimidazolium diethyl phosphate (Deyab 2018). The authors verified that there is a reduction in double-layer capacitance and increased resistance to charge transfer due to adsorption of ionic liquid on the surface of the lead electrode. The energy density increased from 45 to 83 mAh/g when the ionic liquid was added, and this phenomenon improved the energy storage capability.

A study using three acid imidazolium ionic liquids was carried out to verify their effect on the efficiency of acid-catalyzed biodiesel production from palm oil under

Table 8.2 Summary of recent studies using ionic liquids in green processes

Extraction process	Ionic liquid	Main characteristics	References
Extraction of lithium from salt lake brine using ionic liquid	1-Butyl-3-methyl-imidazolium hexafluorophosphate	Methodology: The ionic liquid, sodium perchlorate, and tributyl phosphate were used as the extraction medium. Main response: The single extraction efficiency of lithium was 87.28% at equilibrium time of 10 min	Shi et al. (2015)
Microwave-assisted extraction of lipids from microalgae using an ionic liquid solvent	1-Butyl-3-methylimidazolium hydrogen sulfate	Methodology: 1 g algae and 5 g ionic liquid were added into microwave-compatible reactor tube (100 mL) for lipid extraction. The temperature of the reactor was 120 °C and hold for 10–60 min. Main response: The total amount of extracted lipid using ionic liquid was 23% (oil g/100 g dry algae)	Pan et al. (2016)
Extraction of aromatic hydrocarbons from pyrolysis gasoline using tetrathiocyanatocobaltate-based ionic liquids	Bis(1-ethyl-3-ethylimidazolium) tetrathiocyanatocobaltate and bis(1-butyl-3-methylimidazolium) tetrathiocyanatocobaltate	Methodology: Hydrocarbons and ionic liquids were gravimetrically added to 8 mL vials. The liquid-liquid equilibrium between both phases was reached for 5 h at 313.2 K. Main response: The aromatic content in the pyrolysis gasoline was reduced from 66.1 to 0.09 wt.%	Larriba et al. (2017)
Ionic liquid-based microwave-assisted extraction of verbascoside from Rehmannia root	1-Benzyl-3-methylimidazolium chloride	The microwave-assisted ionic liquid extraction method was used to extract verbascoside. Hydrophobicity of ionic liquids affected extraction efficiency. The optimized extraction conditions were 30 g/g of liquid-solid ratio, 0.3 mol/L of 1-benzyl-3-methylimidazolium chloride, during 10 s of extraction time with 210 W of microwave irradiation power	Fan et al. (2018)

(continued)

Table 8.2 (continued)

Extraction process	Ionic liquid	Main characteristics	References
Extraction of Ag and Au from chloride electronic waste leach solutions using ionic liquids	[Bmim][Tf2N], [Bmim][PF6], and Cyphos 101	Ionic liquids were used as extractants for Ag and Au from chloride aqua regia leach liquors of printed circuit board samples. Cyphos 101 was the most effective extractant for all metals from chloride solution at room temperature and 1:1 of the ionic liquid/sample ratio. The extractions were 95% Ag, 97% Au, and 99% Cu under these conditions	Masilela et al. (2019)
Ionic liquid solvent bar micro-extraction of $CdCl_n^{(n-2)-}$ species for ultra-trace Cd determination in seawater	Trioctylmethylammonium chloride	Three-phase solvent bar micro-extraction system containing the ionic liquid was used for isolation/pre-concentration of Cd from seawater samples. Under optimum conditions, the samples were pre-concentrated 65 times	Herce-Sesa et al. (2018)

Energy storage	Ionic liquid	Main characteristics	References
Electrochemical energy storage in montmorillonite K10 clay-based composite as supercapacitor using an ionic liquid electrolyte	Tetraethylammonium cation	An electrical double-layer capacitor has its improved performance when the ionic liquid was applied as an electrolyte, increasing the acid-base interaction of K10 that has hydroxyl groups on the surface. This system showed a high energy density of 171 Wh/kg at a power density of 1.98 kW/kg	Maiti et al. (2016)
High-energy green supercapacitor driven by ionic liquid electrolytes as an ultra-high stable next-generation energy storage device	1-Ethyl-3-methylimidazolium bis(trifluoromethylsulfonyl) imid	A supercapacitor was synthesized through carbonized watermelon (600 °C) and activated with KOH solution. Supercapacitors were mounted on a standard cell CR2032 with two symmetrical electrodes filled with ionic liquid. The supercapacitor delivers high energy (174 Wh/kg) and high power (20 kW/kg)	Thangavel et al. (2018)

(continued)

Table 8.2 (continued)

Energy storage	Ionic liquid	Main characteristics	References
Ionic liquid as an electrolyte additive for high-performance lead-acid batteries	1-Ethyl-3-methylimidazolium diethyl phosphate	The performance of different concentrations of ionic liquid on the hydrogen gas evolution during the immersion of Pb electrode in 5.0 M H_2SO_4 solution was examined. The battery energy storage increases from 45 to 83 mAh/g by adding ionic liquid	Deyab (2018)
Electrochemical performance of hard carbon negative electrodes for ionic liquid-based sodium-ion batteries over a wide temperature range	Na[FSA]-[C3C1pyrr][FSA] (FSA = bis(fluorosulfonyl) amide, C3C1pyrr = *N*-methyl-*N*-propylpyrrolidinium)	Use of ionic liquids as an electrolyte for sodium-ion batteries. The battery developed works using hard carbon electrode that has a good conductivity even at high temperatures (90 °C). Electrochemical performance of battery was evaluated at a wide temperature range (−10 to 90 °C). Stable cyclability was observed for the HC electrode at 90 °C (i.e., capacity retention ratio of 84% after 500 cycles)	Ding et al. (2015)
Solidification of ionic liquid redox electrolytes using agarose biopolymer for highly performing dye-sensitized solar cells	Mixture of 1:7 v/v of blend of [C_4C_1im]I / EMimDCA + 4-tert-butyl pyridine (4-TBP)	Ionic liquid-based redox electrolytes were prepared and subsequently solidified using agarose natural polysaccharide. The resulted material was incorporated in dye-sensitized solar cells that presented excellent efficiency, 3.98%	Bidikoudi et al. (2015)
Room-Temperature Aluminum-Sulfur Batteries with a Lithium-Ion-Mediated Ionic Liquid Electrolyte	1-Ethyl-3-methylimidazolium chloride ([EMI]Cl)	A lithium-ion mediation strategy enhances the reversibility of Al-S batteries. Cycle life of Al-S batteries is improved with a lithium-ion-mediated ionic liquid and Li+-mediation mechanism was studied with experimental and computational methods	Yu et al. (2018)

(continued)

Table 8.2 (continued)

Reaction	Ionic liquid	Main characteristics	References
Process intensification of transesterification for biodiesel production from palm oil: Microwave irradiation on transesterification reaction catalyzed by acidic imidazolium ionic liquids	Three acidic imidazolium ionic liquid ([HSO_3-BMIM] HSO_4)	Three acidic imidazolium ionic liquids were synthesized and their catalytic performances were investigated. A maximum yield of 98.93% was obtained	Ding et al. (2018)
Esterification of potato starch by a biocatalyzed reaction in an ionic liquid	1-Butyl-3-methylimidazolium chloride	Potato starch was esterified with oleic acid, using ionic liquid. Effects of the reaction parameters on the degree of substitution, such as the time and the temperature, were also studied. The highest degree of substitution (0.22) was found at 60 °C for 4 h	Zarski et al. (2016)
Uniform acid poly ionic liquid-based large particle and its catalytic application in esterification reaction	Poly(ionic liquid) P(VB-VMS)PW	The catalytic performance of the poly ionic liquid for the esterification reaction was investigated. The optimum reaction conditions are 5 wt.% catalyst, 95 °C reaction temperature, 3-h reaction time, and methacrylic acid-to-methanol ratio of 1:1.2. The yield of methyl methacrylate is 100%	Zhang et al. (2015)
Pyridinium-based ionic liquid: A pretreatment solvent and reaction medium for catalytic conversion of cellulose to total reducing sugars	1-Butyl-1-methylpyrrolidinium chloride	Pyridinium-based ionic liquids were synthesized and used for the dissolution of cellulose and further conversion to total reducing sugar in the presence of catalysts. 1-Butyl-1-methylpyrrolidinium chloride was the best solvent to dissolve it up to 28% at 110 °C. Micrography results showed disrupted flaky construction for regenerated cellulose. The optimum conditions were observed at 4-h treatment time, 120 °C, and 12% catalyst loading	Saher et al. (2018)
Covalent functionalization of multiwalled carbon nanotubes with imidazolium-based poly(ionic liquid)s by Diels-Alder "click" reaction	Imidazolium-based poly(ionic liquid)s	Imidazolium-based ionic liquid grafted on the surface of multiwalled carbon nanotubes was prepared by Diels-Alder "click" reaction with a high grafting density. The material was characterized by FTIR, Raman spectroscopy, XRD, XPS, and TEM analyses. A very high grafting density was achieved in deep eutectic solvents in a short time	Le et al. (2015)

FTIR Fourier transform infrared spectroscopy, *XRD* X-ray diffraction, *XPS* X-ray photoelectron spectroscopy, *TEM* transmission electron microscopy microwave irradiation (Ding et al. 2018). Compared with experiments using conventional heating, the use of microwave can intensify biodiesel production. Microwave irradiation reduced energy consumption (44%) and reaction time, with a biodiesel yield of 84.76%. The study of Zarski et al. (2016) used an ionic liquid in the esterification of potato starch as a biocatalyst. A reaction at 60 °C for 24 h presented a high degree of substitution (0.22).

8.3.2 Switchable Solvents

Switchable solvents change their properties as soon as a stimulus is applied and are changed again when that stimulus is removed or another stimulus is applied (Jessop 2015). These solvents are commonly used for extraction and purification processes based on their promising characteristics. Therefore, it is also important to verify the scientific production in the last 5 years, until July 2019, related to switchable solvents. The research was done in a similar manner to that one for ionic liquids, using Science Direct platform. However, in this section, only the keyword "switchable solvent" was used for searching the scientific works. A total of 56 articles were found for the last 5 years. Figure 8.4 shows the scientific production over the years.

Since 2015, the highest number of reports related to switchable solvents was observed in 2017. Even with a reduction in the amount of studies in 2018, the number of articles was higher than in 2015 and 2016. Thus, some recent studies found in the scientific literature can be highlighted.

Biodiesel was produced from oil extraction from *Jatropha curcas* L. oilseeds using a switchable solvent (Zeng et al. 2016). *N,N*-Dimethylcyclohexylamine was used with 200 rpm agitation speed, during 60 min at 30 °C. The extraction yield was approximately 82% with a low residual solvent content (1.7%). The

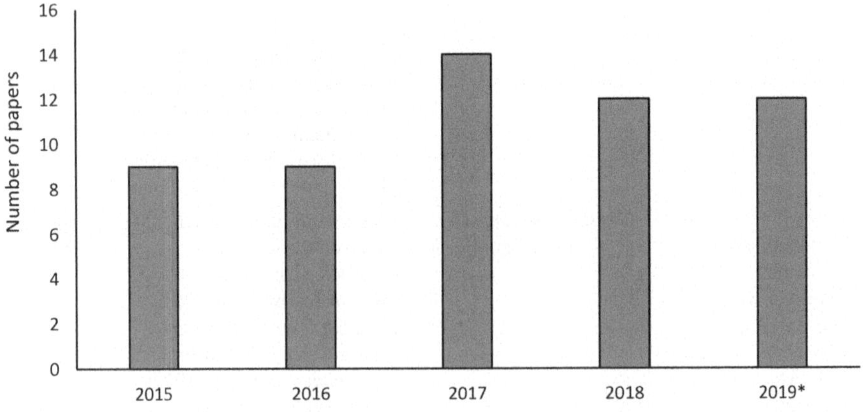

Fig. 8.4 Scientific production in the last 5 years related to studies with switchable solvents; asterisk—data gathered until July 2019

transesterification process using oil extracted with switchable solvent showed a conversion rate of approximately 99.5% (Zeng et al. 2016).

Al-Ameri and Al-Zuhair (2019) proposed the use of *N*,*N*-dimethylcyclohexylamine, *N*-ethyl-butylamine, and dipropylamine for simultaneous wet microalgae cell disruption, oil extraction, and one-step transesterification and biodiesel separation. The best oil yield was 13.6%. The experiment using *N*,*N*-dimethylcyclohexylamine with a 6:1 methanol/oil molar ratio at 35 °C for 1 h yielded the best biodiesel yield of 47.5% (Al-Ameri and Al-Zuhair 2019).

Regarding the use of switchable solvents to capture carbon dioxide, it is possible to highlight the study that uses *N*-ethylpiperidine as solvent (Carrera et al. 2018). A stirred tank reactor was used in the study, with capacity of 2 L of solvent; the absorption experiments were carried out at 22 °C and 98 kPa. The *N*-ethylpiperidine solvent reduced the costs of regeneration by 30% due to the increased gas absorption rate.

Another interesting study applied switchable solvent for preparing fine silica aerogel powder (Bediako et al. 2019). This material is useful for catalysis, pharmaceutical applications, and space research. Bediako et al. (2019) produced this silica material using *N*,*N*-dimethylcyclohexylamine and carbon dioxide as a switchable hydrophilicity solvent. Approximately 70% of pure silica hydrogel water was removed by the instantaneous reaction of *N*,*N*-dimethylcyclohexylamine with carbon dioxide.

8.3.3 Water in Sub/Supercritical State

Subcritical water has very different characteristics from room temperature, 25 °C, water or supercritical water. Subcritical water has been used as a reaction medium (Brunner 2009), as an extraction solvent (Nurdiawati et al. 2019), and in hydrolysis and gasification processes (Cantero et al. 2015; Pedras et al. 2019; Abaide et al. 2019b). Extraction with subcritical water is advantageous due to changes in water properties according to the operation temperature, such as diffusivity, which causes an acceleration of the process (Hartonen and Riekkola 2017).

Hydrolysis with water in subcritical state is also another alternative for using it as a green solvent. This process is also commonly called hydrothermal or liquid hot water treatment (Prado et al. 2016). Methane and hydrogen productions have been gaining prominence through the use of biomass gasification using supercritical water. The great advantage of using this technology stands for the unnecessity to dry the biomass residues that will be subjected to gasification since they can be loaded inside the reactor with high moisture for the production of hydrogen and methane (Okajima and Sako 2014).

Supercritical water is also being studied as a medium for waste destruction as it solubilizes nonpolar organic solvents and oxygen. Waste processing aimed at destroying it is mainly by oxidation with supercritical water. However, the harsh chemical environment and high temperatures lead to problems such as salt precipitation and corrosion (Marrone and Hong 2009).

In a similar way that was taken for the previous solvents, articles related to studies using water at subcritical and/or supercritical states were searched for the last 5 years, until July 2019. The search was done in the Science Direct database analyzing the production of research articles with the following keywords: "sub/supercritical water" and "gasification"; "sub/supercritical water" and "corrosion"; "sub/supercritical water" and "hydrolysis"; "sub/supercritical water" and "oxidation"; "sub/supercritical water" and "synthesis"; and "sub/supercritical" and "liquefaction." The term sub/supercritical refers to the words "subcritical" and "supercritical" individually. There were still searches using the terms "superheat water" and "extraction"; "superheat water" and "hydrolysis"; "superheat water" and "gasification"; and "superheat water" and "liquefaction." The terms pressurized "hot water," "hydrothermal," and "high-pressure water" are also attributed to subcritical water but were not used in this search.

The highest production of scientific articles related to the use of water at subcritical and/or supercritical state is related to gasification processes, followed by researches that studied water in sub/supercritical state as extraction solvent (Fig. 8.5a). Regarding the gasification studies, the years 2016 and 2018 were those with the highest number of articles. For extraction with sub/supercritical water, the highest scientific production was verified in 2018. A trend of increasing research in this area is observed when evaluating the works published in the last 3 years, 2016–2018. The smallest number of researches related to the use of water at subcritical and/or supercritical states refers to liquefaction processes and processes that use water as a solvent in the synthesis medium.

A total of 477 of scientific articles were found in the last 5 years that studied the potential use of water at subcritical and/or supercritical state related to the keywords mentioned in this document. Overall, some different areas have been intensively studied (Fig. 8.5b). Researches that used sub/supercritical water in the gasification process represented 40% of the surveys evaluated in the last years. The second, third, and fourth most studied topics were related to extraction (19%), oxidation (13%), and hydrolysis (11%) processes, respectively.

The main products obtained by sub/supercritical water gasification were hydrogen and methane. Also, the main type of biomass used in recent studies was waste generated by agro-food industry (Okajima and Sako 2014). Regarding the use of sub/supercritical water for the extraction process, it is possible to emphasize that the largest studies had the objective of extracting bioactive compounds with antioxidant activity, phenolic compounds, lignin, and oil.

Sub/supercritical water is also used in the oxidation process. The main studies are focused on energy production and on the degradation of unwanted organic compounds from wastewater, industrial effluent, and sewage sludge. Sub/supercritical water hydrolysis process was studied for obtaining fermentable sugars and bioproducts from lignocellulosic biomass. Therefore, some studies using water in sub/supercritical state as a solvent are presented in Table 8.3.

Hydrogen was produced using beet-based distillery wastewater by subcritical water gasification process (Seif et al. 2015). The authors used a reactor with a volume of 160 mL. The process was performed with 90 mL of wastewater/water

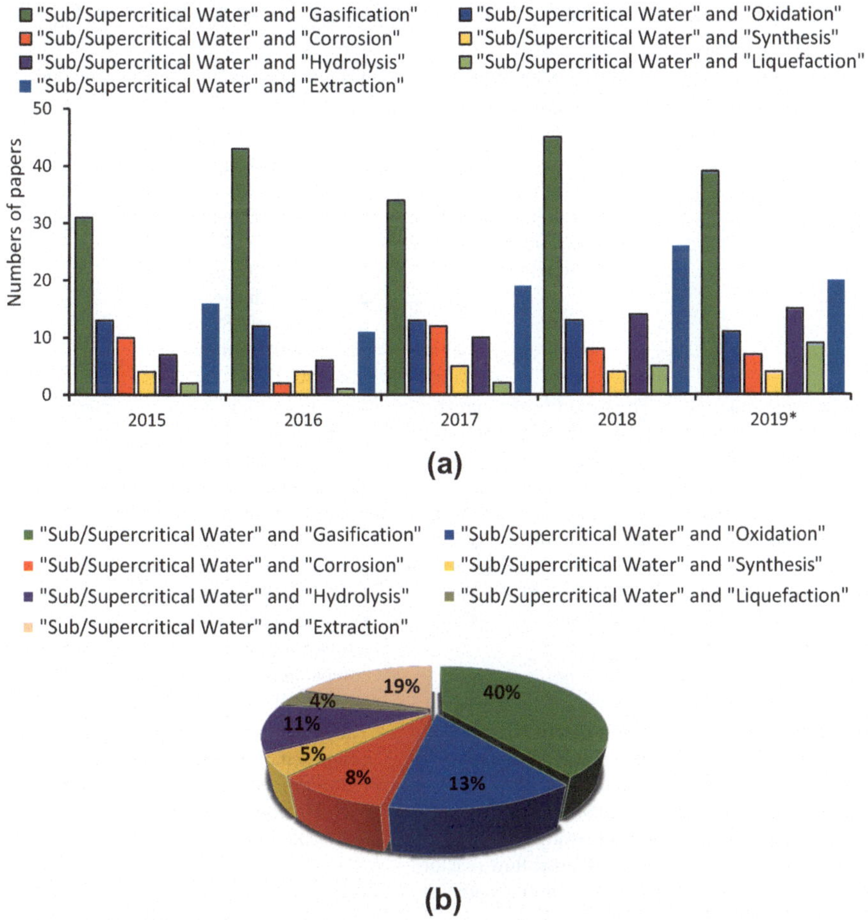

Fig. 8.5 Scientific production in the last 5 years related to studies with sub/supercritical water; asterisk—data gathered until July 2019. Sub/supercritical: subcritical/supercritical

solution, varying feedstock concentrations as 20, 30, and 40 wt.% from 300 to 375 °C at 15, 30, and 45 min. In the best condition, 375 °C, 45 min, and 30 wt.% of feedstock, hydrogen content in gaseous reached 48.8%. Carbon dioxide (48.2 wt.%) and methane (2.3 wt.%) were also produced.

In order to recover energy and to manage pollution, subcritical water gasification of black liquor and wheat straw was studied (Cao et al. 2019). The experiments were performed using an Inconel 625 reactor with a total volume of 10 mL from 500 to 750 °C, during 5–40 min, with a mixture concentration of 5.0–19.1 to wt.% and mixing ratio from 0 to 100%. The highest carbon gasification efficiency of 97.87% was obtained at 750 °C with the reduction of the chemical oxygen demand of the aqueous residues, 672.15 mg/L.

Table 8.3 Summary of recent studies using water in sub/supercritical state as a solvent in green processes

Gasification process	Main characteristics	Main responses	References
Subcritical water gasification of beet-based distillery wastewater for hydrogen production	The process was conducted in a batch reactor at various temperatures (300–375 °C) for different reaction times (15–45 min) and with three feedstock concentrations (20, 30, and 40 wt.%)	The gaseous fraction obtained mainly contained hydrogen and carbon dioxide with fewer amounts of methane and carbon monoxide. Hydrogen mole fraction in gaseous product reached the value of 48.8% at 375 °C after 45-min residence time	Seif et al. (2015)
Subcritical and supercritical water gasification of lignocellulosic biomass impregnated with nickel nanocatalyst for hydrogen production	The nickel-impregnated feedstocks were gasified in sub/supercritical water to study the impacts of temperature (300–500 °C), biomass-to-water ratio (1:5 and 1:10), and residence time (15, 30, and 45 min) at 23–25 MPa	The supercritical water gasification of nickel-impregnated biomass showed high carbon gasification efficiency (19.6–32.6%), higher total gas yields (9.5–16.2 mmol/g), and hydrogen yields (2.8–5.8 mmol/g) when compared to non-catalytic gasification	Nanda et al. (2016)
Supercritical water gasification of microalgae and their constituents in a continuous reactor	The supercritical water gasification process development at 663 °C and 24 MPa. Gasification has been investigated using a 200 mL downflow reactor working continuously	The gas obtained is mainly composed of hydrogen (52.0%), methane (17.9%), and carbon dioxide (23.0%)	Caputo et al. (2016)
System analysis of pulping process coupled with supercritical water gasification of black liquor for combined hydrogen, heat, and power production	Combined hydrogen, power, and low power and medium power steam were produced for the pulping process. The gas product after hydrogen extraction was burned with imported natural gas to supply more heat	Using air 22,604 kW power can be exported after balancing the consumptions and 219 kg of coal equivalent energy can be produced with 1 ton pulp production	Cao et al. (2017)

(continued)

Table 8.3 (continued)

Gasification process	Main characteristics	Main responses	References
Supercritical water gasification of black liquor with wheat straw as the supplementary energy resource	The synergistic effect was observed in the co-gasification due to the addition of wheat straw making better use of the alkali in black liquor. The influences of temperature (500–750 °C), reaction time (5–40 min), mixture concentration (5.0–19.1 wt.%), and mixing ratio (0–100%) were studied	The highest gasification efficiency of 97.87% was obtained at 750 °C. The hydrogen yield increased from 12.29 mol/kg at 500 °C to 46.02 mol/kg	Cao et al. (2019)
Comparative evaluation for catalytic gasification of petroleum coke and asphaltene in subcritical and supercritical water	Comparative evaluation for catalytic gasification of petroleum coke and asphaltene in subcritical and supercritical water	Asphaltene proved to be a better precursor for catalytic hydrothermal gasification leading to 11.97 mmol/g of total gas yield compared to petroleum coke (8.04 mmol/g)	Rana et al. (2019)

Extraction process	Main characteristics	Main responses	References
Simultaneous extraction of oil- and water-soluble phase from sunflower seeds with subcritical water	Extraction kinetics were studied at different temperatures and material/solvent ratios in a batch extractor. Degree of hydrothermal degradation of oils was observed by analyzing the amount of formed free fatty acids and their antioxidant capacities	The highest amount of oil was obtained at 130 °C at a material/solvent ratio of 1/20 g/mL after 30 min of extraction. For all obtained oils, the minimum degree of hydrothermal degradation could be identified	Ravber et al. (2015)
Subcritical water extraction of wild garlic (*Allium ursinum* L.) and process optimization by response surface methodology	The influence of temperature (120–200 °C), extraction time (10–30 min), and added acidifier, hydrochloric acid (0–1.5%), was investigated on the extraction process	The best conditions were defined at a temperature of 180.92 °C, extraction time of 10 min, and added acidifier at 1.09%	Tomšik et al. (2017)
Simultaneous extraction of oil and tea saponin from *Camellia oleifera* Abel. seeds under subcritical water conditions	The process was evaluated at temperature (110–150 °C), extraction time (20–40 min), and solvent-to-material ratios (3:1 5:1–25:1(v/w))	The highest oil extraction yield is 94.07% at the condition of 133.59 °C, 32.03 min, 10.79 mL/g with the tea saponin yield 71.38%, and the highest tea saponin extraction yield is 74.21% with the oil yield 91.27% (121.11 °C, 32.07 min, 8.33 mL/g)	Wu et al. (2018)

(continued)

Table 8.3 (continued)

Extraction process	Main characteristics	Main responses	References
Subcritical water extraction derived lignin for creation of sustainable reinforced composite materials	The extraction process was performed at different temperatures (140, 180, 220 °C), pressure (5.0, 10.0, 15.0 MPa), and flow rate (2, 5, 8 mL/min) during 45 min	The maximum sugar amount was obtained as 310.9 g/kg under the optimized conditions of 15.0 MPa and 150.9 °C and at a flow rate of 8 mL/min for 45 min. Lignin was mixed with epoxy resin at three different ratios to fabricate lignin-based glass fiber-reinforced composite material	Celiktas et al. (2019)
The influence of extraction temperature on polyphenolic profiles and bioactivity of chamomile (*Matricaria chamomilla* L.) subcritical ater extracts	Subcritical water extraction was performed using 10 g of plant material in a reaction vessel. The process of extraction was performed within 30 min upon reaching the desired temperature at 4.5 MPa in a temperature range between 65 and 210 °C	The extracts obtained at 115 °C showed the highest cytotoxicity. Inhibition of α-amylase and α-glucosidase was the highest at 65 and 85 °C, i.e., 0.51 and 4.13 mmolAE/g, respectively. AE/g = acarbose equivalent per gram	Cvetanović et al. (2019)
Liquefaction and hydrolysis	**Main characteristics**	**Main responses**	**References**
Bio-oil production from microalgae via hydrothermal liquefaction technology under subcritical water conditions	This study investigates direct hydrothermal liquefaction of microalgae to produce bio-oil using a high-pressure batch reactor under subcritical water conditions. Varying water concentrations (1:6, 1:7, 1:8, 1:9, and 1:10 ratio) at certain temperature range (200–320 °C), pressure (6.0 MPa), and reaction time (30 min)	The highest bio-oil yield achieved with *Scenedesmus quadricauda* was 18 wt.% at 1:9 ratio	Kumar et al. (2018)
Liquefaction of palm kernel shell to bio-oil using sub- and supercritical water: An overall kinetic study	The liquefaction process was performed at 330–390 °C and 25 MPa for different reaction times	In this study, the constant (k) increased from 0.43 to 0.49 s^{-1} with reaction temperature from 330 to 390 °C	Chan et al. (2019)

(continued)

Table 8.3 (continued)

Liquefaction and hydrolysis	Main characteristics	Main responses	References
Effect of residence time on two-step liquefaction of rice straw in a CO_2 atmosphere: Differences between subcritical water and supercritical ethanol	The process of liquefaction was performed at final reaction temperatures (270–345 °C) and residence times (15 and 30 min)	At 320 °C and 15 min of reaction, solvents have a weak impact on the quality of bio-oil, and higher heating values obtained were 28.57 MJ/kg and 28.62 MJ/kg to subcritical carbon dioxide-supercritical water and subcritical carbon dioxide-supercritical ethanol, respectively	Yang et al. (2017)
Subcritical water hydrolysis of brewer's spent grains: Selective production of hemicellulosic sugars (C-5 sugars)	The total sugar recovery and formation of by-products were evaluated as a function of flow rate (10 and 20 mL/min), reaction temperature (140, 160, 180, and 210 °C), and the solvent/feed ratio (64, 80, and 112)	The best condition obtained 5.1 g/100 g of brewer's spent grains using solvent/feed ratio of 64	Torres-Mayanga et al. (2019)
Obtaining fermentable sugars and bioproducts from rice husks by subcritical water hydrolysis in a semi-continuous mode	The influences of temperature (180, 220, 260 °C) and liquid/solid ratio (7.5 g water/g husks, 15 g water/g husks) on reducing sugar yield and efficiency	The highest yields (18.0 ± 2.9 g/100 g husks) and efficiency (39.5 ± 1.7 g sugars/100 g carbohydrates) were obtained at 220 °C and 7.5 g water/g husks. In such condition, the hydrolyzed solutions presented cellobiose (18.0 g/L), xylose 17.7 g/L), arabinose (3.6 g/L), glucose (1.5 g/L), and levulinic acid (0.7 g/L)	Abaide et al. (2019a)

Regarding some extraction studies that used water in the subcritical state, it is possible to emphasize the study of Wu et al. (2018). A 500 mL reactor was used to extract oil and tea saponin from *Camellia oleifera* Abel. seeds. The process was evaluated at different temperatures, 110–150 °C, during a fixed time interval, 20–40 min, and solvent-to-feed ratios, 3:1 5:1–25:1(v/w). In the best extraction condition, 32 min and 133.6 °C, the oil and tea saponin extraction yields were 94.07% and 71.38%, respectively.

Subcritical water extraction was also used to obtain lignin for preparing reinforced composite (Celiktas et al. 2019). The extraction process for the *Picea orientalis* using water at subcritical state was performed in a semi-continuous mode. The dried *Picea orientalis* was hydrolyzed at 140, 180, and 220 °C for 45 min at 5.0, 10.0, and 15.0 MPa. The maximum concentration of sugar amount was 310.9 g/kg under 15.0 MPa and 150.9 °C during 45 min. The composite material with lignin addition (20%) presented lower strength and brittleness, and the composite with 5% lignin was more ductile (Celiktas et al. 2019).

Rice husks were hydrolyzed using subcritical water for obtaining fermentable sugars and bioproducts (Abaide et al. 2019a). In this study, the semi-continuous mode was used in a 50 mL stainless steel reactor. The influences of temperature, 180–260 °C, and solvent-to-feed ratio, 7.5 and 15 g water/g biomass, on sugar yields and composition were studied. In the best condition, 220 °C and solvent-to-feed ratio of 7.5 g water/g biomass, the yield of reducing sugars was 18 g/100 g biomass, while the concentrations obtained from cellobiose, xylose, arabinose, and glucose were 18.0, 17.7, 3.6, and 1.5 g/L, respectively.

8.3.4 Supercritical Carbon Dioxide

Supercritical carbon dioxide is commonly used in the extraction operations (Conde-Hernández et al. 2017), solubility (Pitchaiah et al. 2018), adsorption (Jedli et al. 2018), and drying techniques (Morbiato et al. 2019; Zabot et al. 2017) as a green solvent. Thus, supercritical carbon dioxide has been constantly studied, which can be observed by a large number of academic publications.

In this sense, it is important to verify the scientific production in the last 5 years, until July 2019, related to supercritical carbon dioxide for a wide kind of applications. The search was done in the Science Direct database analyzing the production of research articles with the following keywords: "supercritical carbon dioxide" and "extraction"; "supercritical carbon dioxide" and "solubility"; "supercritical carbon dioxide" and "adsorption"; and "supercritical carbon dioxide" and "drying."

When analyzing Fig. 8.6a, the highest scientific production in the last 5 years, until July 2019, is related to the use of supercritical carbon dioxide in extraction studies, showing an increase when comparing 2015 and 2016. The studies that used supercritical carbon dioxide for extraction, drying, and adsorption processes have been slightly reduced from 2016 to 2018.

A total of 486 scientific articles were found in the last 5 years, until July 2019, that studied the potential application of supercritical carbon dioxide related to the keywords mentioned in this document (Fig. 8.6b). The researches that used supercritical carbon dioxide for extraction represented 43% of the surveys evaluated. The second and third most studied topics were related to solubility (18%) and drying (5%).

Supercritical carbon dioxide processes used for extraction in the last 5 years, until July 2019, have been focused on oil, metals, and bioactive and phenolic compounds. Regarding studies about the solubility of supercritical carbon dioxide, it is possible to highlight some works that have used this solvent as solvation medium for pharmaceutical compounds. For drying studies using supercritical carbon dioxide as a green solvent, it is possible to present some articles that have evaluated the removal of water from food (Table 8.4).

Supercritical carbon dioxide was used for extracting oil from brewery malt residue (Spinelli et al. 2016). The authors carried out extraction using different extraction conditions. The highest levels of phenolic (0.35 mg/g) and flavonoids

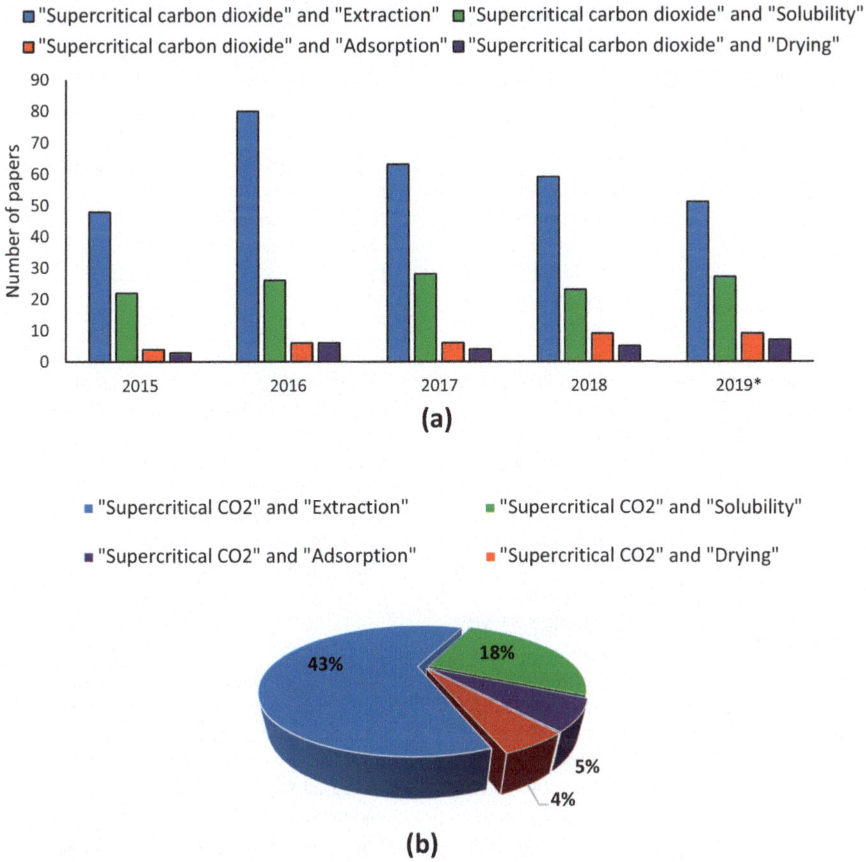

Fig. 8.6 Scientific production in the last 5 years related to studies with supercritical carbon dioxide; asterisk—data gathered until July 2019

(0.22 mg/g) and highest anti-oxidation activity (2.1%) were obtained at 40 °C and 34 MPa for 240 min.

Supercritical carbon dioxide was used to produce a deoiled palm kernel cake (Hossain et al. 2016). The experiments were carried out with the extraction chamber loaded with dried and ground leaves. Extraction conditions were evaluated at 13, 15, and 35 MPa and 40 and 60 °C. A maximum yield of 2.6% was obtained at 35 MPa and 60 °C.

Also related to the use of supercritical carbon dioxide, the study of Argenta et al. (2017) is highlighted. The main purpose was to obtain indium from cell phone liquid crystal screens using supercritical carbon dioxide and co-solvents as an alternative to organic acids. The samples and co-solvents, citric acid and hydrogen peroxide, were inserted in a stainless steel reactor. The pressures of 8, 10, and 15 MPa and temperatures of 60, 80, and 100 °C were used to evaluate the condition

Table 8.4 Summary of recent studies using supercritical carbon dioxide as a solvent in green processes

Extraction process	Main characteristics	Main responses	References
Cesium removal from contaminated sand by supercritical CO_2 extraction	Process parameters' influence such as operating pressure (25–29 MPa) and temperature (40–80 °C), initial cesium/extractant/cationic exchanger amounts (molar ratio between 1:12:12 and 1:100:100), and soil moisture (0–6 wt.%) has been tested	Low temperature (40 °C) and high pressure (29 MPa) allow optimizing processability for cesium removal. Extraction from sand is successful with yields up to 95%	Leybros et al. (2016)
Supercritical carbon dioxide extraction of brewer's spent grain	This study finds proper supercritical fluid extraction conditions to give bioactive compounds from brewer's spent grain. The effects of three factors including pressure (15–35 MPa), temperature (40–60 °C), and carbon dioxide + ethanol (0–60% ethanol concentration, v/v) were investigated	In the best condition (35 MPa, 40 °C, and 60% ethanol) were obtained the maximum concentration of phenolics (0.35 ± 0.01 mg/g sample), flavonoids (0.22 ± 0.01 mg/g sample), and antioxidant compounds (2.09 ± 0.04%/g sample)	Spinelli et al. (2016)
Extraction yield and kinetic study of *Lippia graveolens* with supercritical CO_2	Supercritical fluid extractions were performed at three pressures (13, 15, and 35 MPa) and two temperatures (40 and 60 °C)	A maximum yield of 2.6% on a dry weight basis was obtained at 35 MPa and 60 °C. The extraction yield increased with temperature and pressure	Soto-Armenta et al. (2019)
Utilization of supercritical carbon dioxide extraction technology to produce deoiled palm kernel cake	Supercritical carbon dioxide is used to produce deoiled palm kernel cake from palm kernel. The influences of temperature, pressure, and extraction time on the yield were investigated	The optimum supercritical carbon dioxide extraction process parameters were temperature 60 °C and pressure 44.6 MPa at 50 min of extraction. The oil yield was 49.2% under these experimental conditions	Hossain et al. (2016)
Supercritical CO_2 extraction of indium present in liquid crystal displays from discarded cell phones using organic acids	An effective method for indium extraction from liquid crystal display screens of cell phones using alternative organic acids was studied. For comparative purposes, leaching tests are performed with supercritical carbon dioxide and co-solvents, as well as under conventional conditions using citric and malic acids	At atmospheric pressure, 76.5% of indium was extracted in 180 min. With supercritical carbon dioxide was extracted 94.6% at 30 min	Argenta et al. (2017)

(continued)

Table 8.4 (continued)

Extraction process	Main characteristics	Main responses	References
Supercritical carbon dioxide extraction of lycopene from tomato processing by-products: Mathematical modeling and optimization	The model was tested to investigate the effects of peel-to-seed ratio (30/70 to 70/30), temperature (40–80 °C), and pressure (30–50 MPa) on the lycopene yield	The maximum lycopene yield was 1.32 mg/kg of raw material at 80 °C, 50 MPa, and peel/seed ratio of 70/30	Hatami et al. (2019)
Solubility and drying	**Main characteristics**	**Main responses**	**References**
Mixed solid phase model using the equation of state based on hole theory for solubility prediction of the pharmaceutical compound in supercritical CO_2	A mixed solid phase model using an equation of state based on hole theory is developed for solubility calculation in supercritical carbon dioxide considering the dissolution of carbon dioxide into solid phase at the high-pressure condition	It is found that the results for the pharmaceutical compounds with S, N, F, Cl, and I atoms using the mixed solid phase model give the calculation performance better than those of the pure solid phase model	Sakabe et al. (2015)
Solvation free energy and solubility of acetaminophen and ibuprofen in supercritical carbon dioxide: Impact of the solvent model	Classical molecular dynamics simulations are used to estimate the solvation free energy of two pharmaceutical solids, namely, ibuprofen and acetaminophen in carbon dioxide, over the density range of interest in supercritical processes. Thus, three popular carbon dioxide models (Zhang, EPM2, and TraPPE) are studied	The value estimated using the TraPPE model is the highest for the solubility of solutes. The EPM2 model is intermediate, and Zhang model is the lowest	Noroozi et al. (2016)
The solubility of fenamate drugs in supercritical carbon dioxide by using a semi-flow apparatus with a continuous solvent-washing step in the depressurization line	Solubility of niflumic and tolfenamic acid in supercritical carbon dioxide has been measured at temperatures between 40 and 70 °C and pressures of 16–40 MPa. The use of a continuous solvent-washing step in the depressurization line avoids the solubility data to be underestimated	The results obtained were compared with previous work, and its values were up to 57–105% smaller than the values obtained in current literature. The best models to describe the experimental data were the Adachi-Lu model for niflumic acid and Valle-Aguilera for tolfenamic acid	Banchero and Manna (2016)

(continued)

Table 8.4 (continued)

Solubility and drying	Main characteristics	Main responses	References
Supercritical carbon dioxide combined with high-power ultrasound as innovate drying process for chicken breast	This study explores the feasibility to apply supercritical carbon dioxide drying alone or in combination with high-power ultrasounds to improve the shelf life and safety of raw chicken meat	An inactivation (6 log CFU/g) of mesophilic bacteria and yeasts and molds was achieved with both the supercritical carbon dioxide processes while oven drying at 75 °C, CFU = colony-forming unit	Morbiato et al. (2019)
Comparison of three types of drying (supercritical CO_2, air, and freeze) on the quality of dried apple—quality index approach	This study develops a quality index and examines the effects of drying apples using three technologies (supercritical carbon dioxide drying, air drying, and freeze drying) during a period of 6 months in ambient conditions	After 6 months, samples dried in supercritical carbon dioxide and freeze-dried apples, both packed in polyethylene-coated aluminum with 100% nitrogen	Djekic et al. (2018)

that yielded higher indium extraction. The extraction yield obtained in the best condition, 30 min, 100 °C, 15 MPa, solid/liquid ratio of 1:20 with 5% hydrogen peroxide, was 94.6%, and the process was selective for indium.

Regarding solubility studies that have been using supercritical carbon dioxide, it is possible to emphasize a study evaluating fenamate drugs on supercritical carbon dioxide (Banchero and Manna 2016). In this work, the drug was charged inside the 50 mL cylindrical vessel, and then it was dispersed in a bed of 0.4–0.6 mm glass spheres. The solubilities of niflumic and tolfenamic acid were evaluated at different temperatures and pressures. The higher solubility of niflumic acid was 6.06×10^{-5} at 70 °C and 40 MPa. For tolfenamic acid, the solubility in supercritical carbon dioxide in term of molar fraction was 3.18×10^{-5} in the same condition observed for niflumic acid.

The effects of supercritical carbon dioxide, air, and freeze drying were studied on dry apple quality (Djekic et al. 2018). Supercritical drying was carried out using supercritical carbon dioxide at 12.5 MPa in the temperature range from 50 to 108 °C for 16 h. An Arcen air dryer type was used for air drying at 60 °C for 8 h. Freeze drying was carried out in a freeze dryer at 20.0 Pa during sublimation and 5.0 Pa during desorption. The main results indicated that apples dried in supercritical carbon dioxide had the best scores. The main results indicated that apples dried in supercritical carbon dioxide maintained the best quality. With respect to shelf life, the quality of some dried apple slices was reduced during the second half of the shelf life (Djekic et al. 2018).

8.4 Future Expectations

Ionic liquids, switchable solvents, supercritical carbon dioxide, and water in subcritical and/or supercritical state are promising solvents for industrial processes, especially for unitary operations as extraction, separation, and purification. The future expectations for green solvents appear as environmentally friendly options to minimize the generation of toxic solvents that pollute the environment. Thus, after a constructive overview, the expansion of recent studies using green solvents in several processes demonstrates a promising path for industrial applications. These solvents are attractive for several applications as they are easily combined and versatile with respect to processing technologies. The high cost of green solvents or technologies that modify their state can be compensated by the reduction of energy consumption, time savings, reduction of purification stages, and no generation of undesirable by-products. Thus, it is interesting to note that there is a tendency for scientific projects and articles to be developed, also evaluating economic aspects with the focus of making future industrial applications feasible.

Acknowledgments The authors are grateful to the Coordination for the Improvement of Higher Education Personnel (CAPES), National Council for Scientific and Technological Development (CNPq: 409583/2018-9), and the Research Support Foundation of the State of Rio Grande do Sul (FAPERGS: 16/2551–0000522–2; 17/2551–0000893–6). M. A. Mazutti, M. V. Tres, F. Castilhos, and G. L. Zabot (304882/2018-6) thank CNPq for the productivity grants.

References

Abaide ER, Ugalde G, Luccio MD, Moreira RFPM, Tres MV, Zabot GL, Mazutti MA (2019a) Obtaining fermentable sugars and bioproducts from rice husks by subcritical water hydrolysis in a semi-continuous mode. Bioresour Technol 272:510–520. https://doi.org/10.1016/j.biortech.2018.10.075

Abaide ER, Mortari SR, Ugalde G, Valerio A, Luccio MD, Moreira RFPM, Kuhn RC, Priamo WL, Tres MV, Zabot GL, Mazutti MA (2019b) Subcritical water hydrolysis of rice straw in a semi-continuous mode. J Clean Prod 209:386–397. https://doi.org/10.1016/j.jclepro.2018.10.259

Akiya N, Savage PE (2002) Roles of water for chemical reactions in high temperature water. Chem Rev 102:2725–2750. https://doi.org/10.1021/cr000668w

Al-Ameri M, Al-Zuhair S (2019) Using switchable solvents for enhanced, simultaneous microalgae oil extraction-reaction for biodiesel production. Biochem Eng J 141:217–224. https://doi.org/10.1016/j.bej.2018.10.017

Amaral GV, Silva EK, Cavalcanti RN, Cappato LP, Guimaraes JT, Alvarenga VO, Esmerino EA, Portela JB, Sant'ana AS, Freitas MQ, Silva MC, Raices RSL, Meireles MAA, Cruz AG (2017) Dairy processing using supercritical carbon dioxide technology: theoretical fundamentals, quality and safety aspects. Trends Food Sci Technol 64:94–101. https://doi.org/10.1016/j.tifs.2017.04.004

Amarasekara AS (2016) Acidic ionic liquids. Chem Rev 116:6133–6183. https://doi.org/10.1021/acs.chemrev.5b00763

Argenta AB, Reis CM, Mello GP, Dotto GL, Tanabe EH, Bertuol DA (2017) Supercritical CO_2 extraction of indium present in liquid crystal displays from discarded cell phones using organic acids. J Supercrit Fluids 120:95–101. https://doi.org/10.1016/j.supflu.2016.10.014

Banchero M, Manna L (2016) Solubility of fenamate drugs in supercritical carbon dioxide by using a semi-flow apparatus with a continuous solvent-washing step in the depressurization line. J Supercrit Fluids 107:400–407. https://doi.org/10.1016/j.supflu.2015.10.008

Bediako BBA, Zhou P, Rugabirwa B, Liu Q, Su Y, Wang H, Li J (2019) A switchable hydrophilicity solvent mediated process to prepare fine silica aerogel powder as an excellent flatting agent. Adv Powder Technol 30:565–571. https://doi.org/10.1016/j.apt.2018.12.009

Bidikoudi M, Perganti D, Karagianni C-S, Falaras P (2015) Solidification of ionic liquid redox electrolytes using agarose biopolymer for highly performing dye-sensitized solar cells. Electrochim Acta 179:228–236. https://doi.org/10.1016/j.electacta.2015.02.122

Brunner G (2005) Supercritical fluids: technology and application to food processing. J Food Eng 24:1152–1170. https://doi.org/10.1016/j.jfoodeng.2004.05.060

Brunner G (2009) Near critical and supercritical water. Part I Hydrolytic and hydrothermal processes. J Supercrit Fluids 47:373–381. https://doi.org/10.1016/j.supflu.2008.09.002

Cantero DA, Vaquerizo L, Mato F, Bermejo MD, Cocero MJ (2015) Energetic approach of biomass hydrolysis in supercritical water. Bioresour Technol 179:136–143. https://doi.org/10.1016/j.biortech.2014.12.006

Cao C, Gui L, Jin H, Cao W, Jia Y, Yao X (2017) System analysis of pulping process coupled with supercritical water gasification of black liquor for combined hydrogen, heat and power production. Energy 132:238–247. https://doi.org/10.1016/j.energy.2017.05.104

Cao C, Zhang Y, Li L, Wei W, Wang G, Bian C (2019) Supercritical water gasification of black liquor with wheat straw as the supplementary energy resource. Int J Hydrogen Energy 44:15737–15745. https://doi.org/10.1016/j.ijhydene.2018.10.006

Caputo G, Dispenza M, Rubio P, Scargiali F, Marotta G, Brucato A (2016) Supercritical water gasification of microalgae and their constituents in a continuous reactor. J Supercrit Fluids 118:163–170. https://doi.org/10.1016/j.supflu.2016.08.007

Carrera M, Gómez-Díaz D, Navaza JM (2018) Switchable hydrophilicity solvents for carbon dioxide chemical absorption. J Ind Eng Chem 59:304–309. https://doi.org/10.1016/j.jiec.2017.10.036

Celiktas MS, Yaglikci M, Maleki FK (2019) Subcritical water extraction derived lignin for creation of sustainable reinforced composite materials. Polym Test 77:105918. https://doi.org/10.1016/j.polymertesting.2019.105918

Chan YH, Quitain AT, Yusup S, Uemura Y, Sasaki M, Kida T (2019) Liquefaction of palm kernel shell to bio-oil using sub- and supercritical water: an overall kinetic study. J Energy Institute 92:535–541. https://doi.org/10.1016/j.joei.2018.04.005

Chemat F, Vian MA, Cravotto G (2012) Green extraction of natural products: concept and principles. Int J Mol Sci 13:8615–8627. https://doi.org/10.3390/ijms13078615

Cocero MJ, Cabeza A, Abad N, Adamovic T, Vaquerizo L, Martínez CM, Pazo-Cepeda MV (2018) Understanding biomass fractionation in subcritical & supercritical water. J Supercrit Fluids 133:550–565. https://doi.org/10.1016/j.supflu.2017.08.01

Conde-Hernández LA, Espinosa-Victoria JR, Trejo A, Guerrero-Beltrán JA (2017) CO_2-supercritical extraction, hydrodistillation and steam distillation of essential oil of rosemary (Rosmarinus officinalis). J Food Eng 200:81–86. https://doi.org/10.1016/j.jfoodeng.2016.12.022

Cvetanović A, Švarc-Gajić J, Zeković Z, Jerković J, Zengin G, Gašić U, Tešić Z, Mašković P, Soares C, Barroso MF, Delerue-Matos C, Đurović S (2019) The influence of the extraction temperature on polyphenolic profiles and bioactivity of chamomile (*Matricaria chamomilla* L.) subcritical water extracts. Food Chem 271:328–337. https://doi.org/10.1016/j.foodchem.2018.07.154

Davis JH (2004) Task-specific ionic liquids. Chem Lett 33:1072–1077. https://doi.org/10.1246/cl.2004.1072

Deyab MA (2018) Ionic liquid as an electrolyte additive for high performance lead-acid batteries. J Power Sources 390:176–180. https://doi.org/10.1016/j.jpowsour.2018.04.053

Ding H, Nohira T, Hagiwara R, Fukunaga A, Sakai S, Nitta K (2015) Electrochemical performance of hard carbon negative electrodes for ionic liquid-based sodium ion batteries over a wide temperature range. Electrochim Acta 176:344–349. https://doi.org/10.1016/j.electacta.2015.07.024

Ding H, Ye W, Wang Y, Wang X, Li L, Liu D, Gui J, Song C, Ji N (2018) Process intensification of transesterification for biodiesel production from palm oil: microwave irradiation on transesterification reaction catalyzed by acidic imidazolium ionic liquids. Energy 144:957–967. https://doi.org/10.1016/j.energy.2017.12.072

Djekic I, Tomic N, Bourdoux S, Spilimbergo S, Smigc N, Udovicki B, Hofland G, Devlieghre F, Rajkovic A (2018) Comparison of three types of drying (supercritical CO_2, air and freeze) on the quality of dried apple – quality index approach. LWT Food Sci Technol 94:64–72. https://doi.org/10.1016/j.lwt.2018.04.029

Endres F, Frank SZ (2006) Air and water stable ionic liquids in physical chemistry. Phys Chem Chem Phys 8:2101–2116. https://doi.org/10.1039/B600519P

Fan Y, Xu C, Ling J, Zhang L, Yang L, Zhou Z, Zhu Y, Zhao D (2018) Ionic liquid-based microwave-assisted extraction of verbascoside from Rehmannia root. Ind Crops Prod 124:59–65. https://doi.org/10.1016/j.indcrop.2018.07.063

Guzonas D, Novotny R, Penttilä SP, Toivonen A, Zheng W (2018) Materials and water chemistry for supercritical water-cooled reactors. Woodhead Publishing, Cambridge, 278p

Halder P, Kundu S, Patel S, Setiawan A, Atkin R, Parthasarthy R, Paz-Ferreiro J, Surapanene A, Shah K (2019) Progress on the pre-treatment of lignocellulosic biomass employing ionic liquids. Renew Sustain Energy Rev 105:268–292. https://doi.org/10.1016/j.rser.2019.01.052

Hartonen K, Riekkola M (2017) Water as the first choice green solvent. In: Pena-Pereira F, Tobiszewski M (eds) Handbook of the application of green solvents in separation processes. Elsevier, Amsterdam, pp 19–15. https://doi.org/10.1016/B978-0-12-805297-6.00002-4

Hatami T, Meireles MAA, Ciftci ON (2019) Supercritical carbon dioxide extraction of lycopene from tomato processing by-products: mathematical modeling and optimization. J Food Eng 241:18–25. https://doi.org/10.1016/j.jfoodeng.2018.07.036

Herce-Sesa B, López-López JA, Moreno C (2018) Ionic liquid solvent bar micro-extraction of $CdCln^{(n-2)-}$ species for ultra-trace Cd determination in seawater. Ind Crops Prod 124:59–65. https://doi.org/10.1016/j.chemosphere.2017.11.004

Herrero M, Mendiola JA, Ibáñe E (2015) Gas expanded liquids and switchable solvents. Curr Opin Green Sustain Chem 5:24–30. https://doi.org/10.1016/j.cogsc.2017.03.008

Hossain DS, Norulaini NA, Naim AYA, Zulkhari ARM, Bennama NAN, Omar AKM (2016) Utilization of the supercritical carbon dioxide extraction technology for the production of deoiled palm kernel cake. J CO_2 Util 16:121–129. https://doi.org/10.1016/j.jcou.2016.06.010

Jedli H, Brahmi J, Hedfi H, Mbarek M, Bouzgarrou S, Slimi K (2018) Adsorption kinetics and thermodynamics properties of supercritical CO_2 on activated clay. J Petrol Sci Eng 166:476–481. https://doi.org/10.1016/j.petrol.2018.03.064

Jessop PG (2015) Switchable solvents as media for synthesis and separations. Aldrichimica Acta 48:18–21

Jessop PG, Heldebrant DJ, Li X, Eckert CA, Liotta CL (2005) Reversible nonpolar-to-polar solvent. Nature 436:1102

Jessop PG, Kozycz L, Rahami ZG, Schoenmakers D, Boyd AR, Wechsler D, Holland AM (2011) Tertiary amine solvents having switchable hydrophilicity. Green Chem 13:619–623. https://doi.org/10.1039/C0GC00806K

Jessop PG, Mercer SM, Heldebrant DJ (2012) CO_2-triggered switchable solvents, surfactants, and other materials. Energ Environ Sci 5:7240–7253. https://doi.org/10.1039/C2EE02912J

Joglekar HG, Rahman I, Babu S, Kulkami BD, Joshi A (2006) Comparative assessment of downstream processing options for lactic acid. Sep Purif Technol 52:1–17. https://doi.org/10.1016/j.seppur.2006.03.015

Kenneth SR (1997) Ionic liquids for clean technology. J Chem Technol Biotechnol 68:351–356. https://doi.org/10.1002/(SICI)1097-4660(199704)68:4<351::AID-JCTB613>3.0.CO;2-4

Kiss AA, Lange JP, Schuur B, Brilman DWF, van der Ham AGJ, SRA K (2016) Separation technology–making a difference in biorefineries. Biomass Bioenergy 95:296–309. https://doi.org/10.1016/j.biombioe.2016.05.021

Knez Z, Pantic M, Cor D, Zoran NZ, Hrncic MK (2019) Are supercritical fluids solvents for the future? Chem Eng Process 141:107532. https://doi.org/10.1016/j.cep.2019.107532

Kumar PK, Krishna SV, Verma K, Pooja K, Bhagawan D, Srilatha K, Himabindu V (2018) Bio oil production from microalgae via hydrothermal liquefaction technology under subcritical water conditions. J Microbiol Methods 153:108–117. https://doi.org/10.1016/j.mimet.2018.09.014

Kuramochi T, Elzen M, Levin K, Peters G, Blanco G, Cantzler J, Hans F, Koberle A, Luna L, Mathur R, Muntean M, Olivier J, Padilla JV, Rich D, Sha F, Soest H (2018) Chapter 2 - Trends and progress towards the Cancun pledges, NDC targets and peaking of emissions in Emissions Gap Report 2018. United Nations Environment Programme, pp 3–15

Larriba M, Navarro P, Delgado-Mellado N, Stanisci V, García J, Rodríguez F (2017) Extraction of aromatic hydrocarbons from pyrolysis gasoline using tetrathiocyanatocobaltate-based ionic liquids: experimental study and simulation. Fuel Process Technol 159:96–110. https://doi.org/10.1016/j.fuproc.2017.01.027

Leybros A, Grandjean A, Segond N, Messalier M, Boutin O (2016) Cesium removal from contaminated sand by supercritical CO_2 extraction. J Environ Chem Eng 4:1076–1080. https://doi.org/10.1016/j.jece.2016.01.009

MacFarlane DR, Forsyth M, Howlett PC, Kar M, Passerini S, Pringle JM, Ohno H, Watanabe M, Yan F, Zheng W, Zhang S, Zhang J (2016) Ionic liquids and their solid-state analogues as materials for energy generation and storage. Nat Rev Mater 1:15005

Maiti S, Pramanik A, Chattopadhyay S, Mahanty GDS (2016) Electrochemical energy storage in montmorillonite K10 clay based composite as supercapacitor using ionic liquid electrolyte. J Colloid Interface Sci 464:73–82. https://doi.org/10.1016/j.jcis.2015.11.010

Marıa PD (2014) Recent trends in (ligno)cellulose dissolution using neoteric solvents: switchable, distillable and bio-based ionic liquids. J Chem Technol Biotechnol 89:11–18. https://doi.org/10.1002/jctb.4201

Marrone PA, Hong GT (2009) Corrosion control methods in supercritical water oxidation and gasification processes. J Supercrit Fluids 51:83–103. https://doi.org/10.1016/j.supflu.2009.08.001

Masilela M, Ndlovu S (2019) Extraction of Ag and Au from chloride electronic waste leach solutions using ionic liquids. J Environ Chem Eng 7:102810. https://doi.org/10.1016/j.jece.2018.11.054

Mercer SM, Robert T, Dixon DV, Jessop PG (2010) "Switchable water": aqueous solutions of switchable ionic strength. ChemSusChem 3:467–470. https://doi.org/10.1002/cssc.201000001

Montoya MF, Sánchez FA, Pereda S (2019) Equilibrium in pressurized systems (sub and supercritical). In: Pereira CG (ed) Handbook of thermodynamics of phase equilibria in food engineering. Academic, Cambridge, pp 385–418. https://doi.org/10.1016/B978-0-12-811556-5.00009-0

Morbiato G, Zambon A, Toffoletto M, Poloniato G, Dall'Acqua S, Bernard M, Spilimbergo S (2019) Supercritical carbon dioxide combined with high power ultrasound as innovate drying process for chicken breast. J Supercrit Fluids 147:24–32. https://doi.org/10.1016/j.supflu.2019.02.004

Nanda S, Reddy SN, Dalai AK, Kozinski JA (2016) Subcritical and supercritical water gasification of lignocellulosic biomass impregnated with nickel nanocatalyst for hydrogen production. Int J Hydrogen Energy 41:4907–4921. https://doi.org/10.1016/j.ijhydene.2015.10.060

NIST (2018) National Institute of Standards and Technology (NIST)—Thermophysical properties of fluids. Available in: https://webbook.nist.gov/chemistry/fluid/. Accessed 21 May 2019

Noroozi J, Ghotbi C, Sardroodi JJ, Karimi-Sabet J, Robert MA (2016) Solvation free energy and solubility of acetaminophen and ibuprofen in supercritical carbon dioxide: Impact of the solvent model. J Supercrit Fluids 109:166–176. https://doi.org/10.1016/j.supflu.2015.11.009

Nurdiawati A, Zaini IN, Irhamna AR, Sasongko D, Aziz M (2019) Novel configuration of supercritical water gasification and chemical looping for highly-efficient hydrogen produc-

tion from microalgae. Renew Sustain Energy Rev 112:369–381. https://doi.org/10.1016/j.rser.2019.05.054
Okajima I, Sako T (2014) Energy conversion of biomass with supercritical and subcritical water using large-scale plants. J Biosci Bioeng 117:1–9. https://doi.org/10.1016/j.jbiosc.2013.06.010
Pan J, Muppaneni T, Sun Y, Reddy HK, Fu J, Lu X, Deng S (2016) Microwave-assisted extraction of lipids from microalgae using an ionic liquid solvent [BMIM][HSO4]. Fuel 178:49–55. https://doi.org/10.1016/j.fuel.2016.03.037
Pedras BM, Nascimento M, Sá-Nogueira I, Simões P, Paiva A, Barreiros S (2019) Semi-continuous extraction/hydrolysis of spent coffee grounds with subcritical water. J Ind Eng Chem 72:453–456. https://doi.org/10.1016/j.jiec.2019.01.001
Pitchaiah KC, Lamba N, Sivaraman N, Madras G (2018) Solubility of trioctylmethylammonium chloride in supercritical carbon dioxide and the influence of co-solvents on the solubility behavior. J Supercrit Fluids 138:102–114. https://doi.org/10.1016/j.supflu.2018.04.002
Pollet P, Davey EA, Ureña-Benavides EE, Eckert CA, Liotta CL (2014) Solvents for sustainable chemical processes. Green Chem 16:1034–1055. https://doi.org/10.1039/C3GC42302F
Pouliot Y, Conway V, Leclerc P (2014) Chapter 3 - Separation and concentration technologies in food processing. In: Clark S, Jung S, Lamsal B (eds) Food processing: principles and applications. Publisher: John Wiley & Sons, 2ed, pp 33–60
Prado JM, Lachos-Perez D, Foster-Carneiro T, Rostagno A (2016) Sub- and supercritical water hydrolysis of agricultural and food industry residues for the production of fermentable sugars: a review. Food Bioprod Process 98:95–123. https://doi.org/10.1016/j.fbp.2015.11.004
Promdej C, Matsumura Y (2011) Temperature effect on hydrothermal decomposition of glucose in sub- and supercritical water. Ind Eng Chem Res 50:8492–8497. https://doi.org/10.1021/ie200298c
Rana R, Nanda S, Maclennan A, Hu Y, Kozinski JA, Dalai AK (2019) Comparative evaluation for catalytic gasification of petroleum coke and asphaltene in subcritical and supercritical water. J Energy Chem 31:107–118. https://doi.org/10.1016/j.jechem.2018.05.012
Ravber M, Knez Z, Skerget M (2015) Simultaneous extraction of oil- and water-soluble phase from sunflower seeds with subcritical water. Food Chem 166:316–323. https://doi.org/10.1016/j.foodchem.2014.06.025
Ritchie H, Roser M (2017) CO_2 and other greenhouse gas emissions. Published online at OurWorldInData.org. Retrieved from: https://ourworldindata.org/co2-and-other-greenhouse-gas-emissions. Accessed 5 August 2019
Sakabe J, Uchida H, Shimoyama Y (2015) Mixed solid phase model using equation of state based on hole-theory for solubility prediction of pharmaceutical compound in supercritical CO_2. J Supercrit Fluids 100:26–33. https://doi.org/10.1016/j.supflu.2015.02.002
Seif S, Tavakoli O, Fatemi S, Bahmanyar H (2015) Subcritical water gasification of beet-based distillery wastewater for hydrogen production. J Supercrit Fluids 104:212–220. https://doi.org/10.1016/j.supflu.2015.06.014
Shi C, Jia Y, Zhang C, Liu H, Jing Y (2015) Extraction of lithium from salt lake brine using room temperature ionic liquid in tributyl phosphate. Fusion Eng Des 90:1–6. https://doi.org/10.1016/j.fusengdes.2014.09.021
Shukla SK, Pandey S, Pandey S (2018) Applications of ionic liquids in biphasic separation: aqueous biphasic systems and liquid–liquid equilibria. J Chromatogr A 1559:44–61. https://doi.org/10.1016/j.chroma.2017.10.019
Sindhu R, Binod P, Pandey A (2016) Biological pre-treatment of lignocellulosic biomass - an overview. Bioresour Technol 199:76–82. https://doi.org/10.1016/j.biortech.2015.08.030
Soto-Armenta LC, Sacramento-Rivero JC, Ruiz-Mercado CA, Lope-Navarrete MC, Rocha-Uribe JA (2019) Extraction yield and kinetic study of *Lippia graveolens* with supercritical CO_2. J Supercrit Fluids 145:205–210. https://doi.org/10.1016/j.supflu.2018.12.018
Spinelli S, Conte A, Lecce L, Padalino L, Nobile MAD (2016) Supercritical carbon dioxide extraction of brewer's spent grain. J Supercrit Fluids 107:69–74. https://doi.org/10.1016/j.supflu.2015.08.017

Sunol AK, Sunol SG (2014) Safer solvents and processes: ionic liquid. In: Wypych G (ed) Handbook of solvents: use, health, and environment, ChemTech Publising, vol 2, pp 635–677. https://doi.org/10.1016/B978-1-895198-65-2.50010-7

Thangavel R, Kannan AG, Ponraj R, Thangavel V, Kim D, Lee Y (2018) High-energy green supercapacitor driven by ionic liquid electrolytes as a ultra-high stable next-generation energy storage device. J Power Sources 383:102–109. https://doi.org/10.1016/j.jpowsour.2018.02.037

Tobiszewski M, Mechlińska A, Zygmunt B, Namieśnik J (2009) Green analytical chemistry in sample preparation for determination of trace organic pollutants. TrAC Trends Anal Chem 28:943–951. https://doi.org/10.1016/j.trac.2009.06.001

Tomšik A, Pavlić B, Vladić J, Cindrić M, Jovanov P, Sakač AM, Vidović S (2017) Subcritical water extraction of wild garlic (*Allium ursinum* L.) and process optimization by response surface methodology. J Supercrit Fluids 128:79–88. https://doi.org/10.1016/j.supflu.2017.05.012

Torres-Mayanga PC, Azambuja SPH, Tyufekchiev M, Tompsett GA, Timko MT, Goldbeck R, Rostagno MA, Foster-Carneiro T (2019) Subcritical water hydrolysis of brewer's spent grains: selective production of hemicellulosic sugars (C-5 sugars). J Supercrit Fluids 145:19–30. https://doi.org/10.1016/j.supflu.2018.11.019

Vekariya RL (2017) A review of ionic liquids: applications towards catalytic organic transformations. J Mol Liq 227:44–60. https://doi.org/10.1016/j.molliq.2016.11.123

Wai LT, Lim JS, Hashim H, Mustaffa AA, Ho WS (2016) Review of pre-combustion capture and ionic liquid in carbon capture and storage. Appl Energy 183:1633–1663. https://doi.org/10.1016/j.apenergy.2016.09.103

Wu H, Li C, Liu R, Zhang A, Xiao Z, Ma L, Li J, Deng S (2018) Simultaneous extraction of oil and tea saponin from *Camellia oleifera* Abel. seeds under subcritical water conditions. Fuel Process Technol 174:88–94. https://doi.org/10.1016/j.fuproc.2018.02.014

Yang T, Wang J, Li B, Kai X, Li R (2017) Effect of residence time on two-step liquefaction of rice straw in a CO_2 atmosphere: differences between subcritical water and supercritical ethanol. Bioresour Technol 229:143–151. https://doi.org/10.1016/j.biortech.2016.12.110

Yu X, Boyer MJ, Hwang GS, Manthiram A (2018) Room-Temperature Aluminum-Sulfur Batteries with a Lithium-Ion-Mediated Ionic Liquid Electrolyte. Chem 4:586–598. https://doi.org/10.1016/j.chempr.2017.12.029.

Zabot GLZ, Bitencourte IP, Tres MV, Meireles MAA (2017) Process intensification for producing powdered extracts rich in bioactive compounds: an economic approach. J Supercrit Fluids 119:261–273. https://doi.org/10.1016/j.supflu.2016.10.003

Zabot GLZ, Moraes MN, Meireles MAA (2018) Process integration for producing tocotrienols-rich oil and bixin-rich extract from annatto seeds: a techno-economic approach. Food Bioprod Process 109:122–138. https://doi.org/10.1016/j.fbp.2018.03.007

Zarski A, Ptak S, Siemion P, Kapusniak J (2016) Esterification of potato starch by a biocatalysed reaction in an ionic liquid. Carbohydr Polym 137:657–663. https://doi.org/10.1016/j.carbpol.2015.11.029

Zeng S, Tao C, Parnas R, Liang WJB, Liu Y, Lu H (2016) *Jatropha curcas* L. oil extracted by switchable solvent N, N-dimethylcyclohexylamine for biodiesel production. Chin J Chem Eng 24:1640–1646. https://doi.org/10.1016/j.cjche.2016.04.023

Zhang J, Zhang S, Han J, Hu Y, Yan R (2015) Uniform acid poly ionic liquid-based large particle and its catalytic application in esterification reaction. Chem Eng J 271:269–275. https://doi.org/10.1016/j.cej.2015.0

Chapter 9
Applications of Supercritical Carbon Dioxide in the Rubber Industry

Asim Ali Yaqoob, Mohamad Nasir Mohamad Ibrahim, Khalid Umar, Akil Ahmad, and Siti Hamidah Mohd. Setapar

Contents

9.1 Introduction

Increasing interest in different fields such as herbal medicine, natural products, and the rubber industries has shown that the conventional and traditional techniques such as the solvent extraction process and hydrodistillation are not effective because the results are insufficient and unstable. Therefore, the solvent extraction process has mostly used organic solvents that can barely produce an extract. This extract is

A. A. Yaqoob · M. N. M. Ibrahim (✉) · K. Umar
School of Chemical Sciences, Universiti Sains Malaysia, Penang, Malaysia
e-mail: mnm@usm.my

A. Ahmad (✉)
School of Industrial Technology, Universiti Sains Malaysia, Penang, Malaysia
e-mail: akilchem@yahoo.com

S. H. Mohd. Setapar
Department of Chemical Processes, Malaysia-Japan International Institute of Technology, University Technology Malaysia, Kuala Lumpur, Selangor, Malaysia

Inamuddin, A. M. Asiri (eds.), *Advanced Nanotechnology and Application of Supercritical Fluids*, Nanotechnology in the Life Sciences,
https://doi.org/10.1007/978-3-030-44984-1_9

usually not suitable for organoleptic and other health reasons. Organic solvents also provide only inadequate selectivity and have some associated compounds in addition to the significant active compounds. Another traditional process is the distillation process, which allows volatile-based compound separation but with high temperature effects (Mićić et al. 2011). So, the conventional methods have failed currently to fulfil the demands of the modern world. Now, to overcome the problems associated with these conventional techniques, the scientific community has developed a new technique called supercritical fluid extraction (SFE), especially to enhance the selectivity and working efficiency of extraction. Supercritical fluid extraction is a new emerging technology and is still not extensively used, but as a novel method of extraction its use is increasing day by day. SFE is defined as a process that separates one component from another compound by using a supercritical fluid such as CO_2. CO_2 serves as a solvent in SFE that can extract from both solid and liquid matrices. Carbon dioxide has been declared by some researchers to be a most stable and highly used supercritical fluid that can be improved by using cosolvents (methanol or ethanol) by the doping process to enhance its activity. The scientific community has found that CO_2 is very simple to use, cost effective, toxin free, nonflammable, chemically stable, removable from the extract without any effort, and with high affinity to volatile materials. Furthermore, change in temperature and pressure above the critical level of CO_2 carried out a distinct change in dielectric constant and density; that is, the solvent influence of supercritical CO_2 can be attained. Supercritical extraction with CO_2 has some extraction limitations and conditions, such as the requirements of critical temperature and critical pressure, that is, 31 °C and 74 bar, respectively. High temperature and pressure demands make it more costly even than conventional methods. Furthermore, the addition of cosolvents as modifiers may bring minor changes in these parameters. SFE is a very significant and effective technique that is used only for substantial compensation and notable advantage. The CO_2 itself is a nonpolar solvent with limited dissolving stability, so it cannot serve as a solvent in all cases, specifically in cases of polar solutes. Then, modifiers can increase the ability and stability of extracted materials. Food-based modifiers such as ethanol can serve and are found very helpful in the assembly of extracted material. Some benefits are decreased by utilizing a solvent that is in vaporous form at 20–25 °C (Cunha et al. 2018; Kavoura et al. 2019; Melgosa et al. 2019). There are some reported advantages of the supercritical extraction process as compared to conventional extraction methods; for example, SFE dissolving stability is dependent on the conditions of temperature and pressure level, high-boiling substances are removable at lower temperatures, and thus are toxic free without any residue, and thermally labile material can be removable with less damage at a lower temperature range. The supercritical fluid properties can be changed by fluctuating temperature and pressure, permitting selective extraction. For example, extraction of volatile oils from plants can be achieved with lower pressure at 100 bar whereas liquid extraction also eliminates lipids. By using CO_2 as a solvent, lipids can be removed at elevated pressures, and addition of the cosolvent ethanol can also remove phospholipids at higher pressures. Unsaturated fatty acids and polyphenols can be extracted from wine wastes by the same route

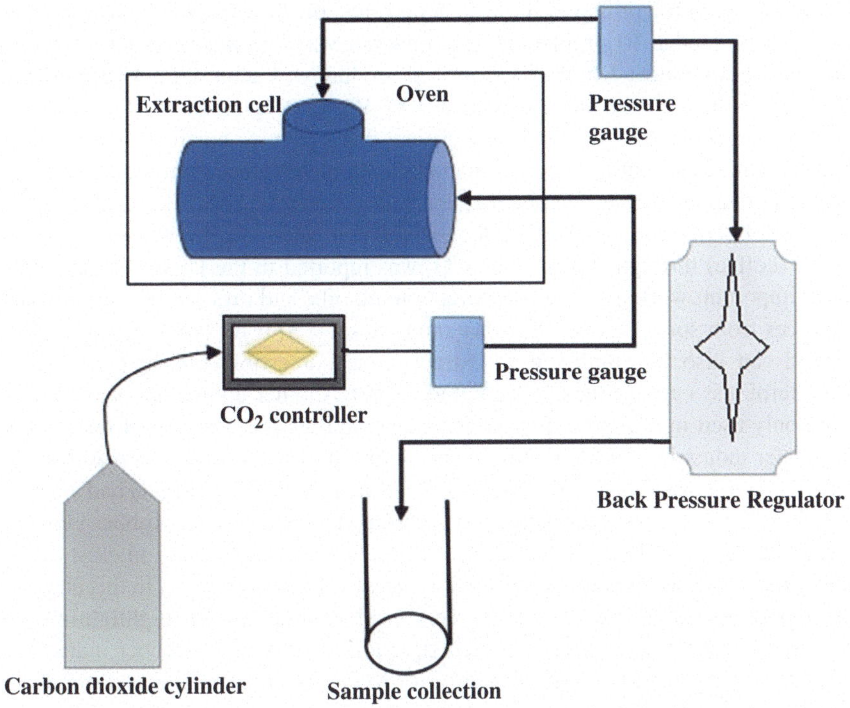

Fig. 9.1 Schematic diagram of supercritical extraction through carbon dioxide

(Aizpurua-Olaizola et al. 2015). In simple words, extraction is a diffusion-based method in which the solvents are very important because a solvent is used to diffuse into the targeted material and to extract material from that targeted material (the matrix) and diffuse it back into the solvent. The diffusivity ability is faster in supercritical fluids than in liquids, so the extraction rate is much faster than in the conventional route. Furthermore, because insignificant viscosities and surface tension are more absent than in liquids, the used solvent can enter into the targeted material (matrix) unreachable by liquids. Extraction using an organic liquid may take longer, but in the case of SFE, a maximum of 1 h might be required (Kinney et al. 2017; Naziri et al. 2016; Özkal and Yener 2016; Azwanida 2015). The basic principles and working routes of the supercritical fluid extraction system are shown in Fig. 9.1. The SFE has a current countercolumn and collection vessel to extract material from the targeted matrix. The liquid-based material is carried out through a packed column, producing a countercurrent among CO_2 mean flow upward and material flow downward. Ultrafiltration technology is used to filter the serum material to eliminate excessive water and reduce the size of particles before use in the SFE method. By adjusting both parameters, temperature and pressure, CO_2 can extract the desired material through selectivity. Formerly, CO_2 proceeded into the extractor and was heated at a definite temperature to attain the supercritical phase and also was forced to attain the required pressure. In the general route, the flow rate of CO_2 (supercritical) was sustained at 30 g/min and serum flow rate was 5 g/min. The total extraction time was set to almost 1 h; the extraction procedure was achieved between 40 and 50 °C while the pressure range was set at 100–300 bar (Razak and Arif 2014).

The distinct characteristics of supercritical fluids provide positive advantages to biochemical separation methods. Some applications have been entirely industrialized and commercialized. The largest and most common application is the decaffeination of coffee and tea. In 1974, the extraction process of caffeine from the main source (coffee) through supercritical CO_2 was reported in the United States (US). Other important working zones of extraction are oils, and different types of aroma resources from spices such as brewery industries use SFE for hop extraction. This method can also be employed to extract some comestible oils and produce a cholesterol-less egg residue (Yousefi et al. 2019). Another unique application very commonly used in Malaysia and other Asian countries is extraction of material in the rubber industries. The scientific name of the rubber tree is *Hevea brasiliensis*, which is commonly found in the Amazon Rainforest. In 2007, Malaysia had approximately 1,229,940 ha of latex plantation according to the Malaysian Rubber Industry Board (Lee et al. 2013). Therefore, SFE through CO_2 earned great interest in the rubber industries for generating different types of valuable products. In this chapter, the importance of SFE with the solvents used is summarized, highlighting the importance of CO_2 as a solvent in the extraction process in the rubber industries. This chapter might be particularly useful for Malaysian readers and researchers to plan future research directions because the scope and relationship with SFE in the presence of CO_2 has been summarized extensively in different rubber industries.

9.2 Rubber Industry

In the modern era, no one can ignore the importance of the rubber industries because rubber products have become an essential factor in the human lifestyle. Rubber is considered to be a vital key material for the survival of modern life on the Earth. Rubber-based products are one better source to fulfil the current demands of the world population. The rubber industry transforms elastomers (rubber) into different valuable products, such as rubber mats, automobile parts, tires, and exercise stretching bands. The rubber industries include latex material, polymer-based material, natural rubber (from *Hevea brasiliensis*), and synthetic rubbers. Rubber actually belongs to isoprene, a natural hydrocarbon polymer, also known as *cis*-1,4-polyisoprene, that is present naturally in the form of a milky latex in several plants. Many other materials such as fatty acids, inorganic and organic materials, proteins, and salts are also present in natural rubber but in trace amounts. Recently, a group of researchers stated that rubber can be produced artificially by using different chemical materials. In simple words, rubber can be well defined as an elastic, sticky solid that when produced in the form of milky latex is called natural rubber or latex obtained from various rubber plants. So, it can be distinguished by various types generally used in production of industrial goods such as tires, medical devices, and shock mounts as well as ordinary products for human use (Hernández Santana et al. 2018). Natural rubber is present in Southeast Asia in very large quantities, especially in Malaysia, Thailand, and Indonesia, whereas another type, synthetic rubber, is frequently artificially produced by different chemical materials and methods in highly industrialized developed countries such as Japan, Germany, the United States, and east and west Europe. Brazil is one of the developing countries producing synthetic rubber at a high industrial scale to improve the country economy and financial condition (Jomo 2019).

9.2.1 Natural Rubber

Natural rubber (NR) is an essential cultivated material used in the manufacture of different valuable products. The rubber industries and their products have a key position in the socioeconomic textile picture of several emerging countries. More than 20 million people work in rubber cultivation to fulfil their basic requirements of survival and livelihood (Shattuck 2018). NR is also known as latex, caucho, gum elastic, Amazonian, or *caoutchouc* rubber. It is defined as an elastic material produced from the latex of trees of the genera *Ficus* and *Hevea*. In terms of chemical constituents, NR consists of the organic compound isoprene and some other minor organic compounds with water. NR is an elastic hydrocarbon in nature because polyisoprene is classified as an elastomer. Presently, NR is obtained from trees of the genera *Ficus* and *Hevea* in the form of latex. The latex material is a sticky, colloid-like milky liquid harvested by making cuts on the tree bark and accumulating the latex fluid. The collected fluid is refined for marketable processing and used for the production of many commercial items. It is possible to produce NR from many different plant species, but the most common and significant source is *Hevea brasiliensis*, which is present in large quantities in tropical America. These tropical trees grow under some special environmental conditions such as temperatures between 24 and 30 °C and rainfall less than 2000 mm per year. These stated conditions indicate the ideal altitude of these tree is about 600 m, and the leading rising area is about 10 °C on the Equator. However, these trees are also present in north Guatemala, China, and Mexico. Sir Henry Alexander Wickham, known as a bio-thief and British explorer, smuggled 70,000 rubber plant seeds into the East Indies (Bekkedahl 1945). After that, rubber cultivation began in parts of Asia, and nowadays also in Malaysia, Thailand, and Indonesia, known as one of the largest NR suppliers to the world. These countries contribute almost 90% of the world's NR. Recently, Africa started rubber plantations to cover their needs. The natural sources for NR latex include especially the genera *Ficus* and *Hevea brasiliensis* (Amazonian rubber tree), which are members of the spurge family, Euphorbiaceae. The species *H. brasiliensis* is the most suitable because it grows under certain environmental conditions that allow significant cultivation. Congo rubber, which earlier was a key source for rubber production, comes from the genus *Landolphia*.

Dandelion is also a good source for NR because of the presence of latex. This latex shows a similar class as the NR from pure rubber trees, gutta percha, and *Manilkara* species, which may also serve as a source agent for the production of NR (Kohjiya 2015). NR industries have been industrialized over the past 100 years. Industrial development has relied on targeted exploration and growth in several areas such as natural science, productivity investigation, chemistry, physics, and a plethora of growth applications. NR has many applications because it is an admirable barrier to pathogens and wastewater, which is why latex is used mostly in surgical gloves and condoms to keep people safe from diseases, with major roles also in the automotive industries, adhesives, etc. Malaysia, Thailand, and Indonesia are considered as the world's largest producers of rubber, followed by China, Sri Lanka, India, Vietnam, and many other countries. According to a published report in 2001, approximately 6.5 million tonnes NR are produced every year worldwide to fulfil the industrial demands (Chua and Oh 2010).

9.2.2 Synthetic Rubber

Synthetic rubber (SR) plants were first used around 1945 in Europe, America, and Japan, and in 1960, the SR demand increased. According to reported information, the world is producing 11.5 million tonnes SR material every year. Currently, the US is one of the top SR producers supplying rubber to other countries (Rodgers 2015; Thomas and Stephen 2010; Erickson et al. 2011). SR is a white-coloured, crumbly, brittle, plastic-based mass that is handled and vulcanized by the same route as NR is treated. In the short term, SR is produced artificially with properties similar to those of NR. Mostly, SR is gained by using the polymerization or poly-condensation method for unsaturated monomers. There are many varieties of SR, reflecting the several diverse applications and mechanical/chemical properties. In World War II, the US was losing its NR sources because of the war crises. So, to meet the rubber industry demands from their country, governments took steps toward rubber plants to produced synthetic rubber by artificial methods and different manufacturing industries (Myhre et al. 2012). In 1941, the SR production was 8000 tons; when SR demand increased suddenly in 1945, its production reached 820,000 tons (Hoisington 1960). After the war, the government analysed and sold SR to different industries for commercial use. Therefore, the US is still known as the largest producer for SR and is supplying SR material to the whole world for commercial uses. SR is produced from several petroleum monomers such as polymers. The most dominant SR is styrene-butadiene rubber, which is derived from 1,3-butadiene and styrene through the copolymerization method. Butyl SR is fabricated from different sources such as chloroprene (2-chloro-1,3-butadiene), isoprene (2-methyl-1,3-butadiene), and isobutylene with a minor ratio of isoprene, and many other types are produced by the same route. The common examples or types of both natural and synthetic rubber are summarized in Table 9.1, with their significant applications. Although SR has diverse characteristics, they have some common

Table 9.1 Different types of rubber with their properties and potential applications

Sample number	Rubber/elastomer	Basic properties	Application	Source (references)
1	Natural rubber	Oil-resistant, inflamed by solvents; weathering via oxygen, UV light properties	Tires, bridge, shock mounts, couplings, drinking seals, footwear, condoms, conveyor belts, hoses, rolls, building bearings, moulded products, gloves, therapeutic devices, adhesives, etc.	Kohjiya (2014)
2	Butyl rubber	High resistance against aging and chemicals, lower gaseous permeability, good abrasion resistance, high mechanical properties	Cable insulation, automotive hoses, tire liners, rubberized fabrics, seals	Abe and Ishiguro (2013)
3	Chloroprene	Excellent resistance to oils, grease, good weathering resistance, good flame retardance, resistant to aging	All technical products, conveyor belts, seats belts, clutches, chains, pneumatic suspension structures	Franta (2012)
4	Acrylonitrile butadiene	Excellent resistance to fuel and oils, excellent abrasion resistance, high temperature properties	Technical products manufacturing, mats, motor vehicle parts production, oil hoses, plates, rollers	Thomas and Stephen (2010)
5	Isoprene	More uniform properties, cleaner, highly transparent rubber	Production of mechanical products, such as heating or cooling hoses, construction sections, effective tires, different foodstuffs, utensils	Franta (2012)
6	Polybutadiene	Flexible at lower temperatures, good elasticity	Used in clutches, drive belts, engine bearings, tires, conveyor belts, water seals	Thomas and Stephen (2010)
7	Styrene butadiene	Improved abrasion resistance, lower elasticity, good aging, admirable electrical insulation material, good heat resistance	Tire industry, conveyor and driver seat belts, water seals, mechanical rubber-based products	Mao et al. (2013)

(continued)

Table 9.1 (continued)

Sample number	Rubber/elastomer	Basic properties	Application	Source (references)
8	Ethylene propylene	Flexibility at low heating, good resistance against weathering and heat, high electrical properties	Moulded products, wires, cables, weather stripping, water seals, roofing, ditches, landfill	Stelescu et al. (2013)
9	Nitrile	Resistant to vegetable oil, fuel, organic solvents	Sealants, gaskets, shoe soles, fuel-resistant adhesives, roll coverings, conveyor belts, gloves, oil-drilling equipment	Franta (2012)
10	Silicone	Excellent stability, oil-resistant, resistant to solvents, chemically inert	Wire insulation, adhesives, water seals, specialty moulded, respirators, gaskets, gas masks, medical tubing, food tubs, surgical instrumentation	Shit and Shah (2013)
11	Polysulfide	Good oil resistance, lower temperature, weathering, lower gas permeability	Roller covering, moulded goods, sealants, hose liner, propellant binder, glass sealants, solid rocket gaskets	Syao and Malysheva (2014)
12	Reclaimed rubber	Shorter polymeric chains, relaxed processing, lower mixing duration, less consumption power, lower cost	Tyres, floor mats, adhesives, inner and outer tubes, machine-driven goods, rubberized asphalt	Myhre et al. (2012)
13	Hypalon rubber	Good heat ageing, high chemical resistance, good lower gaseous permeability, good electrical properties, good oil resistance	Static seals explore humid weather conditions	Sisanth et al. (2017)
14	Epichlorohydrin	Good heat, fuel, petrol resistance, good resistance to alkalis, acids, ozone	Automotive fuel, bladders, diaphragms, modern rollers	Franta (2012)
15	Ethylene propylene diene monomer	Good physical properties and high resistance to polar solution	Good bladders, improved cable connectors, good insulators, gaskets, seals	Nair et al. (2012), Zanchet et al. (2012)
16	Fluoroelastomers	Excellent resistance against ozone, fuel, oil, and chemicals	Improved cable connectors, good insulators, gaskets, seals	Petrova and Portnyagina (2014)

(continued)

Table 9.1 (continued)

Sample number	Rubber/elastomer	Basic properties	Application	Source (references)
17	Fluorosilicones	Good electrical power and modest oil resistance	Aerospace fuel system and their components, gaskets, seals, hose linings, O-rings	Tonge (2009)
18	Hydrogenated nitrile rubber	Good tear and tensile strength, high elongation, good abrasion resistance, high compression	Fuel and oil industries	Bystritskaya et al. (2013)
19	Perfluoro elastomers	Limited at lower temperatures	Core sleeves, fuel and oil industries	Aizawa et al. (2010)
20	Polyacrylic	High resistance to hot hydraulic oil, wreathing, zones, and oxidation	Production of automotive transmissions parts	Maiti et al. (2012)
21	Polyurethanes	High humidity resistance and good resistance to acids	Good abrasion-resistant coatings, haul-off pads, linings, gaskets, diaphragms, hoses, tires, wheels	Clemitson (2015)

properties such as flame retardance, improved abrasion resistance, better resistance to oil, high elasticity, good heat resistance, upgraded aging resistance, good electrical insulation, and flexibility at lower temperatures. SR has provided valuable and significant improvements at the industrial level to produce such items as tires for aircraft, cars, trucks, and motorcycles, driver seatbelts, seals, medical and surgical instruments, moulded parts, floor coverings, and utility hoses, but SR requires some conditions to preserve or transport it. It should be stored in a dry environment; some SR require a cool environment for storage, avoiding direct sunlight, with suggested storage time to be not more than 5 years (Williams 2013). There are also some specified packaging conditions for SR: SR material can be packed in polyethylene plastic sheets, and processed SR can be packed in the form of cardboard boxes wrinkled with polyethylene film to avoid impurities. Flexible bulk packaging also usable for SR.

9.3 Importance of the Rubber Industry in Malaysia

Malaysia is one of the big rubber producers, contributing almost 46% of NR production in the whole world. Malaysia is the fifth in order in producing NR after China, Thailand, Indonesia, and Vietnam and is also a large consumer of NR products. In 1877, natural rubber plantation first began when seed was imported from the Amazon Basin to Malaysia: these seeds were used in a trial experiment to produce rubber. Then, after seedlings were fruitfully planted, they gave unexpected results because of the favourable Malaysian environment and certain other factors such as rainfall and temperature range (Verheye 2010). In the Malaysian region an NR plantation is possible anywhere because of the favourable climate and nature of soil, such as constant temperature between 24 and 30 °C, rainfall under 2000 mm annually without any drought, rich subterranean soils with excellent drainage, preferably inelastic, highly oxidized, and reaction is acidic in nature. A sufficient source of cheap labour is a significant feature for plantations and later for the collection of NR latex from the trees. Most active growth for NR is in the western coastal plains areas of Malaysia because it is easy to export from the coastal areas near ports. The largest production of NR is found in the Johor Bahru state of Malaysia. The rubber industries of Malaysia produce more than 500 items based on latex material such as tires, tire-based products, and various industrial and commercial products. In Malaysia, NR producers include plantations, homegrown smallholders, and multinationals in joint cooperation with other countries such as the US, Europe, and Japan. Latex products are the largest portion of the NR industry, including 125 producers making latex thread, medical gloves, catheters, condoms, etc.: these products constitute 81% of the total worth of NR exports. Malaysia has maintained its position as a leading manufacturer and exporter of such latex-based products. Malaysia has provided 1.07 million hectares for cultivation of NR to improve the economy because China is a big customer for the Malaysian rubber industries. Malaysian industries include 185 companies that are producing an extensive variety of NR products such as belting, tubing, hoses, seals, mountings, automotive sheeting, electronic machinery, electrical equipment, and construction equipment. The current need for the rubber industries of Malaysia is to diversify more, highlighting value-added products, and to develop new technology for making rubber-based products that can offer opportunities in different mechanical, engineering, marine, and medical applications (Ali et al. 2018; Sethuraj and Mathew 2012; Yusoff et al. 2019; Kawano 2019). The government is encouraging the growth of NR resource industries to expand the country's sources and is providing reasonable policies to the rubber industries. The export ratio in Malaysia increased more in 2019 than in 2018, that is, to 49,797 tonnes (Fig. 9.2). The Malaysian rubber industries have become an important source for providing top-class glove material to the world for different applications. The produced latex material consumption ratio at industrial level is also shown in Fig. 9.2. Currently, Malaysians are facing some challenges in the progress of the NR industry such as competing against low-priced products from cheaper producers, specifically China and India. So, the Malaysian Rubber Industrial Board should focus on quality and high-productivity products through low-cost methods.

9.4 Supercritical Carbon Dioxide in the Rubber Industry

Supercritical fluid extraction (SFE) is the most active and effective mode to extract the significant constituents of botanicals. Some commonly used solvents have been reported, such as CO_2, water, methane, ethane, and acetone (Table 9.2). Their

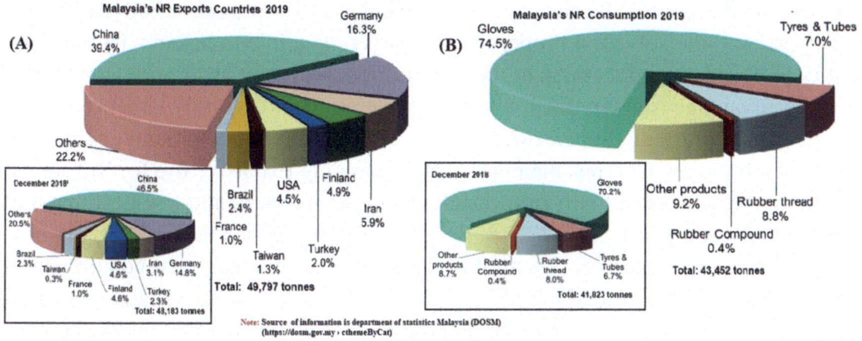

Fig. 9.2 (**a**) Natural rubber (NR) export by countries, 2019. (**b**) Consumption of NR in different sectors. (From Department of Statistics Malaysia (DOSM). https: www//dosm.gov.my)

Table 9.2 Common supercritical fluid extraction (SFE) solvents and their critical properties (Sapkale et al. 2010; Machado et al. 2013)

Sample number	SFE solvent	Chemical formula of solvent	Critical temperature (°C)	Molecular weight of solvent (g/mol)	Critical pressure MPa (atm)
1	Carbon dioxide	CO_2	30.95	44.01	7.38 (72.8)
2	Water	H_2O	373.946	18.015	22.064 (217.755)
3	Ethanol	C_2H_5OH	240.75	46.07	6.14 (60.6)
4	Methanol	CH_3OH	239.45	32.04	8.09 (79.8)
5	Acetone	C_3H_6O	234.95	58.08	4.70 (46.4)
6	Ethylene	C_2H_4	9.25	28.05	5.04 (49.7)
7	Propylene	C_3H_6	91.75	42.08	4.60 (45.4)
8	Ethane	C_2H_6	32.15	30.07	4.87 (48.1)
9	Methane	CH_4	190.4 K	16.04	4.60 (45.4)
10	Propane	C_3H_8	96.65	44.09	4.25 (41.9)
11	Cycrohexanol	$C_6H_{12}O$	356.0	100.158	38.00
12	Monofluoromethane	CH_3F	44.06	34.033	58
13	Cholortrifluoromethane	$CClF_3$	28	104.46	38.7
14	Isobutanol	$C_4H_{10}O$	275.00	74.122	42.00
15	Isopropanol	C_3H_8O	235.00	60.1	47.00
16	Krypton	Kr	−63.04	83.8	54.03
17	Xenon	Xe	16.6	131.29	57.6
18	Nitrous oxide	N_2O	36.05	44.013	71.5
19	Ammonia	NH_3	132.04	17.031	112.5

properties were described by Reid et al. in 1987. However, the scientific community declared that the CO_2 was the best extraction solvent for significant botanicals. Interest developed for the application scope of supercritical CO_2 because the CO_2 solvent had the ambient critical temperature of 31 °C. Therefore, biotic materials can be treated at 35 °C; the density of the solvent (CO_2) at pressure (200 bar) is near to hexane, and the solvation properties are also comparable to hexane. CO_2 has the ability to dissolve triglycerides in the supercritical region at absorptions up to 1%. The key advantage is less reduction in this temperature and a lightly larger reduction rate in critical pressure. Supercritical fluids can yield a product without any solvent residues. Moreover, CO_2 has many properties enabling extensive use for SFE in the rubber industry as it is a toxin-free, tasteless, odourless, inert, and low-cost material (Seghar et al. 2019; Tang et al. 2018; Shi et al. 2019). Reche et al. in studying SFE using CO_2 showed the SFE potential in the presence of *N*-nitrosamines in natural rubbers. Reche et al. studied the SFE parameters when CO served as the extraction fluid throughout the entire process (Reche et al. 2002a); they also described the *N*-nitrosamines in natural latex through sequential SFE and derivatization using CO_2 as the solvent to enhance product selectivity and decrease material handling. Gas chromatography with phosphorus and nitrogen in sensitive exposure ensured product selectivity (Reche et al. 2002b). Kojima et al. studied the segmented behaviour of polyisoprene rubber and its supercritical carbon dioxide effect considering the phase of cross-link-based polyisoprene behaviour in CO_2 at ambient temperatures between 19.85 and 179.85 °C and 0.1–20 MPa pressures (Kojima et al. 2005). Lee et al. studied the extraction of natural rubber from *Hevea brasiliensis*-based seed oil through supercritical CO_2 as solvent and with Soxhlet techniques. The major objective of Lee et al. was to compare the two methods to enhance the value of SFE with CO_2. Experimental observations on the basis of the physical properties of seed oil showed that oil produced by SFE with CO_2 was fresh and pure with an agreeable smell whereas the oil by the Soxhlet process was brownish in colour and less clean. Rubber seed oil extraction by SFE with CO_2 was 33.652% g and with Soxhlet extraction was 7–12% g. SFE also required less time for extraction than Soxhlet extraction. Thus, SFE was declared superior to other methods (Lee et al. 2013). Mangili et al. explored the devulcanization process of different ground tire-based natural rubber in the presence of supercritical CO_2. A 2^4 factorial experimental design was used to explore supercritical CO_2 influence on the devulcanization of ground tires from the life cycle of tires using diphenyl disulfide (DD) as the reagent. Their purpose was to study the effect of pressure, temperature, nature of devulcanizing reagent, handling duration, and their significant interactions. These outcomes are significant because this devulcanization method can take place rapidly and at comparatively lower pressure, with ensuing energy saving (Mangili et al. 2014). Abd Razak et al. investigated vitamin E by using SFE from deproteinised NR serum and analysed by the HPLC method. They clarified that CO_2 usage with SFE is suitable because it is eco-friendly to extract the components with lower costs and no toxicity. Thus, SFE at lower temperature is better for vitamin E extraction because a high temperature can destroy sensitive components (Razak and Arif 2014).

Tessanan et al. studied the preparation of microcellular NR in the supercritical CO_2 method, a promising alternative source to exchange the conventional foaming agent in response to ecological concerns. Some critical parameters for foaming as well as CO_2 saturation time, pressure range, temperature scale, and different cross-linking properties of NR were investigated. The outcomes showed a decrement of regular cell size, size distribution, and growth ratio reliant on improvements in saturation duration and pressure. Furthermore, the addition of saturation temperature increases cell-size distribution ratio and expansion (Tessanan et al. 2019). Zhijun Liu et al. investigated waste tire-based rubber material devulcanized using supercritical CO_2 as the medium in a reaction, by observing the effects of pressure, temperature, reaction duration, and concentration of diphenyl disulfide during devulcanization. The results indicated that the concentration of diphenyl disulfide is the most essential factor. However, change of pressure had no major effect on the devulcanization process (Liu et al. 2015). In 2014, Song et al. prepared silicon synthetic rubber foam by using supercritical CO_2. The sample was pre-treated and fully cured earlier and after the foam process through supercritical CO_2. The results showed that the supercritical CO_2 foaming procedure can be effectively used to fabricate cross-linked rubber by controlling and handling the cross-linked network (Song et al. 2014). Ikeda et al. elaborated the importance of the recycling process through supercritical CO_2 for sulfur-based cross-linked NR. Their main objective was to elaborate the presence of supercritical CO_2 during recycling of sulfur linked with NR (Ikeda 2014). Gao et al. studied graphene fermentation in the NR matrix to make conductive natural rubber through the supercritical CO_2 method, the first work to show that the ultrasound-assisted supercritical CO_2 technique could make graphene-based conductive NR. The NR was completely inflamed with ultrasound-assisted supercritical CO_2 infiltration and rapidly ensuing decompression, and the conductivity of graphene with NR film reached 0.2 S/m. This prepared graphene with NR exhibited application as a strain sensor for fine movement. It was found that supercritical CO_2 is the most biologically and ecologically gentle solvent of those used. The authors believe that the supercritical CO_2 method has become a multipurpose tool that can be functionalized to natural polymers (Gao et al. 2019).

9.5 Application of SFE in the Rubber Industries

SFE with CO_2 is very valuable as a substitute in organic solvents in a number of laboratories and different industrial processes. Water was used very commonly but recently CO_2 has received attention for use as a supercritical fluid at different industrial levels to explore the value of different applications. Industry, especially the rubber industries, prefer to use SFE with CO_2 because this combination offers such advantages as sample recovery, maintaining purity factor, high selectivity in products, and very short processing time, around 10–60 min. SFE with CO_2 reduces product contamination and improves environmental safety. CO_2 as a solvent in different industrial processes and with SFE does not produce any emissions harmful to the environment. SFE technologies in various industrial applications have undergone substantial development in recent years. Several manufacturing sectors are considered: pharmaceutics, energy, materials, food, cosmetics, and wastewater or other waste material treatment industries. Some common applications in rubber industries that depend on SFE with CO_2 are described next, and some other

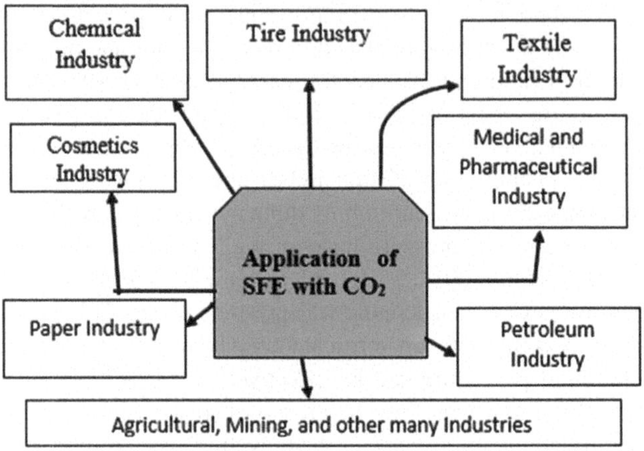

Fig. 9.3 Different industries that use SFE with CO_2 in product extraction from their sources

industries that are not rubber product based have also used SFE with CO_2 in their product manufacturing processes (Fig. 9.3). Supercritical fluid processes are considered the most related, such as formulation, waste treatment, extraction, sterilization, impregnation, cleaning, and energy (Machado et al. 2013).

9.5.1 *Tire Industries*

Rubber-based automobile tire recycling is a major problem for the rubber industries because substantial ecological alarms have been raised, and although many methods for the recycling of scrap tires have been proposed, a significant outcome has not been obtained. Scrap tire recycling can increase further applications in civil engineering to fabricate new novel products and reduce product size. The rubber crumb is an essential form in rubber tire-based recycling industries and the industrial marketplace for rubber growth. Currently, rubber is mostly formed by cryogenic grinding, reclaiming, ambient grinding, and solution grinding. Despite developments, all these methods had some disadvantage such as generation of heat in the rubber, and even at a high scale heat can reduce the rubber and lead to a potential combustion hazard if not cooled properly. In 2004, Tzoganakis et al. described a novel process for rubber crumb devulcanization from used tires supercritical CO_2. Supercritical CO_2 was inoculated in a twin-screw extruder for rubber crumbs to swell them and facilitate rubber extrusion. Scrap rubber can be treated at lower than automatic shear and some forces at several operating conditions carried out to degrees of devulcanization (Tzoganakis and Zhang 2004). Different types of recycled rubber crumbs are considered, and the injection rate of rubber powder flow to the extruder affects the degree of devulcanization with a major impact on the processability and viscosity of the product. Although the concentration of CO_2 is a key factor for extrusion and flow rate, it was verified to have minor effects on devulcanization (Adeola 2018; Bairgi 2015; Latsky et al. 2018; Thomas et al. 2013). Thus, supercritical CO_2 has earned great attention in the tire industries to make it more prolific and valuable.

9.5.2 Cosmetics Industry

Currently, the most demanding ongoing research is on green synthesis and the extraction of natural cosmetic products. The cosmetics industries are using different methods to make their products more valuable. Cosmetics are materials used to improve the outlook, fragrance, face, and surface of the target body. Many antioxidant-like carotenoids, tocopherols, and polyphenols that are present in a large variety of vegetables and fruits, such as beta carotene found in carrots, can be effectively extracted by using supercritical CO_2. Furthermore, extracts produced from rosemary, thyme, chamomile, and lavender are possible from organic solvents. Artificial and some certified organic cosmetics are becoming more popular, but certain chemicals in artificial cosmetic products may cause skin problems such as fungal attack if absorbed into the skin. So, green synthesis of cosmetics is considered to provide safe products, and thus the industry prefers using SFE with CO_2. The increasing trend of natural cosmetics is safer than artificial products (del Valle 2015; Gandhi et al. 2017; Jokić et al. 2017; Kitrytė et al. 2018; Michalak et al. 2017). The herbal components used as cosmetic materials reduce the probabilities of side effects of the product, ultimately increasing the value of natural cosmetics. Many products can be extracted by using SFE such as skin care products, hair products, sunscreen creams, fragrances, makeup, and hair colours.

9.5.3 Medical Rubber Industry

The role of supercritical CO_2 is essential in the medical rubber industry, and extracts produced from different latex plants, diterpenes, antioxidants, triterpenes, and tetraterpenes attract great attention in the medicinal industry. Supercritical fluid extraction used for making fine powders is specific for designing active principles. The Pierre Fabre laboratories in France developed the method to produce fine powders by using supercritical CO_2. Recently, critical pharmaceuticals companies in the United Kingdom provided biodegradable and biocompatible polymers produced by using eco-friendly supercritical CO_2 for various medical and pharmacological applications (Kankala et al. 2017; Muhammadi et al. 2015). The latex-based medical products include condoms, surgical gloves, and many other safety tools. The medical rubber industry has achieved, by substantial study, producing first-class gloves in the presence of chlorine to remove proteins that cause different allergy diseases. The anxiety from HIV-AIDS produced a demand for latex-based gloves. Surgical medical gloves can be manufactured by using nitrile and vinyl but rubber industry experts say they are uncomfortable and more likely to leak. The most common medical-based rubber products are rubber caps that are especially produced for medical applications (Cheng et al. 2019; Mahdi et al. 2015; Ottolina et al. 2016). These caps consist of polychloroprene and natural rubber, rubber tubes for transfusion and infusion sets (a module of infusions suitable for liquid drugs), birth control products (rubber is used for birth control products such as condoms and diaphragms), and rubber medical masks for different purposes: breathing masks, also called oxygen masks, are mass produced by using NR and silicone-based rubber, rubber catheters (medical catheters), and rubber bulb syringes (syringes with a dual cannula rubber bulb and rubber bulb/cannula are offered with different patterned valves; the syringe is employed in infants), and rubber bladders that are commonly used for blood pressure measurement.

9.5.4 Textile Industry

The textile industry has become the largest industry in the world. As concerns clothing globally, the fabric market worth is greater than $400 billion and is still growing every year. Globalization of the fabric trade has caused extremely demanding and developing requirements for fabric outsourcing. A variety of textiles are formed and utilized globally at different applications levels such as technical products, medical equipment, apparel, and household textiles (Fischer and Pascucci 2017; Hasanbeigi and Price 2015; Nica and Potcovaru 2015). The importance of supercritical CO_2 cannot be ignored in the textile industry because, in the dyeing industries, supercritical CO_2 is one source for a solvent that minimizes the environmental and economic problems of traditional dyeing. Con from Holland has had success using supercritical CO_2 in dyeing of textiles to enhance steady business growth. In the tanning of leather for textile leather-based industries, supercritical CO_2 serves to make the leather material more valuable. Tanning of skins into leather material using supercritical CO_2 avoids the use of aqueous-based solutions often problematic to treat before disposal. Chromium salts, a natural tannin, can saturate skin to turn it into leather. In dry cleaning, supercritical CO_2 is used at an industrial scale in the US as an alternate source to chlorinated solvents used traditionally. A wide variety of rubber-based products is employed in the textile industries. Rubber calendaring is a typical process in the production of numerous kinds of fabrics. Many types of SR such as styrene butadiene, and rubber latex specially, are used for finishing purposes (Aladić et al. 2015; Alcázar-Alay et al. 2017; Jantan et al. 2015). Such a type of rubber is produced by using active surface agents applicable on many textiles such as polyester, nylon, and cellulosics. Some others include rubber rollers for fabric printing, pressure hoses in spinning machinery, rubber coverings for curtain screen applications, and rubber moulded for other specific purposes.

9.6 Conclusion and Future Perspectives

Currently, despite all the developments there are still several drawbacks in the supercritical CO_2 technique. Because the fluid viscosity of SFE is less with CO_2, it carries insufficient capacity, and it is probable that blows will become jammed by plugs, resulting in application failure. During the manufacturing process, the fractures may cause pressure to drop quickly, and radius loss is less than the expected radius. CO_2-phase variations are considered too complex; the variations of pressure and temperature of CO_2 fracturing according to time under diverse displacement circumstances, as have been discovered by investigators, may lead to useful conclusions. CO_2 preparation costs have also captured great attention, such as transportation and compression processes, the prospect of gas leakage with enormous CO_2 usage, and mineralization produced by dissolved CO_2 gas in water formation, all of which must be addressed in future. In this chapter different applications were summarized; there also is still a research gap to enhance work stability. Currently, SFE has been functionalized and employed to produce compounds in limited fields such as some latex plants, but there are many rubber-based plants that may be useful in the pharmaceutical and medical fields using the SFE extraction method with CO_2. The new sources can offer novel functionalization and natural antioxidant substances, which allows them to be used in pharmaceutical applications. There is a countless variety of SFE solvents with different chemical composition and properties to increase attention about the sources of bioactive compounds. SFE can provide great development when used in the textile, medical, tire, and pharmaceutical-based industries to make new products by using different solvents. SFE with CO_2 can also give positive outcomes in metal extraction because the complex metallic compounds have good solubility in the presence of supercritical CO_2, which significantly depends on the chemical nature of the compounds found in the process. Many complexing agents are used in heavy metal extraction from water and soil by using SFE. In the modern era, the scientific community has investigated SFE by using different solvents and found CO_2 to be much better because of its properties, and the product is free from toxicity. Researchers have employed this SFE technique in different fields such as medical, food, chemical, cosmetic, textile, biofuel, and rubber-based industries with the prospect of much further progress following increasing research in SFE. As a useful technique it carries many advantages in the environmental arena, as its usage allows reduction in organic solvents, permitting agroindustry deposits by different functional compounds, and removing toxic compounds such as pesticides and heavy metals. Therefore, SFE is an alternative technique that should be employed for extracting and producing different items, especially those frequently used by humans such as new foods, drugs, additives, and medical equipment.

Acknowledgments All authors are thankful to the School of Industrial Technology and Chemical Sciences, Universiti Sains Malaysia, for providing research facilities.

References

Abe Y, Ishiguro H (2013) Chloroprene rubber composition and its application. Google Patents

Adeola A (2018) Fate and toxicity of chlorinated phenols of environmental implications: a review. Med Anal Chem Int J 2(4):000126

Aizawa S et al (2010) Elastomer composition and molded body. Google Patents

Aizpurua-Olaizola O et al (2015) Optimization of supercritical fluid consecutive extractions of fatty acids and polyphenols from *Vitis vinifera* grape wastes. J Food Sci 80(1):E101–E107

Aladić K et al (2015) Supercritical CO_2 extraction of hemp (*Cannabis sativa* L.) seed oil. Ind Crop Prod 76:472–478

Alcázar-Alay SC et al (2017) Polymer modification from semi-defatted annatto seeds using hot pressurized water and supercritical CO_2. J Supercrit Fluid 129:48–55

Ali N et al (2018) Cultivation of *Pleurotus ostreatus* on oil palm fronds mixed with rubber tree sawdust. Chem Eng Trans 63:547–552

Azwanida N (2015) A review on the extraction methods use in medicinal plants, principle, strength and limitation. Med Aromat Plants 4(196):2167–0412.1000196

Bairgi BK (2015) Viability assessment of the use of ground tire rubber in asphalt pavements. Arkansas State University, Jonesboro

Bekkedahl N (1945) Brazil's research for increased rubber production. Sci Mon 61(3):199–209

Bystritskaya EV, Monakhova TV, Ivanov VB (2013) TGA application for optimising the accelerated aging conditions and predictions of thermal aging of rubber. Polym Test 32(2):197–201

Cheng Y et al (2019) Single-port laparoscopic surgery is feasible and safe for hepatic left lateral sectionectomy for benign liver lesions. Gastroenterol Res Pract 2019:1570796

Chua SC, Oh TH (2010) Review on Malaysia's national energy developments: key policies, agencies, programmes and international involvements. Renew Sustain Energy Rev 14(9):2916–2925

Clemitson I (2015) Castable polyurethane elastomers. CRC Press, Boca Raton

Cunha VMB et al (2018) Carbon dioxide use in high-pressure extraction processes. In: Carbon dioxide chemistry, capture and oil recovery. IntechOpen, London, p 211

Erickson B, Singh R, Winters P (2011) Synthetic biology: regulating industry uses of new biotechnologies. Science 333(6047):1254–1256

Fischer A, Pascucci S (2017) Institutional incentives in circular economy transition: the case of material use in the Dutch textile industry. J Clean Prod 155:17–32

Franta I (2012) Elastomers and rubber compounding materials, vol 1. Elsevier, New York

Gandhi K, Arora S, Kumar A (2017) Industrial applications of supercritical fluid extraction: a review. Int J Chem Stud 5:336–340

Gao H et al (2019) Infusion of graphene in natural rubber matrix to prepare conductive rubber by ultrasound-assisted supercritical CO_2 method. Chem Eng J 368:1013–1021

Hasanbeigi A, Price L (2015) A technical review of emerging technologies for energy and water efficiency and pollution reduction in the textile industry. J Clean Prod 95:30–44

Hernández Santana M et al (2018) Routes to make natural rubber heal: a review. Polym Rev 58(4):585–609

Hoisington FR (1960) Commodities: high finance in rubber. Financ Anal J 16(3):105–116

Ikeda Y (2014) Recycling of sulfur cross-linked natural rubber (NR) using supercritical carbon dioxide. In: Chemistry, manufacture and applications of natural rubber. Elsevier, Cambridge, pp 436–451

Jantan SA et al (2015) Solubility determination of tamarind seeds extracts by using supercritical CO_2 extraction. In: Applied mechanics and materials. Trans Tech Publications, Zurich

Jokić S et al (2017) Supercritical CO_2 extraction–a new perspective in the utilisation of food industry by-products. Akademija Tehničkih Znanosti Hrvatske (HATZ) 12(1):7–12

Jomo KS (2019) Southeast Asia's misunderstood miracle. Industrial policy and economic development in Thailand, Malaysia and Indonesia. Routledge, London

Kankala RK et al (2017) Supercritical fluid technology: an emphasis on drug delivery and related biomedical applications. Adv Healthc Mater 6(16):1700433

Kavoura D et al (2019) Supercritical CO_2 extraction of *Salvia fruticosa*. J Supercrit Fluid 146:159–164
Kawano M (2019) Changing resource-based manufacturing industry: the case of the rubber industry in Malaysia and Thailand. In: Emerging states at crossroads. Springer, Berlin, pp 145–162
Kinney LF et al (2017) Apparatus and method for extracting organic compounds from plant material using carbon dioxide. Google Patents
Kitrytė V, Bagdonaitė D, Venskutonis PR (2018) Biorefining of industrial hemp (*Cannabis sativa* L.) threshing residues into cannabinoid and antioxidant fractions by supercritical carbon dioxide, pressurized liquid and enzyme-assisted extractions. Food Chem 267:420–429
Kohjiya S (2014) Chemistry, manufacture and applications of natural rubber. Elsevier, Cambridge
Kohjiya S (2015) Natural rubber. Smithers Rapra, Shropshire
Kojima M et al (2005) Phase behavior of crosslinked polyisoprene rubber and supercritical carbon dioxide. J Supercrit Fluid 35(3):175–181
Latsky C, Kouakou A, Schwarz C (2018) Phase equilibria of CO_2 with components in the light naphtha cut of tyre derived oil. J Supercrit Fluid 131:58–65
Lee N-Y et al (2013) Extraction of rubber (*Hevea brasiliensis*) seed oil using supercritical carbon dioxide and soxhlet extraction. Res J Chem Environ 17:46–52
Liu Z et al (2015) Devulcanizaiton of waste tread rubber in supercritical carbon dioxide: operating parameters and product characterization. Polym Degrad Stab 119:198–207
Machado BAS et al (2013) Supercritical fluid extraction using CO_2: main applications and future perspectives. Sep Sci Technol 48(18):2741–2760
Mahdi BD et al (2015) Single port laparoscopic orchidopexy in children using surgical glove port and conventional rigid instruments. Korean J Urol 56(11):781–784
Maiti M et al (2012) Microcellular foam from ethylene vinyl acetate/polybutadiene rubber (EVA/BR) based thermoplastic elastomers for footwear applications. Ind Eng Chem Res 51(32):10607–10612
Mangili I et al (2014) Full factorial experimental design to study the devulcanization of ground tire rubber in supercritical carbon dioxide. J Supercrit Fluid 92:249–256
Mao Y et al (2013) High performance graphene oxide based rubber composites. Sci Rep 3:2508
Melgosa R et al (2019) Supercritical CO_2 assisted synthesis and concentration of monoacylglycerides rich in omega-3 polyunsaturated fatty acids. J CO2 Util 31:65–74
Michalak I, Chojnacka K, Saeid A (2017) Plant growth biostimulants, dietary feed supplements and cosmetics formulated with supercritical CO_2 algal extracts. Molecules 22(1):66
Mićić V et al (2011) Supercritical fluid extraction with carbon dioxide at different pressures. Contemp Mater 1:85–87
Muhammadi S et al (2015) Bacterial polyhydroxyalkanoates-eco-friendly next generation plastic: production, biocompatibility, biodegradation, physical properties and applications. Green Chem Lett Rev 8(3–4):56–77
Myhre M et al (2012) Rubber recycling: chemistry, processing, and applications. Rubber Chem Technol 85(3):408–449
Nair AB, Kurian P, Joseph R (2012) Ethylene–propylene–diene terpolymer/hexa fluoropropylene–vinylidinefluoride dipolymer rubber blends: thermal and mechanical properties. Mater Des (1980–2015) 36:767–778
Naziri E et al (2016) Advantages of supercritical fluid extraction for recovery of squalene from wine lees. J Supercrit Fluid 107:560–565
Nica E, Potcovaru A-M (2015) Labor management and dress culture in the Victorian textile industry. Econ Manag Finance Mark 10(2):96
Ottolina J et al (2016) Surgical training for ovarian non-excisional laparoscopic laser surgery: fiber CO_2 laser and free-beam CO_2 laser compared. J Minim Invasive Gynecol 23(7):S200
Özkal SG, Yener ME (2016) Supercritical carbon dioxide extraction of flaxseed oil: effect of extraction parameters and mass transfer modeling. J Supercrit Fluid 112:76–80
Petrova NN, Portnyagina VV (2014) Wear-resistant rubber based on propylene oxide rubber and ultrafine polytetrafluoroethylene. Google Patents

Razak NHA, Arif SAM (2014) Supercritical fluid extraction of vitamin E from de-proteinised natural rubber (DPNR) serum. Adv Environ Biol 8(8):2689–2695

Reche F et al (2002a) Optimization of parameters for the supercritical fluid extraction in the determination of N-nitrosamines in rubbers. J Chromatogr A 963(1–2):419–426

Reche F et al (2002b) Determination of N-nitrosamines in latex by sequential supercritical fluid extraction and derivatization. J Chromatogr A 976(1-2):301–307

Rodgers B (2015) Rubber compounding: chemistry and applications. CRC Press, Boca Raton

Sapkale G et al (2010) Supercritical fluid extraction. Int J Chem Sci 8(2):729–743

Seghar S et al (2019) Thermo-mechanical devulcanization and recycling of rubber industry waste. Resour Conserv Recycl 144:180–186

Sethuraj MR, Mathew NT (2012) Natural rubber: biology, cultivation and technology, vol 23. Elsevier, New York

Shattuck W (2018) Appropriate-ing rubber: scalar translations in Southern Thailand. Singap J Trop Geogr 39(1):132–148

Shi Y et al (2019) Uncovering the driving forces of carbon dioxide emissions in Chinese manufacturing industry: an intersectoral analysis. Environ Sci Pollut Res 26(30):31434–31448

Shit SC, Shah P (2013) A review on silicone rubber. Natl Acad Sci Lett 36(4):355–365

Sisanth K et al (2017) General introduction to rubber compounding. In: Progress in rubber nanocomposites. Elsevier, New York, pp 1–39

Song L et al (2014) Preparation of silicone rubber foam using supercritical carbon dioxide. Mater Lett 121:126–128

Stelescu MD, Manaila E, Craciun G (2013) Vulcanization of ethylene-propylene–terpolymer-based rubber mixtures by radiation processing. J Appl Polym Sci 128(4):2325–2336

Syao O, Malysheva G (2014) Properties and application of rubber-based sealants. Polym Sci Ser D 7(3):222–227

Tang W et al (2018) Carbon nanotube-reinforced silicone rubber nanocomposites and the foaming behavior in supercritical carbon dioxide. J Supercrit Fluid 141:78–87

Tessanan W et al (2019) Microcellular natural rubber using supercritical CO_2 technology. J Supercrit Fluid 149:70–78

Thomas S, Stephen R (2010) Rubber nanocomposites: preparation, properties, and applications. Wiley, New York

Thomas JF, West BH, Huff SP (2013) Effects of high-octane ethanol blends on four Legacy flex-fuel vehicles, and a turbocharged GDI vehicle. ORNL/TM-2015/116

Tonge L (2009) Fluorocarbon elastomer silicone vulcanizates. Google Patents

Tzoganakis C, Zhang Q (2004) Devulcanization of recycled tire rubber using supercritical carbon dioxide. In: ANTEC-conference proceedings. Wiley, Hoboken

del Valle JM (2015) Extraction of natural compounds using supercritical CO_2: going from the laboratory to the industrial application. J Supercrit Fluid 96:180–199

Verheye W (2010) Growth and production of rubber. In: Land use, land cover and soil sciences. UNESCO-EOLSS, Oxford

Williams PT (2013) Pyrolysis of waste tyres: a review. Waste Manag 33(8):1714–1728

Yousefi M et al (2019) Supercritical fluid extraction of essential oils. Trends Anal Chem 118:182–193

Yusoff S, Mohamed Z, Ahmad AZ (2019) Environmental impact evaluation of rubber cultivation and industry in Malaysia. In: Climate change and agriculture. IntechOpen, London

Zanchet A et al (2012) Use of styrene butadiene rubber industrial waste devulcanized by microwave in rubber composites for automotive application. Mater Des 39:437–443

Chapter 10
Compressed Fluids for Food By-product Biorefinery

Mónica Bueno

Contents

10.1 Introduction

Nowadays, the society is facing different challenges such as sustainable food and energy supply, the reduction of the use of fossil fuels, human health, environmental degradation and climate change. One of the critical points for a sustainable development is the reduction or even total elimination of wastes (Herrero and Ibáñez 2018). Approximately one-third of food produced for human consumption is lost or wasted globally, which amounts to about 1.3 billion tons per year (FAO 2011). For instance, fruit by-products such as bagasse, peels, trimmings, stems, shells, bran and seeds account for more than 50% of fresh fruit (Torres-León et al. 2018).

These food residues can be used as a source of (1) cellulose and hemicellulose that can be hydrolysed into fermentable sugars for bioethanol production; (2) lignin, whose combustion could replace fossil fuels; and (3) bioactive compounds that can be used as additives for food and cosmetic products or in nutraceutical formulations. In this context, food by-products are the perfect biomass for developing biorefinery approaches (Fig. 10.1), since the concept of biorefinery is based on the continuous processing of biomass into bioenergy, biofuels, biochemicals and other added-value co-products (Cherubini 2010).

M. Bueno (✉)
Laboratory of Foodomics, Institute of Food Science Research (CIAL-CSIC), Madrid, Spain
e-mail: monica.bueno@csic.es

Inamuddin, A. M. Asiri (eds.), *Advanced Nanotechnology and Application of Supercritical Fluids*, Nanotechnology in the Life Sciences,
https://doi.org/10.1007/978-3-030-44984-1_10

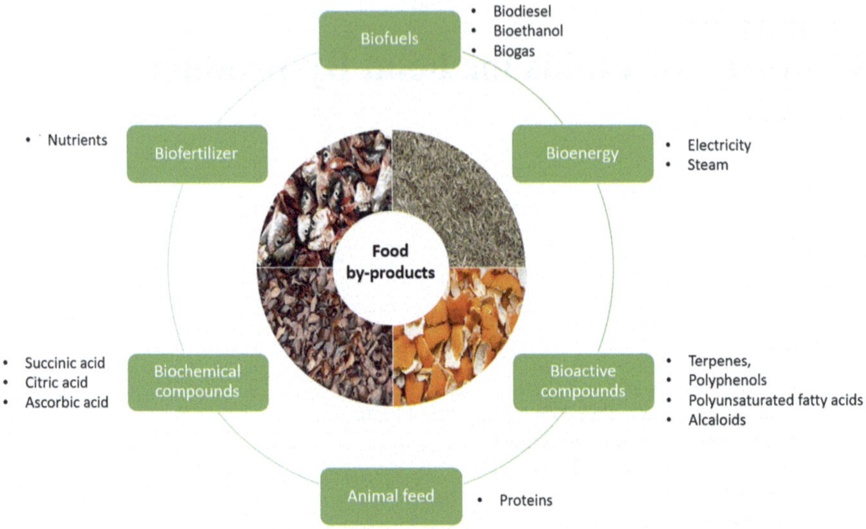

Fig. 10.1 Possible uses of food by-products

Nevertheless, to prevent the 'green character' and profitability of the extraction and recovery of added-value products from food residues, new extraction processes involving less time, an enhancement of extraction efficiencies and a reduction in the environmental impact are requested. Compressed fluid technologies can meet such requirements when using Generally Recognized As Safe (GRAS) solvents. The GRAS solvents most commonly used are water, ethanol, ethyl acetate, D-limonene or CO_2. Thus, much attention has been paid to valorize by-products from agri-food industries through biorefinery approaches based on compressed fluid technologies in recent years.

For this reason, the objective of this chapter is to highlight the use of compressed fluid technologies as environmentally safe procedures for the extraction of a variety of compounds from food by-products and more specifically for the extraction of bioactive compounds that are of interest for food, cosmetics and pharmaceutical industries. Recovery of other products used for the production of bioethanol and bioenergy will be also commented.

10.2 Extraction Processes Based on Compressed Fluids

The common denominator of compressed fluid processes is that they operate under medium-high pressures, which confers a greater efficiency in contrast to conventional atmospheric pressure methods. These operational conditions allow the solvents to acquire physical-chemical properties that may not be attainable otherwise (Sánchez-Camargo et al. 2017). These techniques are supercritical fluid extraction

(SFE), gas-expanded liquids (GXL), pressurized liquid extraction (PLE) and subcritical water extraction (SWE).

A basic system used to perform compressed fluid extractions consists of a reservoir for the extraction solvent, a pump to pressurize the fluid, an oven containing the extraction vessel, a restrictor to control the high extraction pressure and a trapping vessel. Nevertheless, the equipment used could show varying degrees of sophistication according to the process needs. For example, a modifier pump could be added if change in solvent polarity due to online solvent mixtures is needed.

Furthermore, in particular circumstances, these techniques satisfy the green chemistry principles and, therefore, they turn into one of the most efficient ways to achieve sustainable development. In addition, medium-to-high operational pressures are easily tuneable as well as temperatures. This fact leads to selective and fast extractions with controlled temperature, which is an advantage for the extraction of thermosensitive compounds. Consequently, compressed fluid technologies using green solvents have been widely employed as an alternative to organic solvents for the extraction of bioactive compounds from different biomasses. Depending on the polarity of the target compound, different green compressed fluids and technologies can be used, as can be seen in Fig. 10.2. Mainly, compounds presenting high and medium-high polarity are conveniently extracted using PLE or SWE, whereas compounds with low polarity are satisfactory extracted by SFE..

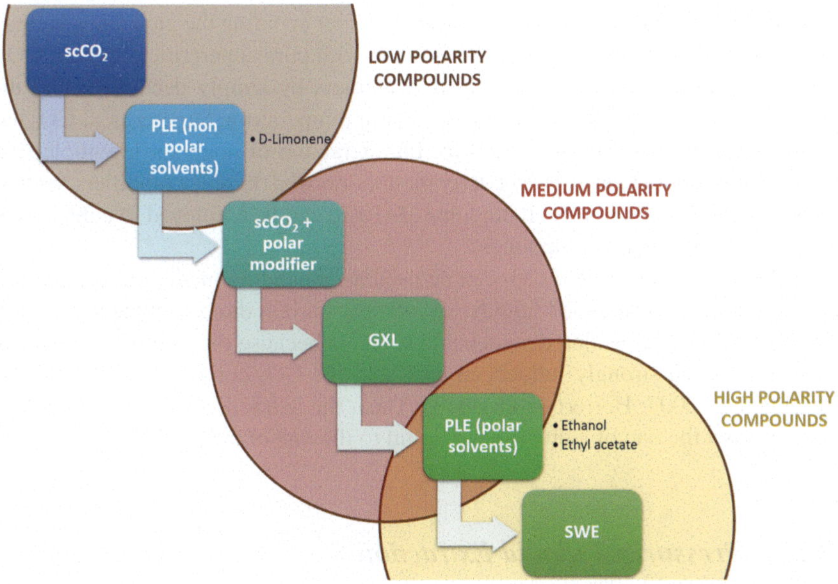

Fig. 10.2 Compressed fluid techniques and green solvents used to extract bioactive compounds from food by-products. *scCO*$_2$ supercritical CO_2, *PLE* pressurized liquid extraction, *GXL* gas-expanded liquid, *SWE* subcritical water extraction

10.2.1 Supercritical Fluid Extraction

SFE is based on the use of solvents at pressures and temperatures above their critical points. One characteristic property of a supercritical fluid is its low viscosity producing a fast diffusivity. In this way a fast mass transfer is attained which is fundamental for a good separation process. At supercritical conditions, the solvating power of the fluid can be easily modified by small changes in temperature and pressure favouring the extraction of target compounds. For that reason, pressure and temperature are parameters that must be optimized for each application, as well as flow rate, extraction time and solvent-to-feed ratio. The operating mode usually involves dynamic extraction; nevertheless, sometimes a previous static period is carried out, followed by a prolonged dynamic extraction.

Considering the absence of toxicity, water can be suggested as an interesting green supercritical fluid. However, a great amount of energy is consumed to achieve its critical point (Tc = 374 °C, Pc = 22.1 MPa); then supercritical water is not the most popular supercritical solvent. Notwithstanding, supercritical water provides fast hydrolysis of biomass causing rapid degradation of the polymeric structures as, for example, the breakdown of polysaccharides into fermentable sugars, as will be discussed in Sect. 10.3.

On the other hand, carbon dioxide is the most commonly used supercritical fluid due to its easily attainable critical point (Tc = 31 °C, Pc = 7.4 MPa). Moreover, CO_2 is non-flammable and non-toxic. It is also considered a GRAS solvent and environmentally friendly due to its recycled waste character avoiding the greenhouse effect. Another important feature of extractions made with pure supercritical CO_2 ($scCO_2$) is the possibility of attaining solvent-free extracts by simply decompressing the system because carbon dioxide is a gas at room temperature. Supercritical CO_2 has low polarity and therefore is limited for the extraction of nonpolar low-molecular weight compounds. However, to overcome this drawback, polar modifiers such as ethanol or ethyl acetate in small amounts (1–10%) can be employed to enhance the solubility of more polar compounds.

Furthermore, compressed CO_2 can be used as expanding media for organic solvents to obtain gas-expanded liquids, specifically carbon dioxide-expanded liquids (CXLs). CXLs are considered switchable solvents because their physical-chemical properties can be strongly influenced by rising the CO_2 composition (Jessop and Subramaniam 2007; Herrero et al. 2017). Thus, the extraction behaviour of CXLs can vary from the one of a pressurized liquid to the one of neat $scCO_2$.

10.2.2 Pressurized Liquid Extraction

PLE is based on the use of solvents at high temperatures and pressures, but always below their critical point. Usually, temperatures from 20 to 250 °C and pressures of 3.5–20 MPa are employed to preserve the solvents always in the liquid state during the whole extraction procedure. By applying those conditions, mass transfer rates are enhanced. Simultaneously, solvent viscosity and surface tension are decreased. Accordingly, higher extraction yields are obtained in faster extraction processes that use lower volumes of organic solvents.

The versatility of PLE relies on the use of a wide variety of solvents and a combination of them. Solvent selection must take into account the nature of target analytes. Moreover, if GRAS solvents, such as ethanol, ethyl acetate, ethyl lactate, D-limonene or water, are employed, this technique is considered as environmentally friendly. When water is used as extraction solvent, this technique is commonly called subcritical water extraction (SWE).

As already mentioned for SFE, PLE method development also involves the optimization of different factors. In this case, temperature is the parameter of utmost importance. Apparently, an increase in the temperature favours the extraction yield; however, the bioactivity of some thermolabile compounds could be negatively affected by high temperatures. On the other hand, several papers revealed that extraction pressure is not significant once it is high enough to keep the solvent in a liquid state (Herrero et al. 2013; Plaza and Turner 2015). Extraction mode is a crucial parameter that significantly influences extraction yield and time. In the static mode, sample and solvent are maintained for a specific time at constant temperature and pressure, while in the dynamic mode, the solvent continuously flows through the sample. In the latter, thermal degradation is prevented, as a result of faster extraction procedures.

10.3 Applications

As has been mentioned before, food wastes coming from different industries (fruits, vegetables, grains, fish, among others) still contain different bioactives and other compounds that can be extracted and transformed into new biochemicals and bioenergy, what constitutes a key point for the valorization of food by-products, using the biorefining concept. For a more comprehensive information about biorefinery of food by-products, other recent reviews are also available in the valorization of cereal residues (ElMekawy et al. 2013; Celiktas et al. 2014; Alonso 2018), winemaking-related wastes (Zacharof 2017), coffee residues (Aristizabal-Marulanda et al. 2017; Karmee 2018; Zabaniotou and Kamaterou 2019) and fruit by-products (de la Torre et al. 2019).

Common bioactives from food by-products are carotenoids (Mustafa et al. 2012; Ndayishimiye and Chun 2017), phenolic compounds (Espinosa-Pardo et al. 2017; Ersan et al. 2018), lipids (Akgun et al. 2014), phytosterols (Vagi et al. 2007) and tocopherols (Ibáñez et al. 2000), among others. These bioactive compounds have different health-promoting properties such as immunomodulatory (Talhaoui et al. 2016), anti-cancer (Lee et al. 2009), antioxidant (Araujo et al. 2015), anti-diabetic (Diaz-de-Cerio et al. 2016) or anti-inflammatory (Cavalcanti de Albuquerque et al. 2019) which indicates their potential in food, cosmetic and pharmaceutical industries. To recover these bioactive compounds from food residues, only the combined uses of environment-friendly processes, such as PLE or SFE using green solvents, are discussed in this section. Table 10.1 summarizes the extraction conditions employed in the most remarkable biorefinery applications. Obviously, extraction conditions will vary according to the target compounds and the matrix. The effect of process parameters, such as temperature, pressure, time or solvent composition, has been widely evaluated by means of response surface methodology (RSM) (Da Porto et al. 2015; Vigano et al. 2016b; Mackela et al. 2017). Furthermore, when considering a bioactivity as one of the response variables, we can talk about bio-based directed methodologies (Brazdauskas et al. 2016).

In general, the food by-product matrices are very complex, and many processes have been evolved coupling or integrating different techniques. For instance, Da Porto et al. (2015) coupled an ultrasound-assisted extraction (UAE) with a SFE in sequence for the extraction of polyphenols from grape marc. The polyphenol extraction was carried out after the extraction of lipids using $scCO_2$, making the whole extraction more efficient. The total polyphenol recovery as well as the antioxidant activity obtained by the combined UAE+SFE was significantly higher (28% and 60%, respectively, $p < 0.05$) than the ones obtained by SFE alone. Ultrasound has been described as a versatile tool due to its easy simultaneous coupling with different compressed fluid techniques for the extraction of tocopherols and polyphenols, as an example, from rice bran or passion fruit seeds by UAE-SFE (Barrales et al. 2015; Soares et al. 2018) and from pomegranate peels by UAE-PLE (Sumere et al. 2018) or UAE-GXL (Santos et al. 2019). In summary, the use of ultrasounds not only improves the extraction recoveries but also allows a reduction of extraction time or cycles in the case of PLE (Sumere et al. 2018), thanks to the vibration effect promoted by the ultrasonic waves at the interfaces between the solid matrix and solvent.

The usefulness of enzyme treatments has been proved as advantageous for bioactive extraction. These enzymes are capable to break down strong cell walls of some samples. For instance, Mackela et al. (2017) applied an enzyme-assisted extraction (EAE) as part of a multiple-step PLE for a better recovery of polyphenolic compounds. They found an increment in the solubilization of the constituents (approximately 4–5 times) which might be strongly bound to the polymeric matrix of buckwheat hulls.

As already stated, SFE is appropriate for recovering low-polar compounds, and particularly neat $scCO_2$ is used for the extraction of lipids (Da Porto et al. 2015; Kryzeviciute et al. 2016; Kühn and Temelli 2017; Mazzutti et al. 2018). Among those lipids, long-chain omega-3 polyunsaturated fatty acids (PUFAs) should be highlighted because they are present in many marine processing by-products. A special mention should be done on docosahexaenoic acid and eicosapentaenoic acid which are recommended in the prevention of cardiovascular diseases in patients with coronary heart diseases (Siscovick et al. 2017). However, for each fish residue, it is necessary to perform a specific optimization for the SFE process for PUFA extraction, using wide ranges of pressure and temperature (Gallego et al. 2019a). Some of the major benefits of PUFA extraction employing SFE are the selective

Table 10.1 Some representative biorefinery applications involving the use of compressed fluids for the extraction of bioactive compounds from food-related by-products

Matrix	Target compounds	Technique	Solvent	*T* (°C)/P (MPa)	Mode of extraction	Extraction time (min)	Flow rate (mL min^{-1})	References
Tomato peel and seeds	Carotenoids (lycopene, β-carotene)	SFE	CO_2	80/40	Dynamic	120	4 #	Kehili et al. (2016)
	Proteins (amino acids)							
	Sugars and carboxylic acids; cellulose and lignin	SWH	Water	160/5	Static	30		
Grape seed	Triglycerides	SFE	CO_2 + hexane	60/50	Dynamic	–	1 #	Manna et al. (2015)
	Polyphenols	SFE	CO_2 + 10% ethanol	60/50	Dynamic	–	1 #	
Grape marc	Lipids	SFE	CO_2	45/28	Dynamic	180	167 #	Da Porto et al. (2015)
	Polyphenols (proanthocyanidins)	UAE + SFE	CO_2 + 10% ethanol	40/8	Dynamic	300	100 #	
Grape marc	Monomeric anthocyanins	PLE	Acidified ethanol/water (50:50)	40/10	Dynamic	40	5 #	Vitor Pereira et al. (2019)
	Other phenolic compounds	PLE	Ethanol/water (50:50)	100/10	Static	180		
Orange peel	α-Terpineol and D-limonene	SFE	CO_2	45/35	Dynamic	–	10.8 #	Barrales et al. (2018)
	Phenolic compounds (glycosylated flavonoids and hydroxybenzoic acids)	PLE	Ethanol/water (75:25)	65/10	Dynamic	20	2.37 #	

(continued)

Table 10.1 (continued)

Matrix	Target compounds	Technique	Solvent	*T* (°C)/P (MPa)	Mode of extraction	Extraction time (min)	Flow rate (mL min^{-1})	References
Tangerine peel	Oil (D-limonene)	SFE	CO_2	35/10	Dynamic	30	2	Tsitsagi et al. (2018)
	Carotenoids (β-carotene)	SFE	CO_2 + 7% acetone	40/15	Dynamic	60	2	
	Flavanones (hesperidin)	SFE	CO_2 + 7% methanol	60/25	Dynamic	30	2	
	Heteropolysaccharides (pectin)	UAE						
Passion fruit seeds	Tocopherols and tocotrienols	SFE	CO_2	60/17	Dynamic	–	20.64 #	Vigano et al. (2016b)
	Fatty acids	SFE	CO_2	50/17	Dynamic	–	20.64 #	
	Carotenoids	SFE	CO_2	60/26	Dynamic	–	20.64 #	
	Phenolic compounds (piceatannol and scirpusin B)	PLE	Ethanol/water (50:50)	70/10	Dynamic	120	30 #	Vigano et al. (2016a)
Mango peel	Nonpolar flavonoids and carotenoids	SFE	CO_2	40/30	Dynamic	450	1100	Garcia-Mendoza et al. (2015)
	Polyphenols	PLE	Ethanol	40/30	Dynamic	330	1100	
Olive pomace	Oil	SFE	CO_2	68/28	Dynamic	60	20 #	Kazan et al. (2015)
	Proteins	SWE	Water	88/22	Dynamic	30	2.26	
	Lignin and fermentable sugars	SWH	Water	200/22	Dynamic	30	2	
	Lignin and fermentable sugars	SWH-EAE	Water	50/22		4080		
Olive mill waste	Squalene, mono- and polyunsaturated fatty acids	SFE	CO_2	70/25	Dynamic	420	1.33 #	Schievano et al. (2015)
	Polyphenols, squalene, mono- and polyunsaturated fatty acids	SFE	CO_2 + 0.25% ethanol	70/25	Dynamic	480	1.33 #	

Cranberry pomace	Lipids	SFE	CO_2	50/40	Dynamic	90	1000	Kühn and Temelli (2017)
	Phenolic compounds	SFE	CO_2/ethanol/water	50/40	Dynamic	90	1000	
Raspberry pomace	Lipids	SFE	CO_2	60/45	Static + dynamic	10 + 110	2000	Kryzeviciute et al. (2016)
	Polyphenols	PLE	Ethanol/water (50:50)	80/10.3	Static (3 cycles)	5 × 3		
Black chokeberry pomace	Lipids	SFE	CO_2 + 10% ethanol	40/40	Static + dynamic	15 + 150	2000	Grunovaite et al. (2016)
	Polyphenols (phenolic acids and anthocyanins)	PLE	Ethanol	70/10	Static	45		
Black chokeberry pomace	Tocopherols	SFE	CO_2	40/40	Static + dynamic	15 + 150	2000	Grunovaite et al. (2016)
	Phenols (anthocyanins, proanthocyanidins, flavonoids and phenolic acids)	PLE	Water/ethanol/formic acid (52.2:46:1.8)	165/10	Static	20		Brazdauskas et al. (2016)
Sea buckthorn pomace and seeds	Tocopherols	SFE	CO_2	60/35	Static + dynamic	30 + 180	2000–3000	Kitryte et al. (2017)
	Sugar acids, phenolics and triterpenes	PLE	Ethanol	70/10	Static (three cycles)	5 × 3		
	Glucose, fructose and maltose	EAE						
Cocoa bean hulls	Fatty acids	SFE	CO_2	40/20	Dynamic	120	11 #	Mazzutti et al. (2018)
	Phenolic compounds and alkaloids (caffeine and theobromine)	PLE	Ethanol	70/10	Static	20		
Juçara residues	Anthocyanins	SWE	Acidified water	40/10	Dynamic	–	1.5	Garcia-Mendoza et al. (2017)
	Non-anthocyanic phenolic compounds	PLE	Acidified ethanol/water (50:50)	80/10	Dynamic	–	1.5	

(continued)

Table 10.1 (continued)

Matrix	Target compounds	Technique	Solvent	*T* (°C)/P (MPa)	Mode of extraction	Extraction time (min)	Flow rate (mL min^{-1})	References
Pressed palm residues	Carotenoids	SFE	CO_2	45/15	Static + dynamic	10 + –	3.36 #	Cardenas-Toro et al. (2014)
	Fermentable sugars	SWH	Water	250/15	Dynamic	–	–	
Hemp residue	Cannabinoids	SFE	CO_2	70/46.5	Static + dynamic	10 + 120	2000–3000	Kitryte et al. (2018)
	Flavonoids	PLE	Acetone	100/10.3	Static (3 cycles)	15 × 3		
	Flavonoids	PLE	Ethanol/water (80:20)	100/10.3	Static (3 cycles)	15 × 3		
	Mono and disaccharides	EAE						
Buckwheat hulls	Lipophilic substances (phytosterols)	SFE	CO_2 + 5% ethanol	60/55	Static + dynamic	10 + 90	2	Mackela et al. (2017)
	Phenolic compounds and antioxidants	PLE + EAE	Acetone	70/10.3	Static	15		
	Phenolic compounds and antioxidants	PLE + EAE	Ethanol/water (70:30)	70/10.3	Static	15		
	Phenolic compounds and antioxidants	SWE + EAE	Water	140/10.3	Static	15		

T temperature, *P* pressure, # CO_2 flow rate expressed min g min^{-1}, *SFE* supercritical fluid extraction, *PLE* pressurized liquid extraction, *SWE* subcritical water extraction, *UAE* ultrasound-assisted extraction, *EAE* enzyme-assisted extraction, *SWH* subcritical water hydrolysis enrichment of docosahexaenoic and eicosapentaenoic acids due to a non-oxygen environment with very low lipid oxidation, a better quality of the oil obtained in terms of viscosity and colour and a substantial decrease of the extraction of some organic derivatives containing heavy metals in comparison with conventional methods (Nguyen et al. 2017; Gallego et al. 2019a). In accordance with the zero-waste concept, in addition to omega-3-rich oil, various valuable by-products might also be isolated from the seafood or fish wastes. For instance, seafood collagen is used for functional food products or nutraceuticals (Pal and Suresh 2016), other fish proteins are used as fishmeal, and glycerol, saturated fatty acids and short-chain unsaturated fatty acids are used as liquid biofuel (Fiori et al. 2017).

A possible way to increase further the recovery of target bioactives from tomato by-products has been recently explored taking carotenoids as predominant compounds (Lenucci et al. 2013). Firstly, the lycopene $scCO_2$ extraction was performed (65–70 °C, 45 MPa, with a CO_2 flow rate of 300–330 g min^{-1} for 180 min) in a pilot plant scale, followed by the enzymatic hydrolysis of the cell wall polysaccharides that still remain in the $scCO_2$-exhausted matrix. They also report the fermentation of the obtained sugars into bioethanol, which demonstrates that bioethanol production might be an alternative for the valorization of the carbohydrates present in tomato pomace. Moreover, when nothing else can be extracted, Toscano et al. (2015) encourage to carry out the torrefaction of agro-industrial residues with poor energy content, what would turn tomato residue in a solid biofuel with energy properties similar to coal.

However, Kehili et al. (2016) even includes another step in the cascade processing of tomato residues that leads to multiple products: lycopene, proteins with good nutritional quality and lignocellulosic matter for bioethanol production. The difference between this work and the previous described is the way to carry out the hydrolysis to fermentable sugars. In this case, subcritical water hydrolysis (SWH) was preferred due to several benefits. On one hand, conventional acid or base treatments are not yet satisfactory due to the ecological waste issues. On the other hand, enzymatic hydrolysis cannot break down completely the lignocellulosic material. Compared with these treatments, pressurized water technologies have demonstrated obvious advantages: they are extremely fast due to its much higher reaction rate, they use a non-toxic solvent, they do not require additional catalyst and they are scalable (Zhao et al. 2009; Kilpelainen et al. 2014; Prado et al. 2016). SWH to obtain fermentable sugars has been also used with success in other matrices such as coffee waste residues (Mayanga-Torres et al. 2017), rice husks (Requena et al. 2019), durian seeds (Purnomo et al. 2016) or pressed palm fibres (Cardenas-Toro et al. 2014). Nevertheless, it must be taken into consideration that high temperatures may break down the sugars producing 5-hydroxymethylfurfural, furfural and organic acids which are known to be inhibitors for ethanol fermentation because they are toxic to the microorganisms (Miyafuji et al. 2005). Consequently, an optimization of the process must be done for maximizing the yields of fermentable sugars, while degradation products are minimized (Mayanga-Torres et al. 2017). Contrary to what can be thought, this problem is partially solved working with supercritical water. Around the critical point of water, the hydrolysis rate of

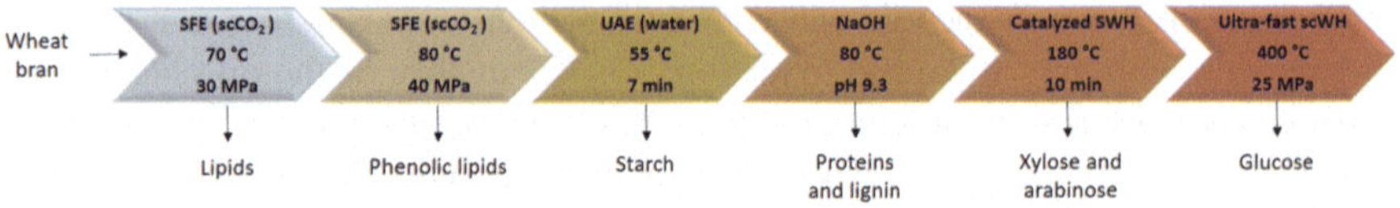

Fig. 10.3 Simplified scheme of the biorefinery cascade processing of wheat bran based on high pressure extraction and hydrolysis proposed by Alonso (2018). (Adapted from Alonso (2018) with permission from Elsevier, Copyright 2018). *SFE* supercritical fluid extraction, *$scCO_2$* supercritical carbon dioxide, *UAE* ultrasound-assisted extraction, *SWH* subcritical water hydrolysis, *scWH* supercritical water hydrolysis

cellulose rises sharply and becomes much faster than the glucose decomposition (Sasaki et al. 1998). Going one step further, Cantero et al. (2015) corroborated a minimal production 5-hydroxymethylfurfural and furfural (0.5% w/w) employing supercritical water (400 °C, 25 MPa) in ultra-fast reactors (0.1–0.7 s) for wheat bran hydrolysis. This ultra-fast supercritical water hydrolysis (ultra-fast scWH) has been proposed as part of a multi-step wheat-bran biorefinery based on high pressure extractions and hydrolysis processes (Alonso 2018), as shown in Fig. 10.3.

Regarding other uses of pressurized water, protein extraction from sustainable low-cost alternatives, such as by-products of oilseeds, fruits and vegetable-processing industries, has concentrated research efforts due to the increased industrial demand of protein sources in recent years. In fact, protein obtained may be used as animal feed and possibly extended to human consumption in the future, although clearly higher dietary requirements will be demanded (Chiesa and Gnansounou 2011). Kazan et al. (2015) described the SWE of proteins from an olive pomace biorefinery by applying 22 MPa under continuous water flow of 2.26 mL min^{-1} at 88 °C for 30 min. In this biorefinery approach, residual oil was obtained by SFE with neat $scCO_2$, proteins by SWE and fermentable sugars and lignin by SWH with or without enzymatic assistance (Table 10.1). In conclusion, this study highlights the feasibility of the high pressure cascade fractionation of the valuable compounds from olive industrial by-products.

More complex approaches are being increasingly developed based on process intensification, in other words, using the same equipment. As previously pointed out, the selectivity of the extraction process depends on solvent and temperature for SFE and PLE and also pressure only in the case of SFE. Some authors have increased the polarity of the supercritical fluid using different solvents as modifier in stepwise extraction methods (Schievano et al. 2015; Tsitsagi et al. 2018). Others have developed a downstream platform process intensification built upon the particular extraction of pure $scCO_2$ depending on its different densities obtained by diverse combination of the extraction parameters (Vigano et al. 2016b). Hence, different fractions of tocols, fatty acids and carotenoids can be extracted (Table 10.1). Therefore, such type of proposals based on tuning pressures and temperatures shows the relevant role of intensified processes as a way to optimize time, capital, wastes and energy.

Another excellent example of a multipurpose system for a process intensification was developed by Mazzutti et al. (2018). This sequential extraction of bioactive compounds from cocoa been hulls comprised the SFE using pure $scCO_2$ with the subsequent PLE using ethanol. Fatty acids were recovered in the first step, whereas mainly caffeine and theobromine were extracted by PLE. The same sequential extraction order and solvents were used in the intensified process for the extraction of nonpolar flavonoids and carotenoids on one hand and polyphenols on the other hand from mango peel (Garcia-Mendoza et al. 2015). Interestingly, the constant values of temperature and pressure would save a lot of time if this biorefinery approach was scaled up. As a final note, it can be said that intensified processes coupling different techniques are a hot topic because a complete residue valorization may be achieved using the same equipment.

10.4 Economic Evaluation

The purpose of biorefinery is to optimize the use of resources and minimize waste, thus maximizing the benefits and profitability (Ferreira 2017). Considering this objective, the concept of advanced biorefineries, also called integrated biorefineries, has been developed (Gallego et al. 2019b). These advanced biorefineries take into consideration multiple feedstocks, products including electricity or biofuels and several platforms. On this wise, bioethanol facilities might be useful examples. Bioethanol has been described as an alternative choice as a substitute for liquid fossil fuels due to its eco-friendly characteristics and relatively low production cost when compared to other bio-based fuels (Prado et al. 2016). The increment demand of bioethanol has led to an increase in the number of ethanol plants, which major raw materials are sugarcane, corn and wheat, which are food sources too. To avoid the fuel versus food dilemma, different solutions are proposed: (1) a greater use of these raw materials through biorefinery perspectives, (2) second-generation bioethanol production (some examples have been already mentioned in the previous section) and (3) combination of both in advanced biorefineries. Figure 10.4 shows a simplified scheme of the extraction of sugarcane wax from the sugarcane biorefinery proposed by Albarelli et al. (2018b). Sugarcane crude wax is usually obtained with a heptane extraction and is rich in fatty acids, phytosterols or long-chain alcohols as policosanol. The greener alternative proposed, a $scCO_2$ extraction, benefits from the electricity and CO_2 generated during the biorefinery in which sugar, bioethanol and biogas are also obtained.

Another interesting example is the case of dried distillers grains with solubles (DDGS), a by-product from the grain bioethanol industry. Until now, this DDGS has been utilized as livestock feed; however, this starch-free material is rich in proteins, lipids and carbohydrates. For this reason, Temelli and Ciftci (2015) have presented DDGS as a potential feedstock for a $scCO_2$-based biorefinery. Minor lipids could be extracted using the recycled CO_2 generated as a result of fermentation and then use them to produce biodiesel in a $scCO_2$ bioreactor. Both examples show the

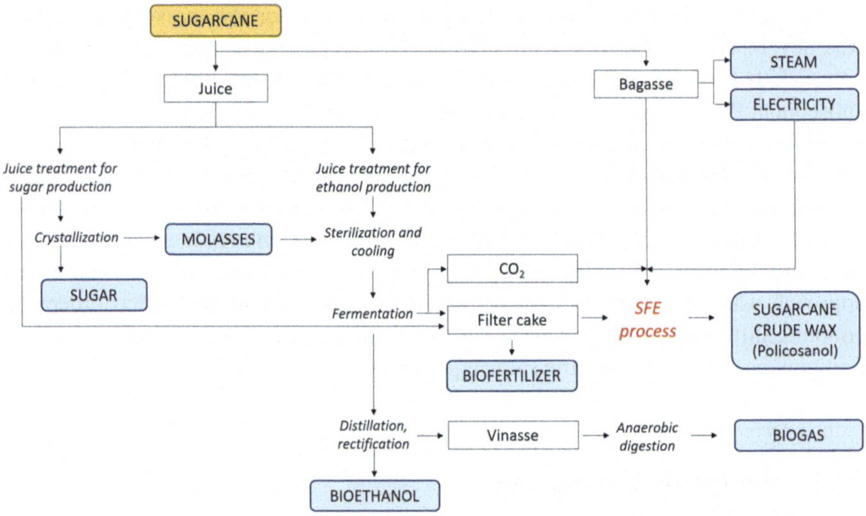

Fig. 10.4 Simplified scheme of the sugarcane biorefinery proposed by Albarelli et al. (2018b). (Adapted from Albarelli et al. (2018b) with permission from Elsevier, Copyright 2018). *SFE* supercritical fluid extraction

potential of assembling a SFE plant inside or in close proximity to a bioethanol fermentation facility, as it had already been postulated by King (2014). The cost of the process when using an SFE plant not only implies the initial investment but also the maintenance costs. Therefore, the SFE plant could reduce logistics and operating costs using directly the high purity CO_2 from the fermentation, ethanol as modifier if required, heat and electricity (Santos et al. 2014; Albarelli et al. 2016, 2018a).

Another important topic in biorefining is evaluating the economic feasibility, which is extremely important to understand the energy, environmental burdens and costs of any production/conversion system giving insight into its sustainability (Gallego et al. 2019b). For that purpose, different process simulation tools such as Aspen Plus (de Melo et al. 2014; Albarelli et al. 2018a, b), Matlab (Albarelli et al. 2016) or SuperPro Designer (Vigano et al. 2017) have been used to perform the technical and economic assessments. For example, Fiori et al. (2017) carried out the simulation of the omega-3 oil extraction performed by $scCO_2$ which is integrated in a biorefinery trout platform using Aspen Plus™ software for the calculation of mass and energy balances, CO_2 equivalent emissions and also the equipment and operating cost analysis. They based their study in an Italian trout processing company whose initial investment should cost about 3.34 M€ (millions of euros). On the other hand, the estimated production of omega-3-rich oil and fish proteins would allow a yearly income of 1.3 M€ and 0.27 M€, respectively. Results also demonstrated that the production of biofuel from the by-products will provide the total electricity needs and cover more than 45% of the thermal energy consumption, thus reducing the utility cost of the plant of about 0.12 M€/year while avoiding the emission of 416 tons of CO_2 each year. This work illustrates the economic feasibility of using compressed fluid-based technologies incorporated into biorefinery approaches when developed on a large scale.

10.5 Conclusions

Throughout this chapter, it has been demonstrated that compressed fluid-based biorefineries of food by-products can play a key role in moving towards a sustainable bioeconomy by increasing process efficiencies while reducing wastes and energy requirements. Moreover, the use of green solvents as part of compressed fluid technologies helps achieve a more favourable world for the human beings and the planet. In this sense, the cascade processing of food wastes using SFE and PLE for a better recovery of bioactive compounds, combined with pressurized water for protein extraction and the hydrolysis of polysaccharides for bioethanol production, seems to be the best way to take advantage of these low-cost raw materials. Nevertheless, further studies should be carried out to understand the behaviour of new types of green solvents (ionic liquids, natural deep eutectic solvents, GXLs, switchable solvents) under pressurized conditions. The alteration in the physicochemical properties of the solvents under pressure conditions causes the modification of the solvent-solute interactions. In this regard, a broader utilization of prediction tools such as Hansen solubility parameters (HSP) could facilitate the solvent selection. In addition, forming associations with industries that produce vast amounts of CO_2 can be a win-win situation for recycling of CO_2, thus reducing the overall CO_2 emissions and environmental impact. However, many approaches have been applied at laboratory scale or 'theoretically' designed for industrial scale. Therefore, future efforts must be focused on the transfer from the laboratory to 'real' large industrial scale.

Acknowledgements M.B. would like to acknowledge the Ministry of Economy and Competitiveness for the "Juan de la Cierva" postdoctoral grant FJCI-2016-30902 and project the support from the AGL2017-89417-R project.

References

Akgun NA, Bulut H, Kikic I, Solinas D (2014) Extraction behavior of lipids obtained from spent coffee grounds using supercritical carbon dioxide. Chem Eng Technol 37:1975–1981. https://doi.org/10.1002/ceat.201400237

Albarelli JQ, Santos DT, Cocero MJ, Meireles MAA (2016) Economic analysis of an integrated annatto seeds-sugarcane biorefinery using supercritical CO2 extraction as a first step. Materials 9:494. https://doi.org/10.3390/ma9060494

Albarelli JQ, Santos DT, Cocero MJ, Meireles MAA (2018a) Perspectives on the integration of a supercritical fluid extraction plant to a sugarcane biorefinery: thermo-economical evaluation of CO2 recycle systems. Food Sci Technol 38:13–18. https://doi.org/10.1590/1678-457x.33516

Albarelli JQ, Santos DT, Meireles MAA (2018b) Thermo-economic evaluation of a new approach to extract sugarcane wax integrated to a first and second generation biorefinery. Biomass Bioenergy 119:69–74. https://doi.org/10.1016/j.biombioe.2018.09.018

Alonso E (2018) The role of supercritical fluids in the fractionation pretreatments of a wheat bran-based biorefinery. J Supercrit Fluid 133:603–614. https://doi.org/10.1016/j.supflu.2017.09.010

Araujo M, Pimentel FB, Alves RC, Oliveira MBPP (2015) Phenolic compounds from olive mill wastes: health effects, analytical approach and application as food antioxidants. Trends Food Sci Technol 45:200–211. https://doi.org/10.1016/j.tifs.2015.06.010

Aristizabal-Marulanda V, Chacon-Perez Y, Cardona Alzate CA (2017) The biorefinery concept for the industrial valorization of coffee processing by-products. In: Galanakis CM (ed) Handbook of coffee processing by-products: sustainable applications. Elsevier, Amsterdam, pp 63–92

Barrales FM, Rezende CA, Martinez J (2015) Supercritical CO2 extraction of passion fruit (Passiflora edulis sp.) seed oil assisted by ultrasound. J Supercrit Fluid 104:183–192. https://doi.org/10.1016/j.supflu.2015.06.006

Barrales FM, Silveira P, Menezes Barbosa PP, Ruviaro AR, Paulino BN, Pastore GM, Macedo GA, Martinez J (2018) Recovery of phenolic compounds from citrus by-products using pressurized liquids - an application to orange peel. Food Bioprod Process 112:9–21. https://doi.org/10.1016/j.fbp.2018.08.006

Brazdauskas T, Montero L, Venskutonis PR, Ibanez E, Herrero M (2016) Downstream valorization and comprehensive two-dimensional liquid chromatography-based chemical characterization of bioactives from black chokeberries (Aronia melanocarpa) pomace. J Chromatogr A 1468:126–135. https://doi.org/10.1016/j.chroma.2016.09.033

Cantero DA, Martinez C, Bermejo MD, Cocero MJ (2015) Simultaneous and selective recovery of cellulose and hemicellulose fractions from wheat bran by supercritical water hydrolysis. Green Chem 17:610–618. https://doi.org/10.1039/c4gc01359j

Cardenas-Toro FP, Forster-Carneiro T, Rostagno MA, Petenate AJ, Maugeri Filho F, Meireles MAA (2014) Integrated supercritical fluid extraction and subcritical water hydrolysis for the recovery of bioactive compounds from pressed palm fiber. J Supercrit Fluid 93:42–48. https://doi.org/10.1016/j.supflu.2014.02.009

Cavalcanti de Albuquerque MA, Levit R, Beres C, Bedani R, de Moreno de LeBlanc A, Saad SMI, Guy LeBlanc J (2019) Tropical fruit by-products water extracts as sources of soluble fibres and phenolic compounds with potential antioxidant, anti-inflammatory, and functional properties. J Funct Foods 52:724–733. https://doi.org/10.1016/j.jff.2018.12.002

Celiktas MS, Kirsch C, Smirnova I (2014) Cascade processing of wheat bran through a biorefinery approach. Energ Conver Manage 84:633–639. https://doi.org/10.1016/j.enconman.2014.04.039

Cherubini F (2010) The biorefinery concept: using biomass instead of oil for producing energy and chemicals. Energ Conver Manage 51:1412–1421. https://doi.org/10.1016/j.enconman.2010.01.015

Chiesa S, Gnansounou E (2011) Protein extraction from biomass in a bioethanol refinery - possible dietary applications: use as animal feed and potential extension to human consumption. Bioresour Technol 102:427–436. https://doi.org/10.1016/j.biortech.2010.07.125

Da Porto C, Lino AN, Decorti D (2015) The combined extraction of polyphenols from grape marc: ultrasound assisted extraction followed by supercritical CO2 extraction of ultrasound-raffinate. LWT Food Sci Technol 61:98–104. https://doi.org/10.1016/j.lwt.2014.11.027

Diaz-de-Cerio E, Verardo V, Maria Gomez-Caravaca A, Fernandez-Gutierrez A, Segura-Carretero A (2016) Exploratory characterization of phenolic compounds with demonstrated anti-diabetic activity in guava leaves at different oxidation states. Int J Mol Sci 17:699. https://doi.org/10.3390/ijms17050699

ElMekawy A, Diels L, De Wever H, Pant D (2013) Valorization of cereal based biorefinery byproducts: reality and expectations. Environ Sci Technol 47:9014–9027. https://doi.org/10.1021/es402395g

Ersan S, Ustundag OG, Carle R, Schweiggert RM (2018) Subcritical water extraction of phenolic and antioxidant constituents from pistachio (Pistacia vera L.) hulls. Food Chem 253:46–54. https://doi.org/10.1016/j.foodchem.2018.01.116

Espinosa-Pardo FA, Nakajima VM, Macedo GA, Macedo JA, Martinez J (2017) Extraction of phenolic compounds from dry and fermented orange pomace using supercritical CO2 and cosolvents. Food Bioprod Process 101:1–10. https://doi.org/10.1016/j.fbp.2016.10.002

FAO (2011) Global food losses and food waste – extent, causes and prevention. Food and Agriculture Organization of the United Nations (FAO), Rome

Ferreira AF (2017) Biorefinery concept. In: Rabaçal M, Ferreira AF, Silva CAM, Costa M (eds) Biorefineries: targeting energy, high value products and waste valorisation. Springer, New York, pp 1–20

Fiori L, Volpe M, Lucian M, Anesi A, Manfrini M, Guella G (2017) From fish waste to omega-3 concentrates in a biorefinery concept. Waste Biomass Valorization 8:2609–2620. https://doi.org/10.1007/s12649-017-9893-1

Gallego R, Bueno M, Herrero M (2019a) Sub- and supercritical fluid extraction of bioactive compounds from plants, food-by-products, seaweeds and microalgae - an update. TrAC Trend Anal Chem 116:198–213. https://doi.org/10.1016/j.trac.2019.04.030

Gallego R, Bueno M, Mendiola JA (2019b) Algae and microalgae biorefinery. In: Torres Gómez MD, Domínguez González H (eds) Valorising seaweed by-products. Nova Science, New York, pp 29–39

Garcia-Mendoza MP, Paula JT, Paviani LC, Cabral FA, Martinez-Correa HA (2015) Extracts from mango peel by-product obtained by supercritical CO2 and pressurized solvent processes. LWT Food Sci Technol 62:131–137. https://doi.org/10.1016/j.lwt.2015.01.026

Garcia-Mendoza MP, Espinosa-Pardo FA, Baseggio AM, Fernandez Barbero G, Marostica Junior MR, Rostagno MA, Martinez J (2017) Extraction of phenolic compounds and anthocyanins from jucara (euterpe edulis mart.) residues using pressurized liquids and supercritical fluids. J Supercrit Fluid 119:9–16. https://doi.org/10.1016/j.supflu.2016.08.014

Grunovaite L, Pukalskiene M, Pukalskas A, Venskutonis PR (2016) Fractionation of black chokeberry pomace into functional ingredients using high pressure extraction methods and evaluation of their antioxidant capacity and chemical composition. J Funct Foods 24:85–96. https://doi.org/10.1016/j.jff.2016.03.018

Herrero M, Ibáñez E (2018) Green extraction processes, biorefineries and sustainability: recovery of high added-value products from natural sources. J Supercrit Fluid 134:252–259. https://doi.org/10.1016/j.supflu.2017.12.002

Herrero M, Castro-Puyana M, Mendiola JA, Ibáñez E (2013) Compressed fluids for the extraction of bioactive compounds. TrAC Trend Anal Chem 43:67–83. https://doi.org/10.1016/j.trac.2012.12.008

Herrero M, Mendiola JA, Ibáñez E (2017) Gas expanded liquids and switchable solvents. Curr Opin Green Sustain Chem 5:24–30. https://doi.org/10.1016/j.cogsc.2017.03.008

Ibáñez E, Palacios J, Senorans FJ, Santa-Maria G, Tabera J, Reglero G (2000) Isolation and separation of tocopherols from olive by-products with supercritical fluids. J Am Oil Chem Soc 77:187–190. https://doi.org/10.1007/s11746-000-0030-8

Jessop PG, Subramaniam B (2007) Gas-expanded liquids. Chem Rev 107:2666–2694. https://doi.org/10.1021/cr040199o

Karmee SK (2018) A spent coffee grounds based biorefinery for the production of biofuels, biopolymers, antioxidants and biocomposites. Waste Manag 72:240–254. https://doi.org/10.1016/j.wasman.2017.10.042

Kazan A, Celiktas MS, Sargin S, Yesil-Celiktas O (2015) Bio-based fractions by hydrothermal treatment of olive pomace: process optimization and evaluation. Energ Conver Manage 103:366–373. https://doi.org/10.1016/j.enconman.2015.06.084

Kehili M, Schmidt LM, Reynolds W, Zammel A, Zetzl C, Smirnova I, Allouche N, Sayadi S (2016) Biorefinery cascade processing for creating added value on tomato industrial by-products from Tunisia. Biotechnol Biofuels 9:261. https://doi.org/10.1186/s13068-016-0676-x

Kilpelainen PO, Hautala SS, Byman OO, Tanner LJ, Korpinen RI, Lillandt MKJ, Pranovich AV, Kitunen VH, Willfor SM, Ilvesniemi HS (2014) Pressurized hot water flow-through extraction system scale up from the laboratory to the pilot scale. Green Chem 16:3186–3194. https://doi.org/10.1039/c4gc00274a

King JW (2014) Modern supercritical fluid technology for food applications. Annu Rev Food Sci Technol 5:215–238. https://doi.org/10.1146/annurev-food-030713-092447

Kitryte V, Povilaitis D, Kraujaliene V, Sulniute V, Pukalskas A, Venskutonis PR (2017) Fractionation of sea buckthorn pomace and seeds into valuable components by using high pressure and enzyme-assisted extraction methods. LWT Food Sci Technol 85:534–538. https://doi.org/10.1016/j.lwt.2017.02.041

Kitryte V, Bagdonaite D, Venskutonis PR (2018) Biorefining of industrial hemp (Cannabis sativa L.) threshing residues into cannabinoid and antioxidant fractions by supercritical carbon dioxide, pressurized liquid and enzyme-assisted extractions. Food Chem 267:420–429. https://doi.org/10.1016/j.foodchem.2017.09.080

Kryzeviciute N, Kraujalis P, Venskutonis PR (2016) Optimization of high pressure extraction processes for the separation of raspberry pomace into lipophilic and hydrophilic fractions. J Supercrit Fluid 108:61–68. https://doi.org/10.1016/j.supflu.2015.10.025

Kühn S, Temelli F (2017) Recovery of bioactive compounds from cranberry pomace using ternary mixtures of CO2 + ethanol plus water. J Supercrit Fluid 130:147–155. https://doi.org/10.1016/j.supflu.2017.07.028

Lee SD, Yoo G, Chae HJ, In M-J, Oh N-S, Hwang YK, Hwang WI, Kim DC (2009) Lipid-soluble extracts as the main source of anticancer activity in ginseng and ginseng marc. J Am Oil Chem Soc 86:1065–1071. https://doi.org/10.1007/s11746-009-1460-x

Lenucci MS, Durante M, Anna M, Dalessandro G, Piro G (2013) Possible use of the carbohydrates present in tomato pomace and in byproducts of the supercritical carbon dioxide lycopene extraction process as biomass for bioethanol production. J Agric Food Chem 61:3683–3692. https://doi.org/10.1021/jf4005059

Mackela I, Andriekus T, Venskutonis PR (2017) Biorefining of buckwheat (fagopyrum esculentum) hulls by using supercritical fluid, soxhlet, pressurized liquid and enzyme-assisted extraction methods. J Food Eng 213:38–46. https://doi.org/10.1016/j.jfoodeng.2017.04.029

Manna L, Bugnone CA, Banchero M (2015) Valorization of hazelnut, coffee and grape wastes through supercritical fluid extraction of triglycerides and polyphenols. J Supercrit Fluid 104:204–211. https://doi.org/10.1016/j.supflu.2015.06.012

Mayanga-Torres PC, Lachos-Perez D, Rezende CA, Prado JM, Ma Z, Tompsett GT, Timko MT, Forster-Carneiro T (2017) Valorization of coffee industry residues by subcritical water hydrolysis: recovery of sugars and phenolic compounds. J Supercrit Fluid 120:75–85. https://doi.org/10.1016/j.supflu.2016.10.015

Mazzutti S, Goncalves Rodrigues LG, Mezzomo N, Venturi V, Salvador Ferreira SR (2018) Integrated green-based processes using supercritical CO2 and pressurized ethanol applied to recover antioxidant compounds from cocoa (Theobroma cacao) bean hulls. J Supercrit Fluid 135:52–59. https://doi.org/10.1016/j.supflu.2017.12.039

de Melo MMR, Barbosa HMA, Passos CP, Silva CM (2014) Supercritical fluid extraction of spent coffee grounds: measurement of extraction curves, oil characterization and economic analysis. J Supercrit Fluid 86:150–159. https://doi.org/10.1016/j.supflu.2013.12.016

Miyafuji H, Nakata T, Ehara K, Saka S (2005) Fermentability of water-soluble portion to ethanol obtained by supercritical water treatment of lignocellulosics. Appl Biochem Biotechnol 121:963–971

Mustafa A, Trevino LM, Turner C (2012) Pressurized hot ethanol extraction of carotenoids from carrot by-products. Molecules 17:1809–1818. https://doi.org/10.3390/molecules17021809

Ndayishimiye J, Chun BS (2017) Optimization of carotenoids and antioxidant activity of oils obtained from a co-extraction of citrus (yuzu ichandrin) by-products using supercritical carbon dioxide. Biomass Bioenergy 106:1–7. https://doi.org/10.1016/j.biombioe.2017.08.014

Nguyen TT, Barber AR, Corbin K, Zhang W (2017) Lobster processing by-products as valuable bioresource of marine functional ingredients, nutraceuticals, and pharmaceuticals. Bioresour Bioprocess 4:27. https://doi.org/10.1186/s40643-017-0157-5

Pal GK, Suresh PV (2016) Sustainable valorisation of seafood by-products: recovery of collagen and development of collagen-based novel functional food ingredients. Innov Food Sci Emerg Technol 37:201–215. https://doi.org/10.1016/j.ifset.2016.03.015

Plaza M, Turner C (2015) Pressurized hot water extraction of bioactives. TrAC Trend Anal Chem 71:39–54. https://doi.org/10.1016/j.trac.2015.02.022

Prado JM, Lachos-Perez D, Forster-Carneiro T, Rostagno MA (2016) Sub- and supercritical water hydrolysis of agricultural and food industry residues for the production of fermentable sugars: a review. Food Bioprod Process 98:95–123. https://doi.org/10.1016/j.fbp.2015.11.004

Purnomo A, Yudiantoro YAW, Putro JN, Nugraha AT, Irawaty W, Ismadji S (2016) Subcritical water hydrolysis of durian seeds waste for bioethanol production. Int J Ind Chem 7:29–37. https://doi.org/10.1007/s40090-015-0059-3

Requena R, Jimenez-Quero A, Vargas M, Moriana R, Chiralt A, Vilaplana F (2019) Integral fractionation of rice husks into bioactive arabinoxylans, cellulose nanocrystals, and silica particles. ACS Sustain Chem Eng 7:6275–6286. https://doi.org/10.1021/acssuschemeng.8b06692

Sánchez-Camargo AP, Ibáñez E, Cifuentes A, Herrero M (2017) Bioactives obtained from plants, seaweeds, microalgae and food by-products using pressurized liquid extraction and supercritical fluid extraction. In: Ibáñez E, Cifuentes A (eds) Green extraction techniques: principles, advances and applications. Elsevier, Amsterdam, pp 27–51

Santos DT, Albarelli JQ, Rostagno MA, Ensinas AV, Marechal F, Meireles MAA (2014) New proposal for production of bioactive compounds by supercritical technology integrated to a sugarcane biorefinery. Clean Technol Environ Policy 16:1455–1468. https://doi.org/10.1007/s10098-014-0760-5

Santos MP, Souza MC, Sumere BR, da Silva LC, Cunha DT, Bezerra RMN, Rostagno MA (2019) Extraction of bioactive compounds from pomegranate peel (Punica granatum L.) with pressurized liquids assisted by ultrasound combined with an expansion gas. Ultrason Sonochem 54:11–17. https://doi.org/10.1016/j.ultsonch.2019.02.021

Sasaki M, Kabyemela B, Malaluan R, Hirose S, Takeda N, Adschiri T, Arai K (1998) Cellulose hydrolysis in subcritical and supercritical water. J Supercrit Fluid 13:261–268. https://doi.org/10.1016/s0896-8446(98)00060-6

Schievano A, Adani F, Buessing L, Botto A, Casoliba EN, Rossoni M, Goldfarb JL (2015) An integrated biorefinery concept for olive mill waste management: supercritical CO2 extraction and energy recovery. Green Chem 17:2874–2887. https://doi.org/10.1039/c5gc00076a

Siscovick DS, Barringer TA, Fretts AM, Wu JHY, Lichtenstein AH, Costello RB, Kris-Etherton PM, Jacobson TA, Engler MB, Alger HM, Appel LJ, Mozaffarian D, Council Lifestyle Cardiometabolic Health, Council on Epidemiology and Prevention, Council on Cardiovascular Disease in the Young, Council on Cardiovascular and Stroke Nursing, Council on Clinical Cardiology (2017) Omega-3 polyunsaturated fatty acid (fish oil) supplementation and the prevention of clinical cardiovascular disease: a science advisory from the american heart association. Circulation 135:e867–e884. https://doi.org/10.1161/cir.0000000000000482

Soares JF, Dal Pra V, Barrales FM, dos Santos P, Kuhn RC, Rezende CA, Martinez J, Mazutti MA (2018) Extraction of rice bran oil using supercritical CO2 combined with ultrasound. Braz J Chem Eng 35:785–794. https://doi.org/10.1590/0104-6632.20180352s20160447

Sumere BR, de Souza MC, dos Santos MP, Neves Bezerra RM, da Cunha DT, Martinez J, Rostagno MA (2018) Combining pressurized liquids with ultrasound to improve the extraction of phenolic compounds from pomegranate peel (Punica granatum L.). Ultrason Sonochem 48:151–162. https://doi.org/10.1016/j.ultsonch.2018.05.028

Talhaoui N, Vezza T, Maria Gomez-Caravaca A, Fernandez-Gutierrez A, Galvez J, Segura-Carretero A (2016) Phenolic compounds and in vitro immunomodulatory properties of three Andalusian olive leaf extracts. J Funct Foods 22:270–277. https://doi.org/10.1016/j.jff.2016.01.037

Temelli F, Ciftci ON (2015) Developing an integrated supercritical fluid biorefinery for the processing of grains. J Supercrit Fluid 96:77–85. https://doi.org/10.1016/j.supflu.2014.09.028
de la Torre I, Martin-Dominguez V, Acedos MG, Esteban J, Santos VE, Ladero M (2019) Utilisation/upgrading of orange peel waste from a biological biorefinery perspective. Appl Microbiol Biotechnol 103:5975–5991. https://doi.org/10.1007/s00253-019-09929-2
Torres-León C, Ramírez-Guzman N, Londoño-Hernandez L, Martinez-Medina GA, Díaz-Herrera R, Navarro-Macias V, Alvarez-Pérez OB, Picazo B, Villarreal-Vázquez M, Ascacio-Valdes J, Aguilar CN (2018) Food waste and byproducts: an opportunity to minimize malnutrition and hunger in developing countries. Front Sustain Food Syst 2:52. https://doi.org/10.3389/fsufs.2018.00052
Toscano G, Pizzi A, Pedretti EF, Rossini G, Ciceri G, Martignon G, Duca D (2015) Torrefaction of tomato industry residues. Fuel 143:89–97. https://doi.org/10.1016/j.fuel.2014.11.039
Tsitsagi M, Ebralidze K, Chkhaidze M, Rubashvili I, Tsitsishvili V (2018) Sequential extraction of bioactive compounds from tangerine (Citrus Unshiu) peel. Ann Agrar Sci 16:236–241. https://doi.org/10.1016/j.aasci.2018.02.007
Vagi E, Simandi B, Vasarhelyine KP, Daood H, Kery A, Doleschall F, Nagy B (2007) Supercritical carbon dioxide extraction of carotenoids, tocopherols and sitosterols from industrial tomato by-products. J Supercrit Fluid 40:218–226. https://doi.org/10.1016/j.supflu.2006.05.009
Vigano J, Aguiar AC, Moraes DR, Jara JLP, Eberlin MN, Cazarin CBB, Marostica Junior MR, Martinez J (2016a) Sequential high pressure extractions applied to recover piceatannol and scirpusin B from passion fruit bagasse. Food Res Int 85:51–58. https://doi.org/10.1016/j.foodres.2016.04.015
Vigano J, Coutinho JP, Souza DS, Baroni NAF, Godoy HT, Macedo JA, Martinez J (2016b) Exploring the selectivity of supercritical CO2 to obtain nonpolar fractions of passion fruit bagasse extracts. J Supercrit Fluid 110:1–10. https://doi.org/10.1016/j.supflu.2015.12.001
Vigano J, Zabot GL, Martinez J (2017) Supercritical fluid and pressurized liquid extractions of phytonutrients from passion fruit by-products: economic evaluation of sequential multi-stage and single-stage processes. J Supercrit Fluid 122:88–98. https://doi.org/10.1016/j.supflu.2016.12.006
Vitor Pereira DT, Tarone AG, Betim Cazarin CB, Fernandez Barbero G, Martinez J (2019) Pressurized liquid extraction of bioactive compounds from grape marc. J Food Eng 240:105–113. https://doi.org/10.1016/j.jfoodeng.2018.07.019
Zabaniotou A, Kamaterou P (2019) Food waste valorization advocating circular bioeconomy - a critical review of potentialities and perspectives of spent coffee grounds biorefinery. J Clean Prod 211:1553–1566. https://doi.org/10.1016/j.jclepro.2018.11.230
Zacharof M-P (2017) Grape winery waste as feedstock for bioconversions: applying the biorefinery concept. Waste Biomass Valorization 8:1011–1025. https://doi.org/10.1007/s12649-016-9674-2
Zhao Y, Lu W-J, Wang H-T, Yang J-L (2009) Fermentable hexose production from corn stalks and wheat straw with combined supercritical and subcritical hydrothermal technology. Bioresour Technol 100:5884–5889. https://doi.org/10.1016/j.biortech.2009.06.079

Index

Inamuddin, A. M. Asiri (eds.), *Advanced Nanotechnology and Application of Supercritical Fluids*, Nanotechnology in the Life Sciences,
https://doi.org/10.1007/978-3-030-44984-1

The manufacturer's authorised representative in the EU is Springer Nature Customer Service Centre GmbH, Europaplatz 3, 69115 Heidelberg, Germany. If you have any concerns regarding our products, please contact ProductSafety@springernature.com

Printed and bound by CPI Group (UK) Ltd, Croydon, CR0 4YY
30/09/2026
02225559-0010